Computing, Communication and Intelligence

A proceeding of ICCTCCI – 2024

T0323798

First edition published 2024
by CRC Press
4 Park Square, Milton Park, Abingdon, Oxon, OX14 4RN

and by CRC Press
2385 NW Executive Center Drive, Suite 320, Boca Raton FL 33431

CRC Press is an imprint of Informa UK Limited

British Library Cataloguing-in-Publication Data
A catalogue record for this book is available from the British Library

ISBN: 9781032946825 (pbk)
ISBN: 9781003581215 (ebk)

DOI: 10.1201/9781003581215

Font in Sabon LT Std
Typeset by Ozone Publishing Services

Computing, Communication and Intelligence

A proceeding of ICCTCCI – 2024

Edited by:

Srinivas Sethi
Bibhudatta Sahoo
Deepak Tosh
Suvendra Kumar Jayasingh
Sourav Kumar Bhoi

CRC Press
Taylor & Francis Group

Contents

List of Figures

List of Tables

Preface

This Taylor & Francis, CRC Press volume contains the papers presented at the International Conference on Cutting Edge Technology in Computing, Communication and Intelligence (ICCTCCI – 2024) held during 23rd and 24th March 2024 organized by Institute of Management and Information Technology (IMIT), Cuttack, Odisha, India. A lot of challenges at us and no words of appreciation is enough for the organizing committee who could still pull it off successfully.

The conference draws the excellent technical keynote talk and many papers. The keynote talk by Dr. Sridhara Nayak, Senion Scientist, Japan Meteorological Corporation, Osaka, Japan is worth mentioning. We are grateful to all the speakers for accepting our invitation and sparing their time to deliver the talks.

We received 136 full paper submissions and we accepted only 63 papers with an acceptance rate of 46%, which is considered very good in any standard. The contributing authors are from different parts of the globe that includes United States, Japan, Jordan, South Korea, United Kingdom, Crotia and India. The conference also received papers from distinguished authors from the length and breadth of the country including 14 states and many premier institutes. All the papers are reviewed by at least three independent reviewers and in some cases by as many as five reviewers. All the papers are also checked for plagiarism and similarity score. It was really a tough job for us to select the best papers out of so many good papers for presentation in the conference. We had to do this unpleasant task, keeping the Taylor & Francis guidelines and approval conditions in view. We take this opportunity to thank all the authors for their excellent work and contributions and also the reviewers who have done an excellent job.

On behalf of the technical committee, we are thankful to Principal Secretary to Govt. of Odisha, Skill Development and Technical Education Smt. Usha Padhee, to accept the invitation and indebted to Late Prof. A. K. Pani, Professor and Dean, XLRI Jamsedpur, General Chair of the Conference, for his timely and valuable advice. We cannot imagine the conference without his active support at all the crossroads of decision-making process. The university management of the host institute, the honorable Vice Chanceller, Prof. A. K. Rath, Patron of the conference and the Principal Prof. M. R. Kabat, Organising Chair have extended all possible support for the smooth conduct of the Conference. Our sincere thanks to all of them.

We would also like to place on record our thanks to all the keynote speakers, tutorial speakers, reviewers, session chairs, authors, technical program committee members, various Chairs to handle finance, accommodation and publicity and above all to several volunteers. Our sincere thanks to all press, print and electronic media for their excellent coverage of this conference.

We are also thankful to Taylor & Francis, CRC Press publication house for agreeing to publish the accepted and presented papers.

Best wishes.

March, 2024

Prof. Srinivas Sethi
Prof. Bibhudatta Sahoo
Prof. Deepak Tosh
Prof. Suvendra Kumar Jayasingh
Prof. Sourav Kumar Bhoi

Editors Biography

Srinivas Sethi

Prof. Srinivas Sethi is a professor in Computer Science Engineering & Application, Indira Gandhi Institute of Technology, Sarang (IGIT, Sarang), India, and has been actively involved in teaching and research since 1997. He did his Ph.D. in the area of routing algorithms in mobile ad hoc network and is continuing his research work in the wireless sensor network, cognitive radio network, cloud computing, BCI, and Cognitive Science. He is a member of the Editorial Board for different journal and program committee members for different international conferences/ workshops. He is book editor of 4 international conference proceedings published in Springer and Taylor & Francis. He has published more than 100 research papers in international journals, edited book chapters, and conference proceedings. He also published 2 nos. of Books. He completed 8 numbers of research and consultancy projects funded by different funding agencies such as DRDO, DST, AICTE, NPIU, and local Govt. office.

Bibhudatta Sahoo

Prof. Bibhudatta Sahoo is a computer science professor at NIT Rourkela, India, specializing in Algorithmic Engineering, Distributed Systems, Cloud Computing, Fog/Edge Computing & IoT, and Software Defined Networks/NFV. With a wealth of knowledge and expertise in these areas, Prof. Sahoo is at the forefront of cutting-edge research and innovation. Their research focuses on developing efficient algorithms and designing scalable distributed systems to address the challenges posed by modern computing paradigms such as cloud computing, fog/edge computing, and IoT. Driven by a passion for advancing the field, they have significantly contributed to developing software-defined networks and network function virtualization. Through their research, publications, and mentorship, Prof. Sahoo has inspired and empowered countless students, fostering a new generation of computer scientists equipped with the skills to tackle complex problems in the rapidly evolving world of technology.

Deepak Tosh

Prof. Deepak Tosh (Senior Member, IEEE) is an assistant professor of Computer Science at the University of Texas at El Paso. His research focuses on addressing various multi-disciplinary networking and cybersecurity challenges associated to critical national infrastructures, Industrial Internet of Things, Blockchain, and tactical battlefields. He works closely with researchers from U.S. Air Force Research Laboratory (AFRL), Sandia National Lab, and Army Research Laboratory (ARL) on developing resilient data/process provenance mechanisms for industrial operational technology environments, and military applications. His research has been funded from Department of Energy, National Science Foundation (NSF), and Department of Defense. He has authored/co-authored more than 70 peer-reviewed conference papers, book chapters, and journal papers. Two of his research works on Blockchain were also awarded as "Top 50 Blockchain Papers in 2018" at Block chain Connect Conference, 2019. He is also a recipient of prestigious NSF CAREER award, 2022.

Suvendra Kumar Jayasingh

Prof. Suvendra Kumar Jayasingh is working as Associate Professor and HOD in the Department of Computer Science & Engineering, Institute of Management and Information Technology (IMIT), Cuttack (A Constituent College of BPUT, Govt. of Odisha), India after being selected in OPSC (Orissa Public Service Commission) in 2005. He has obtained his Bachelor of Engineering in the year 2003 from University College of Engineering (UCE), Burla (Now VSSUT). He got his M. Tech. in Computer Science & Engineering in 2007 from RVU, Udaipur and Ph. D. in Computer Science & Engineering in 2020 from North Orissa University, Baripada. He is having 20 years of teaching experience in Computer Science & Engineering and MCA. He has published several articles, book chapters in reputed National and International journals and periodicals including Springer and Taylor & Francis and has presented research papers in National and International Seminars and Conferences. His research interests include Artificial Intelligence, Data Mining, Soft Computing, Machine Learning, Computational Intelligence, Database Management System and Algorithm Analysis and Design. He has published a book on "Introduction to Machine Learning" and a UK patent on "Smart Home Air Quality Monitoring Device". He is a life member of Indian Society for Technical Education (ISTE).

Sourav Kumar Bhoi

Prof. Sourav Kumar Bhoi (Sr. Member, IEEE and Member, ACM) completed his Doctoral (2017) and M. Tech. (2013) degrees from Department of Computer Science and Engineering, National Institute of Technology (NIT), Rourkela, India respectively. He completed his B. Tech (2011) from Dept. of CSE, Veer Surendra Sai University of Technology, Burla India. He completed his PDF (2023) research work from Srinivas University, India in CS&IS programme. He is currently working as Assistant Professor since 2016 in Dept. of CSE, Parala Maharaja Engineering College (a Govt. college affiliated to BPUT), Berhampur, India. His research interest includes Vehicular Networks, Internet of Things, Edge and Fog Computing, Machine Learning, and Information Security. He has nearly 10 years of teaching and research experience. He published nearly 130+ research publications in reputed International Journals and Conferences. He published 09 books/book chapters as author/ co-author and 03 Indian Patents. He received nearly 10+ State/National/International Awards for his research works such as prestigious IET Premium Award 2016, India Book of Record Holder 2022, BPUT-University Faculty Research Award 2021, ISTE-Rajalaxmi Award for Best Engineering College Teacher Award of Odisha State 2022, Institution of Engineers (IEI)-Sadananda Memorial Award 2021, Institution of Engineers (IEI)-JN Panda Memorial Award 2024, PDF Researcher with Good performance Award 2023, etc. He also completed one MHRD, Govt. of India research project on Smart Irrigation & continuing with one Project of AICTE, Govt. of India. He also acts as reviewer for many reputed organizations such as IEEE, IET, Elsevier, Springer, MDPI, Hindawi, Wiley, Taylor and Francis, etc. He guided nearly 80+ B.Tech students, 03 PG students, 01 PhD scholar awarded with Ph.D. degree, and currently guiding 03 Ph.D. scholars in new recent areas of Computer Sci. and Engg. He also acts as Faculty In-charge/Member of many administrative level sections/committees at institute and dept.

Reviewers

Alekha Mishra
Anitha A
Anup Maharana
Anupama Sahu
Ashima Rout
Ashis Das
Atish Nanda
Bichitra Mandal
Biswa Senapati
Chittaranjan Mallick
Debabrata Dansana
Debasis Mohapatra
Deepak Tosh
Srinivas Sethi
Sangharatna Godboley
Sanjaya Kumar Panda
Jui Pattnaik
Gunamani Jena
Hemant Apat
icctcci conference
Janmenjoy Nayak
Jemarani Jaypuria
Jitendra Kumar
Jitendra Rout
Jyoti barik
Kali Rath
Kalyan Jena
Kaushik Mishra
Lakhmi Das
Madhu Panda
Niranjan Panigrahi
Prahallad Sahu
Prajna Paramita Nanda
Prasant Dhal
Raghunandan Swain
Rajendra Nayak
Rakesh Swain
Ramesh Sahoo
Ranumayee Sing
Rashmi Sahoo
Sangita Pal
Sanjib Nayak
Sasmita Behera
Satyasundara Mahapatra
Sipra Swain
Subasish Mohapatra

Sumit Kar
Surya Das
Suvendra Kumar Jayasingh
Umashankar Ghugar

Committee Members

Patron
Prof. Amiya Kumar Rath, Honourable Vice Chancellor, BPUT, Odisha

General Chair
Prof. Ashis Kumar Pani, Dean of XLRI-Xavier School of Management, Jamshedpur

Organising Chair
Prof.(Dr.) Manas Ranjan Kabat, Principal, IMIT, Cuttack

Program Chairs
Prof. Srinivas Sethi, IGIT Sarang, India
Prof. Bibhudatta Sahoo, NIT Rourkela, India
Prof. Deepak Tosh, University of Texsas

Convenor
Prof. Suvendra Kumar Jayasingh, HOD, CSE

Editorial Board
Prof. Srinivas Sethi
Prof. Bibhudatta Sahoo
Prof. Deepak Tosh
Prof. Suvendra Kumar Jayasingh
Prof. Sourav Kumar Bhoi

Organising Committee
Prof. Satyaprakash Swain, IMIT, Cuttack
Prof. Srutipragyan Swain, IMIT, Cuttack
Prof. Sujata Ray, IMIT, Cuttack

Advisor board
Prof. Chittaranjan Tripathy, Ex-VC, BPUT & SU
Prof. Bibhu Prasad Panigrahi, IGIT, Sarang
Prof. Mrutunjaya Bhuyan, University of Malaya

Single Phase Shunt Active Power Filter using AHCC and CHCC

Omkar Tripathy, Sritam Parida, Maheswar Prasad Behera

Electrical Engineering Department, IGIT, Sarang, BPUT Rourkela
E-mails: omkar_tripathy@yahoo.com, sritamparida@igitsarang.ac.in,
mpbehera@igitsarang.ac.in

Abstract

This paper outlines a shunt-type Active Power Filter (SAPF) for single-phase systems, featuring integration with a grid-connected battery source. The objective of the presented system is to compensate reactive power, mitigate harmonics, and supply real power to a nonlinear load. The boundary concerning the grid and a battery source is done using a Voltage Source Converter (VSC) connected in a shunt with a nonlinear load. Hence harmonics in source current are mitigated. Adaptive Hysteresis Current Control (AHCC) and Conventional Hysteresis Current Control (CHCC) methods are used for the generation of switching pulses of the converter. Instantaneous Reactive Power Theory (IRPT) is used for evaluating compensating current to mitigate harmonics and current harmonics in the source. It is found that AHCC shows better compensation than CHCC.

Keywords: Voltage Source Converter (VSC), Adaptive Hysteresis Current Control (AHCC), Conventional Hysteresis Current Control (CHCC), Shunt Active Power Filter (SAPF), Instantaneous Reactive Power Theory (IRPT)

1. Introduction

With the advancements in technology aimed at easing human tasks, numerous power electronics devices have integrated into the power grid. These devices range from domestic appliances like mixers and televisions to industrial equipment used in automation and control, all exhibiting nonlinear characteristics (Y. S. Prabhu & *et. al*).

Due to their nonlinear behaviour, they draw nonlinear currents, leading to the generation of current harmonics, which, in turn, degrade power quality. Moreover, since many loads are inductive, the power factor tends to be low (B. UdayaSri & *et. al*).

To address these issues, the implementation of a passive parallel filter tuned to specific harmonic frequencies can be beneficial. However, passive filters come with drawbacks such as series and/or parallel resonance and larger size (M. kale & *et. al*).

The filtering characteristics of the filter are also influenced by the source impedance. Filters can become overloaded due to the high degree of harmonic current they encounter (M. P. Behera & et.al). Compounding the issue, multiple cleaners are often required for recompense, as each filter can address only a single frequency, leading to increased filter costs (P. S. Magdum & *et. al*).

To mitigate these drawbacks, a SAPF can be employed, offering solutions to harmonics arise from various factors, including moderate switching frequencies and changes in load conditions.

The Instantaneous Reactive Power Theory serves as the basis for obtaining the reference or compensating current (S. Srinath & et.al). Gate pulse generation for the voltage source inverter (VSI) is facilitated by techniques such as CHCC and AHCC. These methods compare reference and actual signals to produce the gate pulse (R. Kanagavel & et.al).

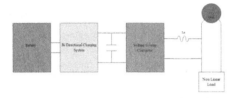

Figure 1: Block diagram of the Simulink Model

A battery is connected to the DC output of VSI. A control design of the bidirectional charging system of the battery is addressed. Hence battery provides real power to the load via VSI at the time of discharging and receives power from the source via VSI (that acts as a rectifier). Hence VSI is also called as dual operator i.e. rectifier and inverter (Omkar Tripathy & et.al).

2. Methodology

The grid-connected battery SAPF setup comprises a shunt voltage source inverter with its DC link interconnected to a battery through a bidirectional charging system. A non-linear load is linked across the grid, and the SAPF connects to the grid through a coupling inductor at the PCC.

The IRPT (PQ Theory) is employed to generate compensating current, while AHCC and CHCC methods are utilized for generating appropriate switching signals for the inverter gates.

The single-phase voltage and current exhibit a phase lag of $\frac{\pi}{2}$. They are then

Figure 2: Block diagram for generating Compensating Current

converted into the alpha-beta stationary reference frame. From this conversion, instantaneous real power and reactive power are derived. While real power comprises both fundamental and harmonic components, the focus for mitigation is primarily on the harmonic component, alongside reactive power. Based on these parameters, compensating current is generated to mitigate harmonic distortions.

$$V_\alpha = V_a = A \tag{1}$$

$$V_\beta = V_a\left(-\frac{\pi}{2}\right) = B \tag{2}$$

$$I_\alpha = I_a = C \tag{3}$$

$$I_\beta = I_a\left(-\frac{\pi}{2}\right) = D \tag{4}$$

$$p = AC + BD \tag{5}$$

$$q = AD - BC \tag{6}$$

$$I_{comp} = \frac{\left\{A\left(-p_{ref}\right) + B\left(q_{ref}\right)\right\}}{A^2 + B^2} \tag{7}$$

Where

$$p_{ref} = p_{loss} - \tilde{p} \tag{8}$$

$$q_{ref} = -q \tag{9}$$

$$p_{loss} = V_{dc} - V_{ref} \tag{10}$$

To uphold a consistent DC bus voltage in the inverter, a slight current draw from the grid (or battery) is necessary for voltage regulation. This current draw is represented by the error signal, denoted as P_{loss}.

The CHCC is integrated into a negative feedback system. Within this setup, the inaccuracy signal, denoted as er(t), is utilised to regulate the buttons of the inverter. This inaccuracy signal is derived from the difference between the looked-for current, Irf(t), and the definite current supplied by the inverter, Iac(t). When the inaccuracy approaches the upper limit, the transistors are activated to reduce the current. Conversely, when the inaccuracy nears the lower edge, the current is increased. It has been observed that the amplitude of the inaccuracy signal directly influences the ripple in the inverter's output current, commonly referred to as the Hysteresis Band. The current is constrained within these restrictions evenif the locus current varies. The specific switching conditions are detailed in Tables 1 and 2.

As the instantaneous compensation current, represented by dic/dt, varies, the AHCC adjusts its HB consequently. Additionally, the voltage Vdc serves to mitigate interference from distortion on the kerbed waveform, thereby optimizing both the converting frequency and the THD. The calculation of the HB for AHCC is outlined below.

$$HB = \frac{0.125 V_{dc}}{f_{sw} L} \left[1 - \frac{4L^2}{V_{dc}} \left(\frac{V_s}{L} + \frac{di_c}{dt} \right)^2 \right] \qquad (11)$$

3. Simulink Model

Several simulation tools exist in the market to validate the proposed method mentioned above. MATLAB stands out as one such tool used for simulation. All simulation work in this research has been exclusively conducted using MATLAB 2023.

Table 1: Switching Condition-1

Condition	Switch Position	State-of Switch
er(t) > HB	Higher	OPEN
	Lower	CLOSED
er(t) < -HB	Higher	CLOSED
	Lower	OPEN

Table 2: Switching Condition-2

Condition	Switch Position	State-of Switch
er(t) ≥ HB	Higher	OPEN
	Lower	CLOSED
er(t) ≤ HB	Higher	CLOSED
	Lower	OPEN

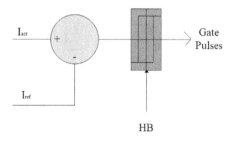

Figure 3: Block diagram to generate gate switching signals

4. Results and Discussion

- **Without any Compensation**

THD of source current when a non-linear load is connected to the grid is 19.96%.

- **Compensation using CHCC**

Figures 6 and 7 shows that the inverter was initially opened i.e. disconnected from the grid. At 0.1s the switch is closed and the inverter gets connected to the grid. From 0.1s to 0.5s, the battery of the inverter gets charged. For 0.5s to 1s, the battery of the inverter gets discharged. Hence shows a Bidirectional Inverter.

From the above simulink result it is noted that the distortion in the source current -

- When compensated while the battery is Charging is 2.81%.
- When compensated while the battery is Discharging is 3.99%.
  - **Compensation using AHCC**

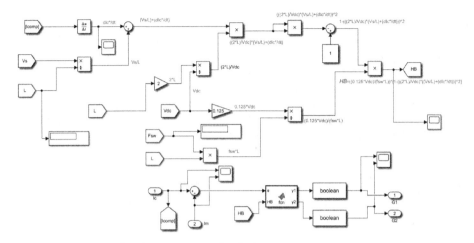

Figure 4: Hysteresis band current control calculation

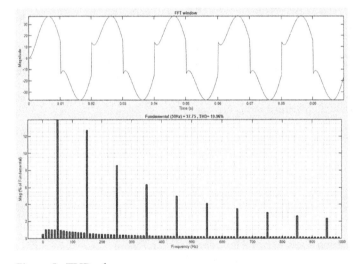

Figure 5: THD of source current

- When compensated while the battery is Charging is 2.81%.
- When compensated while the battery is Discharging is 3.99%.
  - *Compensation using AHCC*

Figures 10 and 11 shows that the inverter was initially opened i.e. disconnected from the grid. At 0.1s the switch is closed and the inverter gets connected to the grid. From 0.1s to 0.5s, the battery of the inverter gets charged. For 0.5s to 1s, the battery of the

inverter gets discharged. Hence shows a Bidirectional Inverter.

From the above Simulink result it has been seen that the THD in the source current -

- When compensated while the battery is Charging at 2.74%.
- When compensated while the battery is Discharging at 3.82%.

Hence AHCC provides better power quality improvement as compared to the CHCC.

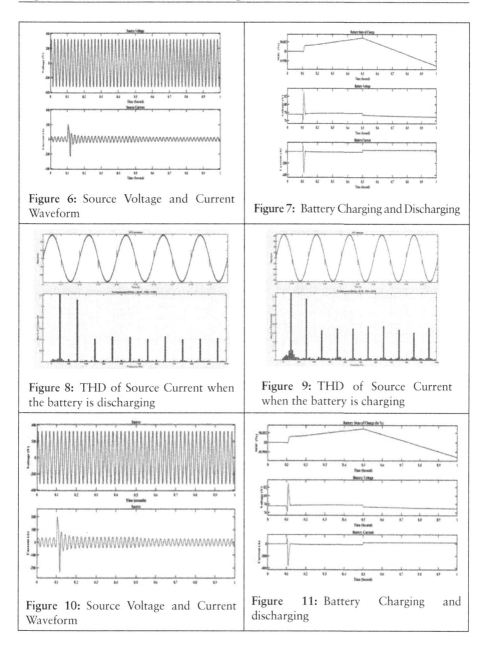

Figure 6: Source Voltage and Current Waveform

Figure 7: Battery Charging and Discharging

Figure 8: THD of Source Current when the battery is discharging

Figure 9: THD of Source Current when the battery is charging

Figure 10: Source Voltage and Current Waveform

Figure 11: Battery Charging and discharging

5. Conclusion

This paper has examined a SAPF integrated with a grid-connected battery source. The investigation reveals successful compensation of reactive power, harmonic mitigation, and provision of real power to a non-linear load. Notably, there is a significant reduction in total harmonic distortion in the source current. Both AHCC and CHCC methods were studied for generating switching pulses in the

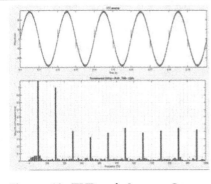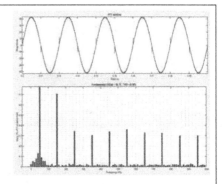

| **Figure 12:** THD of Source Current when Battery is discharging | **Figure 13:** THD of Source Current when Battery is charging |

converter, with AHCC demonstrating superior power quality improvement over CHCC. Additionally, it was observed that the THD of the inverter is higher when the battery is discharged compared to when it is charged.

References

Prabhu, Y. S., Dharme, A., and Talange, D. B. (2014). "A three phase shunt active power filter based on instantaneous reactive power theory." *2014 Annual IEEE India Conference (INDICON)*, Pune, pp. 1–5, doi:10.1109/INDICON.2014.7030398

Udaya Sri, B., Rao, P. A. M., Mohanta, D. K., and Varma, M. P. C. (2016). "Improvement of power quality using PQ-theory shunt-active power filter." *International Conference on Signal Processing, Communication, Power and Embedded System (SCOPES)*, Paralakhemundi, 2016, pp. 20832088.

Kale, M. and Ozdemir, E. (2004). "An Adaptive Hysteresis band current controller for shunt active filter". ELSEVIER. *Electric Power Systems Research*, ISSN: 0378-7796, 73(2), 113–119.

Behera, M. P. and Ray, P. K. (2019). "Three-Phase Grid Connected Bi- Directional Charging System to Control Active and Reactive Power with Harmonic Compensation." *International Journal of Emerging Electric Power Systems*, 20(2), 20180259.

Behera, M. P. and Ray, P. K. (2016). "Three-phase series-connected photovoltaic generator for harmonic and reactive power compensation with battery energy storage device." *Transactions of the Institute of Measurement and Control*, 39(7), 1071–1080.

Behera, M. P., Ray, P. K., and Beng, G. H. (2016). "Single-phase grid-tied photovoltaic inverter to control active and reactive power with battery energy storage device." *IEEE Region 10 Conference (TENCON), Singapore*, 2016, pp. 1900–1904

Magdum, P. S. and Patil, U. T. (2017). "Development of single phase shunt active power filter." *International Conference on Inventive Communication and Computational Technologies (ICICCT), Coimbatore*, 2017, pp. 351–355

Himabindhu, T. (2012). "Performance of Single Phase Shunt Active Power filter Based on P-Q theory Technique Using MATLAB/Simulink." *International Journal of Engineering Research and Technology*, ISSN: 2278-0181, 1(9) Nov., pp. 1–5.

Bhonsle, D. C. and Kelkar, R. B. (2011). "Design and simulation of single phase shunt active power filter using MATLAB." *International Conference On Recent Advancements In Electrcial, Electronics And*

Control Ebgineering, Sivakasi, India, 2011, pp. 237–241, doi: 10.1109/ICONRAEeCE.2011.6129786.

Patel, P., Samal, S., Jena, C., Barik, P. K. (2022). "Shunt active power filter with MSRF-PI-AHCC technique for harmonics mitigation in a hybrid energy system under load changing condition." https://doi.org/10.1080/14488 37X.2022.2114154.

Srinath, S., Prabakaran, S., Mohan, K., Selvan, M. P. "Implementation of Single Phase Shunt Active Filter for Low Voltage Distribution System." *16th National Power Systems Conference, 15th-17th December, 2010,* pp. 295–300.

Kanagavel, R., Vairavasundaram, I., Padmanaban, S. (2019). "Design and prototyping of single-phase shunt active power filter for harmonics elimination using model predictive current control." International transaction of Electrical Energy System. Wiley, https://doi.org/10.1002/2050-7038.12231

Behera, M. P., Tripathy, O., and Ray, P. K. (2020). "UPQC based Grid-Connected Photovoltaic System with Fuzzy Logic Controller." *3rd International Conference on Energy, Power and Environment: Towards Clean Energy Technologies* | 978-1-6654-2536-0/21/$31.00 ©2021 IEEE.

A Comparative Analysis of ANN and RNN Technique for UWB Gaussian Pulse based Indoor Localisation

Sujata Mohanty[1], Aruna Tripathy[2]

[1]Department of ETC, BPUT, Rourkela, India
[2]Department of E & I, OUTR, Bhubaneswar, India
E-mail: Sujatamohanty1988@gmail.com, atripathy@outr.ac.in

Abstract

Localisation is the method that is applied in wireless system in order to upgrade the routing along with build up security. Ultra-wideband (UWB) acts a pivotal role in positioning system due to its varieties of characteristics such as: broad frequency band (7.5GHz), capability of penetrating through the obstacles, high precision (cm), ability of avoiding multipath fading etc. The IEEE 802.15.3a TG is allocated to UWB channel for WPAN, which produces less cost and less consumption power to WPAN devices. There are different UWB Channel models which include: CM-1, CM-2, CM-3, and CM-4. Now a day the Indoor positioning system (IPS) is very important for rapid growth of Artificial intelligence technology along with growth of context aware services in the field of Internet of Things (IoTs). Machine learning (ML) algorithms act as an important tool in the field of system approximator. Here in this paper, the Gaussian monopulse signal is made allowance for IPS. This received pulse undergoes the performance of average and correlation thereupon the originally transmitted pulse in order to calculate the time of arrival (ToA) value for determining the target node location in IPS. It also includes the analysis and evaluation of Artificial Neural Network(ANN), Recurrent Neural network (RNN) ML for UWB localisation in a archetypal indoor positioning system. The MSE of RNN is observed to be minimum during both the training and the testing phase than that of ANN technique.

Keywords: ANN, localisation, ML, RNN, UWB

1. Introduction

Indoor location based system have increased their attention due to the rapid growth in context aware application in area of IoT. The varieties of applications of localisation include automatic warehouse management, surveillance, navigation system, home automation, detecting and locating the items, providing assistance to elderly and disabled people, industrial robots, smart factories, robotic operations, monitoring the equipment location's etc (Bastidas *et al.*, 2022). Ultra-Wideband (UWB) technology is a well-accepted wireless communication technology that employs a broad frequency range to transmit data with unparalleled precision over short distances. UWB has been ingeniously utilised for indoor localisation, enabling the exact determination of the position of the people or objects within indoor environments. With increase of order of derivatives of the Gaussian signal, there

Chaper 2 DOI: 10.1201/9781003581215

is increase of number of zero crossings occurs. That means by taking the derivative, the centre frequency of the pulse also increases (Xianxia *et al.*, 2021). There are different machine learning techniques are used to improve the accuracy of Indoor UWB localisation. The IEEE802.15.3a TG has explicated four different kinds of channel model namely: CM1, CM2, CM3 and CM4. CM1 specify LOS domain where the difference in distance in transmitter and receiver is less than 4m. CM2 is meant for NLOS area with equal separation as CM1. CM3 is meant with NLOS domain having severance in 4 to 10m. CM4 is designed with NLOS surrounding having strong delay dispersion with a result of delay spread 25 ns (Khan *et al.*, 2020). Deep learning (DL), a powerful subset of the machine learning, has been seamlessly integrated into UWB indoor localisation tasks to automatically learn complex patterns from data. However, the suitability of other ML algorithms that have recently been developed to predict the next outcome based on a certain time series dataset is lacking (Li *et al.*, 2019). This paper includes the Gaussian monopulse signal, in order to get the channel domain for positioning in Indoor system. The major goal is to go through the range data for positioning in IPS by taking account of Gaussian monopulse signal in an indoor environment system. The target distance is evaluated by determining the ToF which can be acquired by correlation receiver, then this data is utilised to recognise channel type and device is working in. After this training and testing of ANN and RNN ML techniques has done to find out the better one in terms of MSE for localisation in the field of IPS. The main aim is to find the time, range of operation by using Gaussian monopulse signal and also to investigate the effectiveness of ANN and RNN ML technique in ILS.

2. Literature Survey

UWB technology is an advancing technology in the field of high data rate with low range wireless communication technique. The different propagation time-based techniques includes: TDoA, ToA, and RToF (Dabove *et al.*, 2018). The accuracy of localisation can be increased by considering the machine learning techniques for predicting the target location in the Indoor area system. This ML algorithm learns from the structured data and predicts the output. It also discovers the pattern in that data. ML is the implementation or application of AI, which acknowledges the system to automatically learn and also build on from experiment (Lu *et al.*, 2021).

2.1. Feed forward Neural Network (FNN)

This NN includes one or many layer each of having number of parallel neurons or nodes. These neurons determines the weighted combination of inputs and pass it into non-linear transformation (activation function), hence the neuron can be represented as the generalisation of logistic regression that is called perceptron. The NNs having no feedback are known as feed forward NN (Yin and Lin, 2022). This is also called fully forward Multi-layer Perceptron (MLP). We can back propagate the loss function through the layers for updating the parameters like modify the weight and reduce the error. The sequential data can't be implemented in this ANN. Hence this limitation can be avoided by the use of Recurrent NNs (Krapez *et al.*, 2021).

2.2. Recurrent Neural Network (RNN)

Simple RNN is a type of ANN, Which utilises the sequential data that is time series data for prediction of the target coordinate in the area of localisation. The RNN architecture consists of three layers: an input layer, a hidden layer, output layer. In each step of time, the RNN takes an input vector along with previous hidden state as input values, after this it generates a new hidden state along with an output vector as outputs (Alam *et al.*, 2023). The output vector can be used to make predictions or classify input sequences. The hidden layer of a simple RNN contains a set of neurons, each of which has a weight matrix that determines how input in current time step and that of the previous hidden state are combined (Jha et al., 2019).

3. Proposed Model for Localisation

Considering a rectangular area having size $D_1 \times D_2$. The coordinate index of the target and the receivers are (X,Y), (xi,yi) where i ε {1,2,3}. The distance from transmitter to receiver is given as di.

$$d_i = \sqrt{(X - x_i)^2 + (Y - y_i)^2} \qquad (1)$$

Here, in the Figure 1, the location technique involves two steps; ranging and positioning. In the ranging phase the distance measurement is done between transceivers and in positioning phase the position estimation is done for the target object.

Here the Figure 2 shows, a very short duration pulse is generated through UWB transmitter which is travelled through indoor domain. In receiver part, the UWB receiver finds signal that is combination of both the original transmitted signal along with some addition of noise the transmitted signal along with noise. To avoid noise, the operation of averaging is carried out at receiver section and after this it is entered through correlation part, here correlation is done in between received signal and original transmitted signal. After this output is entered into peak detector, which determines

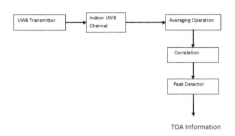

Figure 2: A Block diagram for Simulation in UWB System

values regarding ToA of UWB signal for the purpose of estimating the targetLocation.

3.1. Result and Discussion

Operation of simulation is carried out by MATLAB simulation.

3.2. Simulation Output using Gaussian Monopulsepulse Signal

Here in Figure 3, it shows the Gaussian monopulse signal, where peak position found at 1000 samples as shown in Figure 4, the 1000 samples determines ToA of 16.66ns. This finds ranging of 5m. It corresponds to a channel type primarily of the CM3 (4 to 10mt with NLOS), after this the obtained ranging value is utilised to determine the position value of the target node in IPS.

3.3. Simulation Result for for ANN and RNN

For our work, we utilised the MATLAB Communications Toolbox Library and UWB add-on for estimating the coordinate value of device in accordance with the IEEE® 802.15.4z™ TG.

Here, it is observed that the validation error is found to be 1.85×10^{-6} for ANN technique, as shown in Figure 5. Figure 6 represents the disparity in between trained output and testing output using ANN

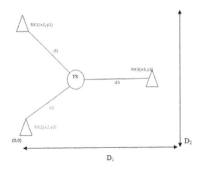

Figure 1: A typical system model

Table 1: Simulation Parameters

Parameters for UWB Simulation	Value
Speed of light	3×10^8 m/s
Pulse shape	Gaussian monopulse
SNR	30 db
Rate of sample	60 GHz
Width of pulse	0.5 ns
Variance	0.2

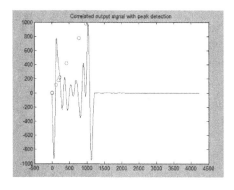

Figure 4: Corelated output signal with peak detection

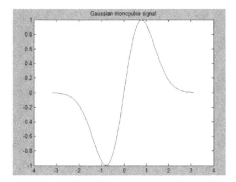

Figure 3: Gaussian monoPulse Signal

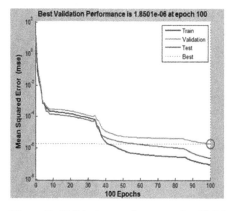

Figure 5: Validation performance for ANN

technique. It is also found that the validation error is found to be 6.11×10^{-8} for RNN technique, as shown in Figure 7. Similarly Figure 8 represents the disparity in between trained output and testing output using RNN technique. Again in final, it is found that after training two techniques (ANN, RNN), the error is high for ANN to that of the RNN technique, since the trained output and the testing output for ANN is found to be more apart than that of the RNN technique.

4. Conclusion and Future Work

This paper surveys the Gaussianmonopulse signal for determining the time information needed for evaluating the location in IPS. It also includes the machine learning techniques: ANN and RNN for UWB Localisation system. The two techniques are compared in terms

validation performance. From the Gaussian monopules output result, it shows a smooth representation of distance estimation in Channel model-3. Now by applying the two machine learning technique, it is found that loss MSE for RNN is less than that of ANN ML algorithm. It is also concluded that by using RNN technique, the actual outputs are very close to the trained output in compare to the ANN method of localisation. Hence the RNN technique is taken into consideration for UWB localisation in IPS.

The future work is to find out localisation accuracy in terms of number of time steps and to investigate the mean localisation error for different DL Algorithms (RNN and CNN) for different number of anchors in IPS.

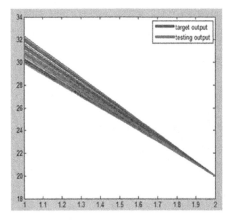

Figure 6: Predicted plot for target and test o/p

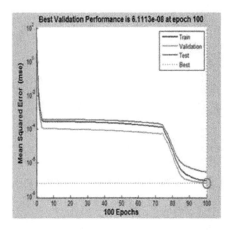

Figure 7: Validation performance for RNN

Acknowledgements

I wish to express my gratitude to Professor A. Tripathy for her evaluation of the manuscript and important comments.

References

Alam, M. S., Mohamed, F. B., Selamat, A., and Hossain, A. B. (2023). A review of recurrent neural network based camera localization for indoor environment. *IEEE Access*, 11: 43985-44009.

Bastidas, S. C., Estivez, M. E., and Quero, J. M. (2022). Review of ultra wideband

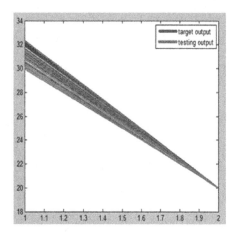

Figure 8: Predicted plot for target and test o/p

for indoor positioning with application to elderly. *Proceeding's of 55th Hawali International Conference on System Science*, pp. 2145-2154.

Dabove, P., Pietra, V. D., Jabar, A. A., Piras, M., and Kazim, S. A. (2018). Indoor positioning using ultra wide band technology: positioning accuracy and sensor performances. *IEEE Conference, Position, Location and Navigation Symposium (PLANS)*, Monterey, CA, USA.

Khan, M. R., Mohapatra, S., and Das, B. (2020). UWB Saleh-Valenzuela model for underwater acquastic sensor network. *International Journal of Information Technology*, Springer, 12: 1073-1083.

Krapez, P, Vidmar, M., and Munih, M. (2021). Distance measurement in UWB radio localization systems corrected with a feed forward neural network model. *Journal of Sensors*, 21(7), 2294.

Li, Z., Xu, K., Zhao, Y., Wang, X., and Shen, M. (2019). Machine learning based positioning: a survey and future direction. *IEEE Network*, 33: 96-101.

Lu, Y. M., Sheu, J. P., and Kuo, Y. C. (2021). Deep Learning for Ultra Wideband Indoor Positioning. *IEEE 23rd Annual International Symposium on Personal Indoor and Mobile Radio Communications (PIMRC)*, Helsinki, Finland.

Jha, B., Koroglu, M. T. and Yilmaz, A. (2019). Trajectory mining for Localization using Recurrent Neural Network. International Conference on Computational science and computational Intelligence (CSCI), IEEE, Las vegas, NV, USA.

Xianjia, Y., Qingqing, L., Queralta, J. P., Heikkonen, J., and Westerlund, T. (2021). Application of UWB networks and positioning to autonomous robots and industrial system. *10th Mediterranean Conference on Embedded Computing (MECO)*, IEEE, pp. 1-9.

Yin, A. and Lin, Z. (2022). Machine Learning aided precise Indoor Positioning. *Journal of Electrical Engineering and system Science, Signal Processing*, Cornel University.

DDOS Attack Detection Using Time Based Features

V. S. A. Chandra Mouli[1], P. Subba Rao[1], Shubhashish Jena[2], G. S. Ganga Prasuna[3], Ch. Sridevi[4], K. K. V. N. Manikantam[4], K. Ajay Babu[4]

[1]BVC Engineering College, Odalarevu, India
[2]IIT Patna, Patna, India
[3]CSE, BVC EC AP, East Godavari, India
[4]Research Scholars, CSE, BVC Engineering College, Odalarevu.
E-mail:principal.bvce@bvcgroup.in,drpsrao.cse@gmail.com,sjena1998@gmail.com

Abstract

One of the most common and expensive cyber security risks nowadays is a distributed denial-of-service (DDoS) assault. Because of their potential to disable network services and cause millions of dollars in damages, businesses and governments must take active measures to detect and mitigate DDoS attacks. While previous studies have shown that shallow and deep learning classifiers are useful for detecting DDoS assaults, nothing has been studied on the use of time-based features and classification for many different kinds of DDoS attacks. In this paper, we propose and analyze the performance of 25 time-based features for binary and multiclass classification of 12 distinct DDoS assault types. In addition, we conducted tests to evaluate the effectiveness of one deep learning classifier vs eight typical machine learning classifiers in two distinct settings. Based to our results, the majority of models were only 70% accurate at identifying different types of DDoS assaults, but they were 99% accurate at detecting DDoS attacks in both the control and time-based tests. Training on this smaller time-based feature subset alone has been shown to effectively cut training time without impacting test accuracy, making it ideal for near-real time applications that use continuous learning.

Keywords: DDoS, CICDDoS2019, SSDP, SYN Flood, PORTMAP, LDAP, CNN

1. Introduction

Existing DDoS classifiers in the literature have a high median training time, leading to poor performance when it comes to identifying DDoS attacks; as a result, we need to introduce a system that can address the problems with the current system. Every day, more than a thousand notable distributed denial-of-service (DDoS) assaults are identified by Arbour Networks' software (Halladay *et al.* 2022). Many of the leading internet service providers (ISPs) in the world are supplied with network security software by this company. These assaults can be directed at everything from individual PCs to the Internet service providers (ISPs) themselves. Due to their simplicity of implementation and invisibility in defence, DDoS attacks are becoming increasingly worrisome. With a distributed denial-of-service (DDoS) assault, the attacker aims to disrupt a network's

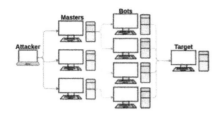

Figure 1: Architecture diagram

normal functioning and prevent authorised users from accessing the system.

As seen in Figure 1, they are carried out by leveraging a botnet, an interconnected group of compromised computers, to send a deluge of requests to the targeted service. A subset of the infected devices, known as "master machines," controls and directs the botnet (Halladay *et al.*, 2022).

2. Methodology

We used SVM, NaiveBayes, Decision Tree Algorithm and Random Forest algorithms. A supervised machine learning approach was used to solve regression and classification issues. However, it is typically applied to classifying data (Priya *et al.*, 2020). When compared to more complex algorithms, the Naive Bayes classifier can be lightning quick. Because the class distributions are no longer entangled, each may be studied separately as a univariate distribution.

2.1. Decision Tree Algorithm

The goal is to develop a simple decision rule-based model from the data attributes that can accurately forecast a target value. This method's benefits include its readability and solvability, as well as its applicability to issues with multiple outputs.

2.2. Random Forest

Each simple decision tree in this model is divided at random, using just some of the characteristics available for that tree's splits. Additionally, the training data used to construct the trees is chosen at random.

In a random forest, each tree takes in information from a different subset of the whole data set during the training phase.

K-Nearest Neighbor (K-NN) is a supervised learning-based machine learning algorithm that is extensively used and among the simplest (Elejla *et al.*, 2019). It was developed earlier than most others. Upon receiving a new data point, the K-Nearest Neighbors algorithm utilises its extensive memory to assign a label. Thus, new data can be efficiently classified as soon as it becomes available by applying the K-NN approach.

The open-source gradient boosted trees technique XGBoost (eXtreme Gradient Boosting) is widely used because of its effectiveness and popularity (Sindian, 2020). Aiming to enhance the prediction of a target variable by averaging the predictions of multiple less complex models, gradient boosting is a supervised learning technique. AdaBoost, or adaptive boosting, is a useful ensemble method for machine learning practitioners. AdaBoost's most popular implementation of an estimator is the one-level decision tree, often known as a tree with a single split.

3. Results and Discussion

All algorithms are trained on the 0 and 1 class labels, and the resulting scenario, named Scenario A or Binary classification, boasts an accuracy of more than 99%. In this scenario, benign is labelled as 0, and malicious attacks are labelled as 1. The accuracy of time-based classification models ranges from 40% to 70%, with the author extracting 25 features and then using all 12 (although in our dataset we identified 10 attacks) attacks to retrain all the models.

4. Conclusion

On the CICDDoS 2019 dataset, we compared the results of eight ML methods and one deep learning model across two

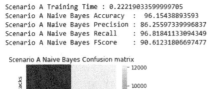

```
Scenario A Training Time : 0.22219033599999705
Scenario A Naive Bayes Accuracy  :  96.15438893593
Scenario A Naive Bayes Precision : 86.25597339996837
Scenario A Naive Bayes Recall    : 96.81841133094349
Scenario A Naive Bayes FScore    : 90.61231806697477
```

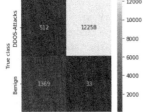

Figure 2: Naive Bayes was trained on the Scenario A dataset, and the resulting accuracy was 96%

```
Scenario B Training Time : 0.0791921579999979
Scenario B Naive Bayes Accuracy  : 15.932825289302851
Scenario B Naive Bayes Precision : 17.53084610402864
Scenario B Naive Bayes Recall    : 13.9413378891148871
Scenario B Naive Bayes FScore    : 9.281948737640382
```

Figure 3: Naive Bayes accuracy as 15% on all attack with 25 features

```
print("Scenario A Training Time = "+str(t_time))
predict = svm_cls.predict(scenario_A_X_test)
calculateMetrics("Scenario A SVM", predict, scenario_A_y_test, ['Benign', 'DDoS-Attacks'])

Scenario A Training Time : 25.559124011200001
Scenario A SVM Accuracy  :  99.8165306556004
Scenario A SVM Precision : 99.3619821806788
Scenario A SVM Recall    : 99.6124577819185B
Scenario A SVM FScore    : 99.4067790031021S
```

Figure 4: Accuracy and other values on scenario 'A' using SVM

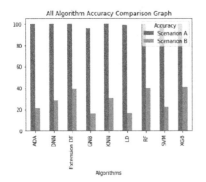

Figure 5: Accuracy graph of all algorithms where blue line represents Scenario

use-cases. Various types of DDoS attacks were classified in both Scenarios A and B. In each scenario, models were initially trained on the 70-feature control dataset. Next, the proposed set of 25 temporal features was used to train the models. For both the temporal and regulating features in Scenario A, the top five models obtained accuracy of more than 98% and more than 99%, respectively. These findings are on par with those of the best DDoS classifiers

0	Naive Bayes Scenario A	86.255973	96.818411	90.612318	96.154389	0.222190
1	Naive Bayes Scenario B	17.530846	13.941338	9.281949	15.932825	0.079192
2	SVM Scenario A	99.361902	96.612488	90.486770	99.816540	25.559125
3	SVM Scenario B	29.378194	20.249062	17.304887	22.360994	7.433814
4	KNN Scenario A	99.543677	99.386317	99.464837	99.809483	32.289757
5	KNN Scenario B	33.584546	34.270966	32.634314	30.600000	0.972629
6	LDA Scenario A	95.714373	98.163196	96.897324	98.863957	0.876817
7	LDA Scenario B	29.278873	17.125568	14.127456	16.680000	0.161966
8	RF Scenario A	99.904984	99.778189	99.841484	99.943591	3.285406
9	RF Scenario B	46.068503	46.229789	44.755900	39.880000	13.765843
10	Ada Boost Scenario A	90.952541	99.889095	99.920792	99.971775	9.215987
11	Ada Boost Scenario B	14.086695	26.549190	16.550250	21.120000	3.011100
12	XGBoost Scenario A	99.889246	99.952509	99.920893	99.971775	13.078317
13	XGBoost Scenario B	48.204841	47.227196	45.496705	41.020000	103.426658
14	DNN Scenario A	99.595879	99.532898	99.564357	99.844764	10.525369
15	DNN Scenario B	41.555822	27.847825	28.180984	28.540000	78.211782
16	Extension Decision Tree Scenario A	100.000000	100.000000	100.000000	100.000000	0.328703
17	Extension Decision Tree Scenario B	45.815878	45.329218	44.209272	39.240000	0.763604

Figure 6: Accuracy, precision, recall, FSCORE and training time for all algorithms and in all algorithms Decision Tree got 100% accuracy

```
#predicting attacks on test data
dt_cls = DecisionTreeClassifier()
dt_cls.fit(scenario_A_X_train, scenario_A_y_train)
testData = pd.read_csv("testDataTestData.csv") #keeping test data

columns = testData.columns
types = testData.dtypes.values
index = 0
for i in range(len(types)):
    name = types[i]
    if name == 'object' and columns[i] != 'Label'
        testData[columns[i]] = pd.Series(label_encoder[index].fit_transform(testData[columns[i]].astype(str)))
        index = index + 1
testData.fillna(0, inplace = True)
tm=testData = testData.values
testData = testData[:,0:testData.shape[1]-1]
test = scaler.transform(testData)
predict = dt_cls.predict(test)#performing prediction on test data
for i in range(len(predict)):
    if predict[i] == 0:
        print("Test Data = "+str(testData[i])+" ----> Predicted AS Normal Traffic")
    else:
        print("Test Data = "+str(testData[i])+" ----> Predicted AS DDOS Attack Traffic")
```

Figure 7: Loading test data and then predicting traffic and normal or attack

```
Test Data = [ 3.9107e+04  1.6000e+01  0.0000e+00  6.2806e+04  0.0000e+00  6.1197e+04
  6.0000e+00  2.6000e+01  1.0000e+00  2.0000e+00  0.0000e+00  0.0000e+00
  0.0000e+00  0.0000e+00  0.0000e+00  0.0000e+00  0.0000e+00  0.0000e+00
  0.0000e+00  0.0000e+00  0.0000e+00  1.0000e+00  0.0000e+00  1.0000e+00
  0.0000e+00  0.0000e+00  1.0000e+00  1.0000e+00  0.0000e+00  0.0000e+00
  1.0000e+00  1.0000e+00  0.0000e+00  0.0000e+00  0.0000e+00  0.0000e+00
  0.0000e+00  0.0000e+00  0.0000e+00  0.0000e+00  0.0000e+00  4.0000e+01
  0.0000e+00  2.0000e+00  0.0000e+00  0.0000e+00  0.0000e+00  0.0000e+00
  0.0000e+00  0.0000e+00  0.0000e+00  0.0000e+00  0.0000e+00  0.0000e+00
  1.0000e+00  0.0000e+00  0.0000e+00  2.0000e+00  0.0000e+00  0.0000e+00
  0.0000e+00  5.8400e+03 -1.0000e+00  0.0000e+00  2.0000e+00  0.0000e+00
  0.0000e+00  0.0000e+00  0.0000e+00  0.0000e+00  0.0000e+00  0.0000e+00
  0.0000e+00  0.0000e+00] ====> Predicted AS Normal Traffic
Test Data = [8.66990000e+04 6.00000000e+00 0.00000000e+00 3.99470000e+04
 0.00000000e+00 6.11970000e+04 6.00000000e+00 2.70000000e+00
 1.00000000e+00 2.00000000e+00 0.00000000e+00 0.00000000e+00
 0.00000000e+00 0.00000000e+00 0.00000000e+00 0.00000000e+00
 0.00000000e+00 0.00000000e+00 0.00000000e+00 0.00000000e+00
```

Figure 8: Square bracket containing test data and then after the symbol we can see prediction as Normal or attack

```
0.00000000e+00 0.00000000e+00 0.00000000e+00 0.00000000e+00
0.00000000e+00 1.00000000e+00 0.00000000e+00 2.00000000e+00
0.00000000e+00 6.40000000e+01 2.43000000e+02 0.00000000e+00
3.20000000e+01 0.00000000e+00 0.00000000e+00 0.00000000e+00
0.00000000e+00 0.00000000e+00 0.00000000e+00 0.00000000e+00
0.00000000e+00 0.00000000e+00] ====> Predicted AS DDOS Attack Traffic
Test Data = [1.40000000e+02 2.90000000e+01 1.00000000e+00 5.04880000e+04
2.00000000e+00 4.43000000e+02 6.00000000e+00 1.00000000e+00
6.08821130e+07 2.70000000e+01 2.70000000e+01 2.46000000e+03
6.61200000e+03 5.81000000e+02 0.00000000e+00 9.11111111e+01
1.97808247e+02 3.01400000e+03 0.00000000e+00 2.44888889e+02
6.82383568e+02 1.49009112e+02 8.86958998e-01 1.14872043e+06
3.19590780e+06 1.00147220e+07 1.00000000e+00 6.08821130e+07
2.34162242e+06 4.32497758e+06 1.01047460e+07 2.00000000e+00
6.07905520e+07 2.33889815e+06 4.33537611e+06 1.01061250e+07
1.00000000e+00 0.00000000e+00 0.00000000e+00 0.00000000e+00
0.00000000e+00 5.64000000e+02 7.00000000e+00 4.43479499e-01
4.34479499e+01 0.00000000e+00 3.01400000e+03 1.64945455e+02
4.99464503e+02 2.49465090e+05 0.00000000e+00 0.00000000e+00
0.00000000e+00 0.00000000e+00 1.00000000e+00 0.00000000e+00
```

Figure 9: Traffic predicted as DDOS

found in the current research. Using solely the recommended time-based features resulted in a modest 1.62% drop in accuracy across the nine models and a sizable 36.28% reduction in median training time. Scenario B's DDoS attacks were categorised by protocol, with the most prevalent ones being SYN Flood, DNS, SNMP, SSDP, LDAP, UDP Flood, MSSQL, PORTMAP, NTP, NETBIOS, TFTP, and UDP-Lag. With accuracy ratings for control features of 74.08% and time-based characteristics of 69.05%, XGB was the best-performing model. Using solely temporal variables, the nine classifiers saw a median drop in accuracy of 7.43% and a drop in training time of 25.29%. Only KNN used hyper-parameter optimisation across all of our tests. Using the CICDDoS2019 dataset, we compared the performance of eight machine learning methods and one deep learning model in two use scenarios. Both Scenarios A and B included descriptions of several DDoS attack types. Models were first trained on the 70-feature control dataset in each case. Subsequently, the suggested collection of twenty-five temporal features was used to train the models. In Scenario A, the top five models achieved over 98% accuracy for the temporal characteristics and over 99% accuracy for the regulating aspects. These findings are on par with those of the best DDoS classifiers found in the current research. Using solely the recommended time-based features resulted in a modest 1.62% drop in accuracy across the nine models and a sizable 36.28% reduction in median training time.

References

Callado, A., Kamienski, C., Szabo, G., Gero, B. P., Kelner, J., Fernandes, S., and Sadok, D. (2009). "A survey on internet traffic identification." *IEEE Communications Surveys and Tutorials*, 11(3), 37– 52, 3rd Quart, doi: 10.1109/SURV.2009.090304.

Cil, A. E., Yildiz, K., and Buldu, A. (2021). "Detection of DDoS attacks with feed forward based deep neural network model." *Expert Systems with Applications*, 169, , Art. no. 114520, doi10.1016/j.eswa.2020.114520.

Halladay, J., Cullen, D., Briner, N., Warren, J., Fye, K., Basnet, R., Bergen, J., and Doleck, T. (2022). "Detection and characterization of DDoS attacks using time-based features." *IEEE Access*, 10, 49794-49807, doi: 10.1109/ACCESS.2022.3173319.

Jena, S., Mohapatra, S., Jena, G., and Jena, S.(2022). "Open-Source Cloud Infrastructure & Application Monitoring and Alerting System." *NeuroQuantology*, 20(10): 7909.

Lashkari, A. H., Gil, G. D., Mamun, M. S. I., and Ghorbani, A. A. (2017). "Characterization of tor traffic using time-based features." In *Proc. 3rd Int. Conf. Inf. Syst. Secur. Privacy*, pp. 253–262, doi: 10.5220/0006105602530262.

Mirsky, Y., Doitshman, T., Elovici, Y., and Shabtai, A. (2018) "Kitsune: An ensemble of autoencoders for online network intrusion detection." In *Proc. Netw. Distrib. Syst. Secur. Symp.*

Mohammed, S. S., Hussain, R., Senko, O., Bimaganbetov, B., Lee, J., Hussain, F., Kerrache, C. A., Barka, E., and Bhuiyan, M. Z. A. (2018). "A new machine learning-based collaborative DDoS mitigation mechanism in software-defined network.." In *Proc. 14th Int. Conf. Wireless Mobile Comput., Netw. Commun. (WiMob)*, Oct., pp. 1–8, doi: 10.1109/WIMOB.2018.8589104.

Murthy, B S N., Srinivas, K., Jena, S., Sandeep, A V L G., Naidu, M S., Ravi, M., and Sudheer, K. (2022). Network intrusion detection using supervised machine learning technique with feature selection. *Mathematical Statistician and Engineering Applications*, 71(4): 5242-5262. https://doi.org/10.17762/msea.v71i4.1115

Sharafaldin, I., Lashkari, A. H., Hakak, S., and Ghorbani, A. A. (2019). "Developing realistic distributed denial of service (DDoS) attack dataset and taxonomy." In *Proc. IEEE 53rd Int. Carnahan Conf. Secur. Technol. (ICCST)*, Oct., pp. 1–8. [Online]. Available: https://ieeexplore.ieee.org/abstract/document/8888419.

Salahuddin, M. A., Bari, M. F., Alameddine, H. A., Pourahmadi, V., and Boutaba, R. (2020). "Time-based anomaly detection using autoencoder." In *Proc. 16th Int. Conf. Netw. Service Manage. (CNSM)*, Nov., pp. 1–9, doi: 10.23919/CNSM50824.2020.9269112.

Ying, X. (2019). "An overview of overfitting and its solutions." *Journal of Physics: Conference Series*, 1168, Feb., Art. no. 022022, doi: 10.1088/1742- 6596/1168/2/022022.

Secure Guard

Deep Shield - A Deep Learning Powered Defense Against DoS Attacks

Gunamani Jena[1], B. S. N. Murthy[2], P. Devabalan[1], G. Aparna[3],
A. L. Prasanna[4], B. S. S. Kiran[5], G. E. V. V. S. Manikanta[6]

[1,2,3]CSE, BVC Engineering College, Odalarevu, India
[3,5,6]Research Scholars, CSE, BVC Engineering College, Odalarevu. Andhra Pradesh.
E-mail: drgjena@gmail.com, bsnmurthy2012@gmail.com, devabalanme@gmail.com

Abstract

A hybrid intrusion detection and prevention system (IDS/IPS) designed specifically to identify and neutralise distributed denial of service (DDoS) attackers using a Deep Learning algorithm is implemented in this research paper, dique. The web server's IDS/ IPS system applies the suggested deep learning model to categorise incoming packets into two groups: benign, which represents regularly occurring traffic, and malignant, which contains probable threats, thereby preventing attacks involving denial-of-service. The Dique's Graphical User Interface (GUI) allows us to toggle between the IDS and IPS modes of operation, as it also provides both a graphic and textual representation of the details of recorded and classified packets. With a multilayered Deep Feed Forward Neural Network trained on the CICD - DoS2019 dataset, our proposed DoS attack categorisation model achieved an accuracy of 99.4%.

Keywords: Dique, hybrid intrusion detection and prevention system, DoS, IDS, FFNN

1. Introduction

Intrusion detection refers to the practise of monitoring a network for signs of malicious activity. The security of the data stored on the computer system as well as its availability could be compromised by such an intrusion. Experts in this industry typically employ a variety of traffic-analysis tools and methods to safeguard data and prevent damage from malicious actors. Cyber crime is growing at an exponential rate as the internet grows and as more people try to secure their computers. As a result, professionals have realised they need more advanced technologies to monitor for intrusion (Jena *et al.*, 2022) and respond to it after the fact rather than before. Typical methods for detecting intrusions include anomaly-based detection, which looks for intruders by continuously observing, collecting, and analyzing network packets that are then categorised as malicious or benign, and signature-based detection, which uses a signature database built from information about previous attacks.

Machine Learning (ML) is one technique used to classify outliers, as ML enables computers to learn without explicit training. Using a large dataset (dataset), it trains an algorithm to identify intrusion attempts (Murthy *et al.*, 2022), keeping the network secure. This algorithm's categorisation skills are continually refined

Figure 1: Architecture

through training on newly processed data (Xin *et al.*, 2018). Although ML appears to be a promising technique, there are a number of limitations that must be considered. For example, it might mistakenly classify malicious data packets as innocuous and allow them to be processed for learning, substantially undermining the algorithm. Additionally, it may incorrectly classify unusual packets as harmless. Another problem is that the sheer amount of data flowing through the network may make it really difficult to analyze the packets in real-time, which may impact the computer system's overall performance. Experts in Artificial Intelligence (AI) are looking into Deep Learning (DL) methods in order to improve the ML techniques for intrusion detection (Apruzzese *et al.*, 2018) because of the pressing necessity to do so. DL is a subarea of ML.

Figure 1 shows that DL algorithm performance increases with data volume, while ML algorithm performance stabilises over time.

2. Methodology

Many CNN-based Deep Learning algorithms have been introduced as a solution to the problem of detecting new attacks, but so far none of them have provided a LIVEGUI for classifying attacks; rather, they are trained to detect on TRAIN and TEST data only. In order to capture LIVE traffic and classify attacks (Unal and Hacibeyoglu, 2018), the author of the proposed project is creating a new GUI-based Dique tool based on Feed Forward Neural Network. The author has designed a multilayer Convolutional Neural Network (CNN) with DENSE and DROPOUT layers to filter out noise in the dataset and

ensure that only useful features are used during model training.

3. Algorithms

Convolutional Neural Networks (CNNs), a type of Deep Learning architecture, is widely used for image categorisation and recognition. Included in its construction are layers of Convolution, Pooling, and Full Connectivity. The input image is filtered by the Convolutional layer so that features can be extracted, the image is down-sampled by the Pooling layer to speed things up, and the fully connected layer provides the prediction. Using back propagation and gradient descent, the network discovers its best filters. The principal component analysis (PCA) features selection approach will select crucial features from the dataset, and the extension CNN will be trained and tested on these features to achieve high accuracy with minimal computational effort.

4. Proposed System

To detect new attacks many Machine Learning algorithms were introduced but they work best on SMALL dataset and unable to detect all new attack variants and to overcome from such issue many CNN based Deep Learning algorithms are introduced but none will provide LIVE GUI to classify such attacks and then trained to detect on TRAIN and TEST data only. So in propose project author is developing new GUI based Dique tool which is developed on Feed Forward Neural Network and can capture LIVE traffic and then classify attacks. To train deep neural network author has design MULTI LAYER CNN with DENSE and DROPOUT layers to remove irrelevant features from the dataset and used only relevant features while training model.

5. Results and Discussion

The proposed CNN and extension CNN with PCA were implemented and the results were compared. The computation times for both were compared. For proposed CNN it is 3.370148 where as for CNN with PCA it was 2.287915. The accuracy, precision graph for CNN with PCA was better. LIVE test data was taken and then using CNN we were able classify records as BENIGN or attack. Using TEST data the model predicted output as DRDOS attack.

In the below Figure 2, x-axis represents the name of attack, where as the y-axis represents count of attack and Figure 3, x-axis represents the Predicted labels, where as y-axis represents the true labels.

6. Conclusion

Deep Learning outperforms conventional ML and DL neural network techniques in recognising DoS assaults (Kasongo and Sun, 2020). This study discovered that the DL techniques of Feed Forward CNN and PCA (Principal Component Analysis) yielded the highest accurate results when training models to identify DoS assaults.

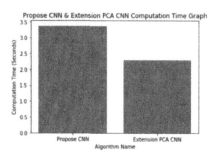

Figure 4: Defining function to calculate accuracy and precision values

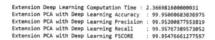

Figure 5: Extension of PCA

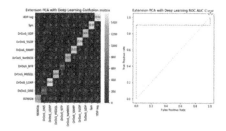

Figure 6: Confusion Matrix and Graphs for extension PCA

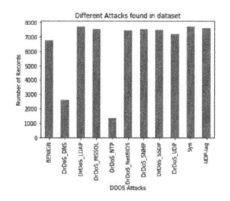

Figure 2: Finding count of each attack in dataset and then plotting graph

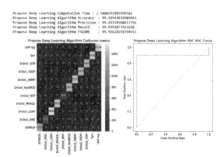

Figure 3: Confusion matrix of our DL model

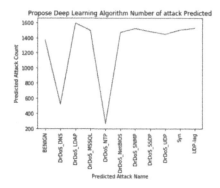

Figure 7: Showing computation time comparison between proposed and extension algorithm

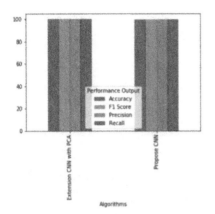

Figure 8: Accuracy, precision graph for both algorithms in different color bars

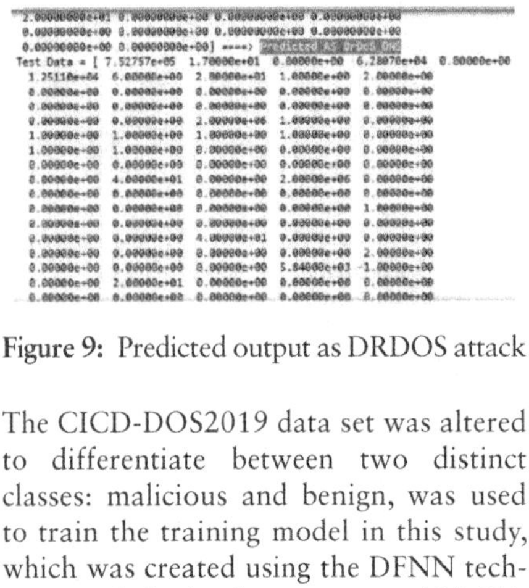

Figure 9: Predicted output as DRDOS attack

The CICD-DOS2019 data set was altered to differentiate between two distinct classes: malicious and benign, was used to train the training model in this study, which was created using the DFNN technique and obtained an accuracy of 99.4%.

References

Amma, B. N. G., and Selvakumar, S. (2019). "Deep radial intelligence with cumulative incarnation approach for detecting denial of service attacks," *Neurocomputing*, 340, 294-308.

Apruzzese, G., Colajanni, M., Ferretti, L., Guido, A., and Marchetti, M. (2018). On the effectiveness of machine and deep learning for cyber security. *Presented at the 10th Int. Conf. CyberConflict.*

Chiba, Z., Abghour, N., Moussaid, K., ElOmri, A., and Rida, M. (2019). "Intelligent approach to build a deep neural network based IDS for cloud environment using combination of machine learning algorithms." *Computer Security*, 86,291-317.

Islam, A. B. M. A. A., and Sabrina, T. (2009). Detection of various denial of service and distributed denial of service attacks using RNN ensemble. *Presented at the 12th Int. Conf. Comput. Inf. Technol.*, Dhaka, Bangladesh.

Jena, S., Mohapatra, S., Jena, G., and Jena, S. (2022). "Open-Source Cloud Infrastructure & Application Monitoring and Alerting System." *NeuroQuantology; Bornova Izmir*, 20(10), 7909-7916. doi:10.14704/nq.2022.20.10. NQ55779.

Kasongo, S. M. and Sun, Y. (2020). "A deep long short-term memory based classifier for wireless intrusion detection system." *ICT Express*, 6(2), 98-103.

Kimand, T. Y. and Cho, S. B. (2018). "Web traffic anomaly detection using C-LSTM neural networks." *Expert Systems with Applications*, 106, 66-76.

Liu, H., Lang, B., Liu, M., and Yanb, H. (2019). "CNN and RNN based payload classification methods for attack detection." *Knowledge-based Systems*, 163, 322-341.

Murthy, B. S. N., Srinivas, K., Jena, S., Sandeep, A V L G., Naidu, M S., Ravi, M., and Sudheer, K. (2022). Network Intrusion Detection using Supervised Machine Learning Technique with Feature Selection. *Mathematical Statistician and Engineering Applications*, 71(4): 5242-5262. https://doi.org/10.17762/msea.v71i4.1115

Priyadarshini, R. and Barik, R. K. (2022). "A deep learning based intelligent framework to mitigate DDoS attack in fog environment." *Journal of King Saud University Computer and Information Sciences*, 34(3), 825-831.

Sharafaldin, I., Lashkari, A. H., Hakak, S., and Ghorbani, A. A. (2019). "Developing realistic distributed denial of service (DDoS) attack dataset and taxonomy." In *Proc. Int. Carnahan Conf. Secur. Technol.*, 1-8.

Thilina, A., Attanayake, S., Samarakoon, S., Nawodya, D., Rupasinghe, L., Pathirage, N., Edirisinghe, T., and Krishnadeva, K. (2016). "Intruder detection using deep learning and association rule mining." *Presented at the IEEE Int. Conf. Comput. Inf. Technol.*

Unal, A. S. and Hacibeyoglu, M. (2018). "Detection of DDOS attacks in network traffic using deep learning." *Presented at the Int. Conf. Adv. Technol., Comput. Eng. Sci. (ICATCES)*.

Xin, Y., Kong, L., Liu, Z., Chen, Y., Li, Y., Zhu, H., Gao, M., Hou, H., and Wang, C. (2018). "Machine learning and deep learning methods for cybersecurity." *IEEE Access*, 6, 35365-35381

Xu, C., Shen, J., Du, X., and Zhang, F. (2018). "An intrusion detection system usinga deep neural network with gated recurrent units." *IEEE Access*, 6, 48697-48707.

Yadav, S. and Subramanian, S. (2016). "Detection of application layer DDoS attack by feature learning using stacked Auto Encoder." *Presented at the Int. Conf. Comput. Techn. Inf. Commun. Technol.(ICCTICT)*.

Zargar, S. T., Joshi, J., and Tipper, D. (2013). "A survey of defense mechanisms against distributed denial of service (DDoS) flooding attacks." *IEEE Communications Surveys and Tutorials*, 15(4), 2046-2069.

Marine Biodiversity Conservation through Computer Vision and Deep Learning

Pradumn Kumar, Praveen Kumar Shukla

Department of Computer Science Engineering, Babu Banarasi Das University, Lucknow, India

E-mail: pradumnyadav18@gmail.com, drpraveenumarshukkla@gmail.com

Abstract

The underwater habitat, home to many fish species, is important for marine biodiversity research. Protecting biodiversity requires accurate fish classification, counting, and localisation. This research examines how computer vision and deep learning can automate these operations and overcome underwater difficulties. The project seeks to simplify underwater environments, occlusions, and lighting. The hybrid model was verified with 95% classification, 97% counting, and.78 mAP scores. The initiative provides tools and insights to examine and protect this endangered ecosystem for future generations.

Keywords: Computer vision, deep learning architecture, classification, counting, localisation, Resnet50, Mobilenet

1. Introduction

Fish species are essential to underwater habitats and the ecological equilibrium of Earth's seas. However, climate change, overfishing, and habitat destruction threaten marine life. Ecological studies need fish species identification, counting, localisation, and segmentation. Automating these tasks requires computer vision and deep learning. This research examines innovative methods for fish identification, counting, and location in underwater images. Fish classification helps researchers identify endangered species, analyze species variation, and track environmental changes on fish populations. Environmental tracking, demographic studies, and resource management require automated counting. The proposed method is evaluated on a Deepfish dataset with diverse marine environments and species. This study advances computer vision technology and benefits marine conservation and research.

2. Literature Review

Smartphone disease diagnosis using segmentation algorithms (Xu *et al.*,2021). With 88.87% accuracy, Random Forest fared well. R-CNNs identify and quantify fish underwater with 87.44% and 80.02% F-Scores (Salman, Ahmad *et al.*, 2020). The hybrid neural network-based (Zhang *et al.*2022) automated offshore salmon farming fish counting method exhibits 95.06% accuracy and 0.99 estimation-data correlation. In occulted river fish species, deep learning-based segmental analysis was used to recognise fish bodies and accurately calculate GIFT fish biomass. YOLOv3 and Mask-RCNN methods were used to test deep learning models for fish detection, categorisation, and tracking using visual audio and video camera data from eight Ocqueoc (Kandimalla *et al.*2022). A robotic eye sensor and a 97.4% accurate Norfair-based YOLOv4 model (Zhang *et al.*, 2020)

Chaper 5 DOI: 10.1201/9781003581215

investigated underwater optical recordings from the Columbia River's Wells Dam fish ladder. Efficient-Net trained on ImageNet and QuT Fish Database (Qiu *et al.*2018) enhances transfer learning prediction accuracy with residual connection architecture and position aware blocks.

3. Methodology

3.1. Dataset

In 2020, DeepFish (Saleh *et al.*) collected 40,000 underwater pictures from 20 tropical Australian marine settings with point-level and segmentation labels to improve fish analysis. For computer vision-based fish analysis, the dataset was preprocessed and updated for uniformity and diversity. 70% of data is split to train, 15% to validate, and 15% to test machine learning models.

3.2. Resnet50

ResNet-50 (Saleh *et al.*, 2020) is a flexible machine vision architecture for image classification and high-level features. The object location, counting, and picture categorisation layers are task specific. ResNet-50 may perform numerous functions as a neural network by merging these paths. Localisation, item counts, and class probability improve with complete results.

3.3. MobilenetV2 Detection and classification Model

MobileNetV2 (Fitrianah *et al.*,2022), an adaptive neural network, detects objects and classifies images because to its depthwise separable convolutions, rapid building, feature extractor convolutional layers, and light It is adaptable, efficient, and compatible with SSD and YOLO object identification frameworks and other detection and classification frameworks.

3.4. Proposed Hybrid Model

The hybrid neural network DenseNet-121 (Lian *et al.*, 2021) categorises, counts, and

Figure 1: Flowchart for proposed Hybrid Model

locates images. Advanced picture feature extraction and layer freezing protect training data in the model. Specialised layers detect visual features and determine categorisation probabilities. We count and localise scene items using convolutional layers and sigmoid activation. A single output layer predicts classes, numbers, and item positions. Ideal loss functions and model compilation allow the model to optimise parameters for various tasks. This hybrid architecture allows multi-task learning in complex visual tasks (Figure 1).

3.4.1. Features are extracted from input photos using the DenseNet-121 architecture

Convolutional Layers: DenseNet-121 uses convolutional layers to generate multi-scale feature maps from input images (I), denoted by (F_{conv}^{i+1}) and the corresponding convolutional operation (Conv).

$$F_{Conv}^{i+1} = Conv(F_{conv}^i) \qquad (1)$$

Dense Connections:

$$F_{Concat}^{i+1} = Concat(F_{concat}^i, F_{Conv}^{i+1}) \qquad (2)$$

High-Level Feature Map:
To identify intricate patterns and subtle representations in images by combining the feature map across all network layers.

$$F_{final} = DenseNet\text{-}121(I) \qquad (3)$$

3.4.2. Layers of the Base Model (Frozen)

The base model's weights are frozen during training to train only necessary layers, preserve feature extraction, and use a pretrained model.

3.4.3. Weight Update during Training(W)

$$W_{new} = W_{old} - \alpha \cdot \frac{\partial Loss}{\partial W_{old}} \qquad (4)$$

Here, Loss is the loss function, α is the learning rate.

3.4.4. Layer Freezing

$$\frac{\partial Loss}{\partial W_i} = 0 \qquad (5)$$

3.4.5. Layers of Customised Classification

Layer for Global Average Pooling (GAP) classification layer:

$$F_{GAP} = 1 / H * W \sum \sum_{j=1}^{n} Ffinal(i,j) \qquad (6)$$

The final feature map's height (H) and width (W) are shown here.

Dense Layer Output:

$$W_{Dense} = ReLU(F_{GAP} + b_{Dense}) \qquad (7)$$

Weights W_dense, bias term b_dense, and Rectified Linear Unit (ReLU) activation function.

Output Layer:

$$P(yi) = \frac{e^{zi}}{\sum_{j=1}^{k} e^{zi}}, \ for \ i1,2,.......k \qquad (8)$$

The input to the softmax function for class i is denoted here as z_i.

3.4.6. Layers of Customised localisation

Layers of Convolution (localisation layer): Base model output to determine object properties.

$$F_{loc} = Conv(F_{final}) \qquad (9)$$

A convolutional layer with a sigmoid activation function predicts item presence using the fundamental model's final feature map, F_final. Equationally,

$$Sigmoid(x) = \frac{1}{1 + e^{-x}} \qquad (10)$$

$F_{loc}, sigmoid$, shows the probability of discovering an object in each area.

3.4.7. Layers of Customised Counting

Flattening converts 2D vector from localisation layer into 1D vector using Flatten function.

$$F_{flattened} = Flatten(F_{loc, sigmoid}) \qquad (11)$$

show object count from localisation output after flattening Figure 2.

$$Count = W_{count} \cdot F_{flattened} + b_{count} \qquad (12)$$

The weights are denoted by W_count, and the bias by b_count.

4. Results

The study compares popular deep learning models ResNet50, MobileNetV2, and DenseNet-121. ResNet50's average precision (mAP) of 0.75 showed robust item detection and categorisation. MobileNetV2 counted and identified objects with 89% accuracy. DenseNet-121 outperformed baselines in object detection, picture categorisation, and counting with 95% and 97% accuracy, respectively. MobileNetV2 can replace ResNet50, and DenseNet-121 is good for item counting (see Table 1 and Figure 3). True Positives (TP), False

Figure 2: Image for fish counting using extra layers (Saleh *et al.*, 2020).

Table 1: Comparison table of different models with their results for classification, counting & mAP(Mean Average Precision)

Architecture	Classification	Counting	mAP
Resnet50	90%	92%	0.75
MobileNet v2	89%	88%	0.72
Proposed Hybrid	95%	97%	0.78

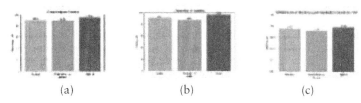

(a)   (b)   (c)

Figure 3: Comparison Graph for (a) classification, (b) counting and (c) mAP

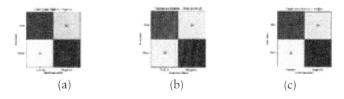

(a)   (b)   (c)

Figure 4: (a) Confusion Matrix for Hybrid Model (b) MobilenetV2 (c) ResNet50

Positives (FP), False Negatives (FN), True Negatives (TN) confusion matrix data are shown Figure 4.

5. Conclusion

Three marine life analysis deep learning models—ResNet50, MobileNetV2, and DenseNet-121—were tested on DeepFish. ResNet50 excelled in species recognition, while MobileNetV2 excelled at photo categorisation and item counting. DenseNet-121 excelled in counting and detecting objects. Future study on fine-tuning, transfer learning, semantic segmentation, and real-time monitoring technologies could improve marine ecosystem monitoring, conservation, and understanding.

References

Abinaya, N. S., D. Susan, and Rakesh Kumar, S.. (2022). "Deep learning-based segmental analysis of fish for biomass estimation in an occulted environment." *Computers and Electronics in Agriculture*, 197: 106985.

Chang, Chin-Chun, Yen-Po Wang, and Shyi-Chyi Cheng. (2021). "Fish segmentation in sonar images by mask R-CNN on feature maps of conditional random fields." *Sensors*, 21(22): 7625.

Ding, Guoqing, Yan Song, Jia Guo, Chen Feng, Guangliang Li, Bo He, and Tianhong Yan. (2017). "Fish recognition using convolutional neural network." In *OCEANS 2017-Anchorage*, pp. 1-4. IEEE.

Fitrianah, Devi, Kristien Margi Suryaningrum, Noviyanti Tri Maretta Sagala, Vina Ayumi, and Siew Mooi Lim. (2022). "Fine-Tuned MobileNetV2 and VGG16 Algorithm for Fish Image Classification." In *2022 International Conference on Informatics, Multimedia, Cyber and Information System (ICIMCIS)*, pp. 384-389. IEEE.

Kandimalla, Vishnu, Matt Richard, Frank Smith, Jean Quirion, Luis Torgo, and Chris Whidden. (2022). "Automated detection, classification and counting of fish in fish passages with deep learning." *Frontiers in Marine Science*, 8: 2049.

Lian, Luya, Tianer Zhu, Fudong Zhu, and Haihua Zhu. (2021). "Deep learning for caries detection and classification." *Diagnostics*, 11(9): 1672.

Lucas, Martyn C. and Etienne Baras. (2000). "Methods for studying spatial behaviour of freshwater fishes in the natural environment." *Fish and Fisheries,* 1(4): 283-316.

Salman, Ahmad, Shoaib Ahmad Siddiqui, Faisal Shafait, Ajmal Mian, Mark R. Shortis, Khawar Khurshid, Adrian Ulges, and Ulrich Schwanecke. (2020). "Automatic fish detection in underwater videos by a deep neural network-based hybrid motion learning system." *ICES Journal of Marine Science*, 77(4): 1295-1307.

Saleh, Alzayat, Issam H. Laradji, Dmitry A. Konovalov, Michael Bradley, David Vazquez, and Marcus Sheaves. (2020). "A realistic fish-habitat dataset to evaluate algorithms for underwater visual analysis." *Scientific Reports*, 10(1): 14671.

Qiu, Chenchen, Shaoyong Zhang, Chao Wang, Zhibin Yu, Haiyong Zheng, and Bing Zheng. (2018). "Improving transfer learning and squeeze-and-excitation networks for small-scale fine-grained fish image classification." *IEEE Access*, 6: 78503-78512.

Xu, Xiaoling, Wensheng Li, and Qingling Duan. (2021). "Transfer learning and SE-ResNet152 networks-based for small-scale unbalanced fish species identification." *Computers and Electronics in Agriculture*, 180: 105878.

Zhang, Song, Xinting Yang, Yizhong Wang, Zhenxi Zhao, Jintao Liu, Yang Liu, Chuanheng Sun, and Chao Zhou. (2020). "Automatic fish population counting by machine vision and a hybrid deep neural network model." *Animals*, 10(2):364.

Comparison of Various Factors Responsible for Task Offloading in Edge Computing Framework

Binayak Sahoo and Alekha Kumar Mishra

Department of Computer Science and Engineering, NIT Jamshedpur, Jamshedpur, India
E-mail: 2020rsca005@nitjsr.ac.in, alekha.cse@nitjsr.ac.in

Abstract

Edge computing is a novel computing paradigm that calls for processing data at the network's edge and has received attention due to the growth of the Internet of Things (IoT) and the adoption of rich cloud services. Computation offloading is a technique for saving time and energy on mobile devices with limited resources by carrying out some activities on other devices or servers. It is viewed as a solution to the limited resource problem of IoT edge devices. The offloading decision problem is influenced by several factors. Due to this, many offloading tasks are unable to fulfil their intended goals. In this paper, we provide a thorough analysis of computational task offloading in edge computing, including offloading schemes and influencing elements. Finally, we have presented the techniques used for task offloading.

Keywords: Bandwidth, edge computing, IoT task offloading, latency

1. Introduction

Due to the limited computational resources of mobile devices, users initially do not experience the same level of satisfaction as users of desktop computers (Lähderanta et al., 2021). The IoT has emerged to enhance human life quality through the collection, processing, and dissemination of sensed and collected data (Nezami et al., 2021). With the emergence of recent applications such as open AI, realistic gaming, etc., a device requires a high-end computational environment (Lin et al., 2023). The cloud has demonstrated its immense capacity of limitless computing capabilities and on-demand resource supply (Guo et al., 2020). This centralised architecture is a limitation for the applications that need instant response. Edge computing emerged with the goal of bringing cloud computing capability to the network edge and addressing the issues that cloud computing by itself is unable to handle, such as bandwidth, latency, and connectivity (Lähderanta et al., 2021). Edge computing technologies are commonly used for computation at the edge.

This work includes the necessity of offloading. First, we discussed the literature survey on task offloading. Then, we have discussed various factors to be considered and various methods for server placement. Finally, this paper concludes.

2. Task Offloading Techniques in Edge Computing

Task offloading makes better use of the resources that are available and prevents

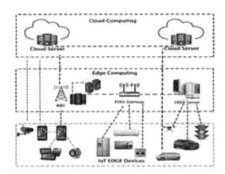

Figure 1: Offloading in edge computing

the overloading of a single node; it improves the performance of the entire system. In this section, we will briefly discuss the task-offloading mechanisms that have recently been documented in the literature. The majority of Mobile-Edge Computing researchers concentrate on finding the best ways to use the server's tremendous computing and storage capabilities (Zhang *et al.*, 2021).

Sabella *et al.* (2016) have proposed that the latency requirements be set to 1 1-ms, which will provide a better service. It makes it possible to host apps at the edge cloud close to the users and thus provides the shortest link between the applications. Zeinab Nezami *et al.* (2021) have presented an approach for decentralised load balancing. They presented a method for effective load balancing over the network and reducing execution costs without violating the deadline. IoT devices request fog nodes for the placement of their service requests. The service request is forwarded to the agent as per the feasibility. Guo *et al.* (2020) introduced a resource provisioning mechanism that dynamically allocates resources based on service usage. Their model incorporates load balancing, migration time, and cost considerations. Load balancing is achieved through a Back Propagation neural network. For application-aware compute offloading in side computing networks, Lin *et al.* (2023) have presented a non-convex branch-and-bound method with stochastic gain. The rules are evaluated, demonstrated to guarantee the necessities of packages and obtain the desired overall performance. By employing the Markov decision process

technique to choose an ideal offloading node, Yang *et al.* (2020) have suggested a way to improve the offloading time in MEC. This strategy uses two vital parameters: available bandwidth and the location of mobile devices. Compared to traditional methods, the findings show how well the suggested strategy reduces offloading time. Zhang *et al.* (2021) have proposed a load balancing and computation offloading (LBCO) scheme for a multiuser edge computing environment. This results in efficient redistribution of users to distribute the load evenly amongst small base stations and lower total communication costs.

3. Offloading Factors

The crucial phase in the offloading process is selecting whether to offload or not. The decision to offload, however, is influenced by a variety of circumstances.

3.1. Applications

Offloading computation benefits various applications like text editors, virus scanners, games, and object/ gesture identification. Managing these variations is challenging due to workload diversity, real-time constraints, and concurrency issues at the data level, impacting application dynamics

3.2. Network

Network quality, pivotal for computation offloading decisions, encompasses bandwidth, interference, link quality, and transmission issues. Parameters like bandwidth, intermittent connectivity, and device processor gauge network fluctuations. Bandwidth, affected by user location and congestion, is unpredictable in unattended, hostile environments. Network interference significantly impacts offloading systems, especially for latency-sensitive applications (Duan *et al.*, 2022).

3.3. Server Capability

When offloading computation, choosing the right server is vital. Factors like

Table 1: A summary of related surveys.

Contributor	Method Used	Solution	Simulator	Limitations
Xu et al. (2022)	Game theory	Combined use of Cloud and edge resources.	Java	Task priority not considered
Huang et al. (2021)	Lagrangian duality	Load balancing using mixed-integer linear programming problem.	CPLEX	Not suitable for different types of tasks
Guo et al. (2020)	Theory			
Lin et al. (2019)	Back Propagation neural network model	Load estimation in Advance	ARIMA model	It may not be fit for complicated task.
Lähderanta et al. (2021)	The power of two choices, theory	Reducing latency Maintain backup server	Latency-aware video analytics platform.	Domain Specific
Duan et al. (2022)				Old data set
Shen et al. (2021)	Euclidean distance for latency	Inter & Intra-Cluster load balancing	R based Software package	May not be suitable
	Cloudlet model	Kuhn–Munkres weighted bipartite graph matching Algorithm	Peer-Sim	for games, etc. Not fit for busy area
	Dynamic Edge server placement		Nanjing traffic Dataset	

processing power, resource availability, distance, and access protocol should be considered. Features such as elastic load balancing and resource management are crucial for efficient offloading. Random-access load balancing ensures users can connect to any available edge server (Lähderanta et al., 2021).

3.4. Time Constraints

The selection of an appropriate server for computation offloading is critical, considering factors like processing power, resource availability, distance, and access protocol. Random-access load balancing ensures user connectivity to any available edge server. Real-time applications, with strict time constraints, necessitate performance assurances, with soft real-time applications allowing some flexibility in task deadlines (Wang et al., 2022).

3.5. Latency

The latency between edge devices and servers is crucial for edge network performance. Operators need to evaluate the hop count and network location of potential edge sites to ensure they meet latency requirements for specific use cases (Sabella et al., 2016).

3.6. Technology used for Task Offloading

In our search across Google Scholar and Scopus using keywords like "offloading IoT Edge" and "task offloading edge computing" among others, we found 100 papers from 2022. It reveals that the majority of researchers favoured the deep reinforcement method for task offloading in their studies. The result is shown in Figure 2.

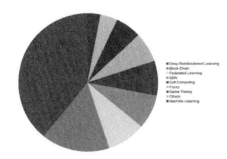

Figure 2: Technology used for task offloading in edge computing

4. Edge Server Placement Methods

4.1. Workload

The workload is calculated by parameters like how much time is taken to process the request and the total number of user connections (Lin *et al.*, 2023). The maximum number of connected users is considered as peak load. From peak load, average load and how much workload can be provided to a specific server can be calculated.

5. Conclusion

As data processing capabilities improve at the network's edge, services are increasingly shifting from the cloud. Offloading computation to the edge reduces the need to send data to the cloud. However, various real-world factors can hinder offloading strategies. Our paper identifies key factors influencing task offloading at edge devices. We found that deep reinforcement learning is particularly effective for optimising offloading in complex scenarios.

References

Duan, Zhenhua, Cong Tian, Nan Zhang, Mengchu Zhou, Bin Yu, Xiaobing Wang, Jiangen Guo, and Ying Wu. (2022). "A novel load balancing scheme for mobile edge computing." *Journal of Systems and Software*, 186 (April): 111195.

Guo, Jingjing, Chunlin Li, Yi Chen, and Youlong Luo. (2020). "On-demand resource provision based on load estimation and service expenditure in edge cloud environment." *Journal of Network and Computer Applications*, 151 (February): 102506.

Huang, Ping-Chun, Tai-Lin Chin, and Tzu-Yi Chuang. (2021). "Server placement and task allocation for load balancing in edge-computing networks." *IEEE Access*, 9: 138200-138208.

Lähderanta, Tero, Teemu Leppänen, Leena Ruha, Lauri Lovén, Erkki Harjula, Mika Ylianttila, Jukka Riekki, and Mikko J. Sillanpää. (2021). "Edge computing server placement with capacitated location allocation." *Journal of Parallel and Distributed Computing*, 153: 130-149.

Lin, Li, Xiaofei Liao, Hai Jin, and Peng Li. (2019). "Computation Offloading toward Edge Computing." *Proceedings of the IEEE*, 107 (8): 1584-1607.

Lin, Rongping, Xuhui Guo, Shan Luo, Yong Xiao, Bill Moran, and Moshe Zukerman. (2023). "Application-aware computation offloading in edge computing networks." *Future Generation Computer Systems*, 146 (September): 86-97.

Nezami, Zeinab, Kamran Zamanifar, Karim Djemame, and Evangelos Pournaras. (2021). "Decentralized edge-to-cloud load balancing: service placement for the internet of things." *IEEE Access*, 9: 64983-5000.

Sabella, D., A. Vaillant, P. Kuure, U. Rauschenbach, and F. Giust. (2016). "Mobile-edge computing architecture: the role of MEC in the internet of things." *IEEE Consumer Electronics Magazine*, 5 (4): 84-91.

Shen, Bowen, Xiaolong Xu, Lianyong Qi, Xuyun Zhang, and Gautam Srivastava. (2021). "Dynamic server placement in edge computing toward internet of vehicles." *Computer Communications*, 178 (October): 114-23.

Wang, Mingzhi, Tao Wu, Tao Ma, Xiaochen Fan, and Mingxing Ke. (2022). "Users' experience matter: delay sensitivity-aware computation offloading in mobile edge computing." *Digital Communications and Networks*, 8 (6), 955-963.

Xu, Fei, Yue Xie, Yongyong Sun, Zengshi Qin, Gaojie Li, and Zhuoya Zhang. 2022. "Two-stage computing offloading algorithm in cloud-edge collaborative scenarios based on game theory." *Computers & Electrical Engineering*, 97 (January): 107624-24.

Yang, Guisong, Ling Hou, Xingyu He, Daojing He, Sammy Chan, and Mohsen Guizani. (2020). "Offloading time optimization via Markov decision process in mobile edge computing." *IEEE Internet of Things Journal*, 1-1.

Zhang, Wei-Zhe, Ibrahim A. Elgendy, Mohamed Hammad, Abdullah M. Iliyasu, Xiaojiang Du, Mohsen Guizani, and Ahmed A. Abd El-Latif. (2021). "Secure and optimized load balancing for multitier IoT and edge-cloud computing systems." *IEEE Internet of Things Journal*, 8 (10): 8119-32.

Load Balancer Model in the Cloud Computing Environment

Ravendra Singh[1], Anupam Singh[2,3], Raj Pal Singh[4], Satyasundara Mahapatra[4]

[1]Department of Computer Science & Engineering, IFTM University, Moradabad, India
[2]Department of Computer Science and Engineering,
Graphic Era Hill University, Dehradun, India
[3]Department of Computer Science and Engineering,
Graphic Era Deemed to be University, Dehradun, India
[4]Department of Computer Science & Engineering,
Pranveer Singh Institute of Technology, Kanpur, India
E-mail: ravendra85@gmail.com, anupamsing.cse@geu.ac.in, raj.1august@gmail.com,
satyasundara123@gmail.com

Abstract

As the state of sophisticated technology evolves, cloud-based services are becoming more crucial. In these situations, the user's needs may determine how much work is done on the suitable server in the open data-driven virtualisation environment. The newest technology, cloud computing, gives users immediate access to computer features without requiring their direct involvement. For this reason, cloud-based enterprises must be scalable to be successful. The goal of this research is to provide an original virtualised cluster architecture that preserves the centralised management of cloud server resources while enabling cloud-based applications to scale flexibly. Resources can be constantly changed through auto-scaling to satisfy various demands. To automate the setting up and balance of virtualised resources, an automatic scaling method based on ongoing implementation sessions will be started. The recommended approach also takes energy costs into account. The suggested study effort has demonstrated that the recommended technique may effectively handle spikes in load demand while retaining higher resource utilisation. Order group measurements offer auto-scaling features that let you automatically add or remove instances based on fluctuations in load from a controlled instance group. To handle cloud services effectively and discover the most practical, ideal solution, this study paper offers an examination of auto-scaling processes in cloud services.

Keywords: Auto-scaling, virtualisation, virtual machine, cloud computing

1. Introduction

In the Internet Interconnected Things era, competition must be addressed by ensuring energy efficiency, limiting data transfer, and preserving operational efficiency while reducing computational complexity. In numerous engineering applications, such as efficiency control, remote monitoring, and infrastructure management, the event-triggering approach is at the forefront of achieving these goals (Liu *et al.* (2021) and Dhar *et al.* (2017). Software as a Service (SaaS), Platform as a Service (PaaS), and Data as a Service (IaaS) are the three distinct kinds of online computing models Singh and Agarwal (2023). Thus, dynamic scalability is one of the key features of cloud computing Singh and Agarwal (2022).

A virtual machine is reduced in an ongoing resizing configuration to better match the needs of the increasing resource requirements Saxena *et al.* (2021) and Shafiq *et al.* (2022). Since virtual computers form the backbone of Amazon EC2 Singh and Agarwal (2022). Their architecture, with its auto-provisioning structure, on-premises network balancer, and cloud-based cluster monitor system, is ideal for virtual web applications Tian *et al.* (2021). For security reasons, if a user's virtual machines are a member of a specific group, they are unable to use resources from other groups Mishra and Tripathi (2018). An optimisation method locates and adjusts the financial system's configuration to meet SLA and budget limitations Chirammal *et al.* (2016). The auto-scaling architecture of the cloud is described below in Figure 1.

The study presents a simulated cluster environment with virtual machines hosting dynamic-scaling applications. A synopsis of the remaining text appears below: In Section 2, an example of a virtual online computing environment is shown. Section 3 of the proposal details the auto-scaling algorithm. The framework and results are explained in Section 4. The findings and next steps are outlined in Section 5.

2. Design for Architecture

Web-based services and concurrently processed applications are taken into consideration in two scenarios in a cloud computing environment. A scalable structure that

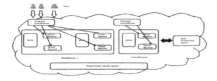

Figure 1: Shows a cloud's auto-scaling architecture

successfully manages these two circumstances is shown in Figure 1. A front-end traffic balancer, an ensemble monitor system, or an auto-provisioning and automatic scaling algorithm are required to achieve this design. The auto-provisioning mechanism adjusts the number of VMs horizontally as the load of the global cluster that the VMs belong to varies.

2.1. Target Audiences and Load Balancing

A web server needs to have a few features. The load balancer is one of them. The load of the user is balanced among several servers using the load balancer.

2.1.1. The AWS Cloud Provides Three Several Kinds of Load Balancers

The following lists the three different kinds of load balancers.

1. Balanced Network Loading
2. The balancing of application load.
3. Standard Load Distribution. **Balanced Network Loading** - TCP and UDP are both capable of transferring data here. This system can handle millions of queries at once and operates at Layer
4. A 504 warning will appear if the server is silent.

The Balance Application Load - At the Layer 7 level, this function mixes HTTP and HTTPS traffic. Requests are moved independently without the aid of algorithms. **Standard Load Distribution** - It served as the initial load balancer for AWS. AWS does not advise utilising this load balancer because it only does fundamental load balancing among servers.

2.1.2. Target Audiences

Each load balancer will have a Target Group designated for it. Target Group frequently distributes traffic across their targets' [servers] using what is known as the Round Robin algorithm. Figure 2 depicts the load balancer and the EC2 instances it is connected to.

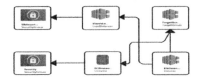

Figure 2: Target groups and the load balancer

2.2. Group for Automatic Scaling (ASG)

Server scaling and administration are automated with Auto Scaling Groups. ASG starts a new server based on the configuration or launch template that is attached. A server's lowest, highest, and desirable capacities can all be set. With ASG, idle servers can be terminated automatically if the load decreases, reducing the burden and saving money.

(a) Scaling policies

ASGs have scaling policies that specify when to launch new servers and when to shut down idle servers according to those policies.

1. Scaling and Target Tracking
2. Step Scaling
3. Basic Scaling

(b) Launch settings

Templates are the foundation of an ASG server configuration.

3. System of Load Balancing

The suggested work architecture distributes web apps through a front-end load balancer. The virtual cluster can handle HTTP requests faster as it grows.

3.1. Algorithm for Auto-Scaling System

Output: An FLB is a collection of front-load balancers.

3.1.1. A Cloud-Based Autoscaling Algorithm

A virtualised cluster monitor can recognise a web application that has excessive

```
Input: n: number of Clusters
Vα: In a Virtual Cluster, the
same computation system   is
run   by   multiple   virtual
machines.
VMβ: In the virtual system, the
number of active sessions
VMMax: The maximum number of
sessions for a virtual machine in μ
Cluster
VMγ1: The upper threshold for a
session
VMγ2: The low-threshold level for a
session
VMζ: Virtual machines above the upper
threshold is recorded in a virtual machine
set
```

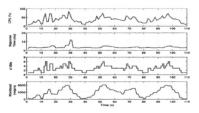

Figure 3: Use of the auto-scaling system

HTTP sessions open. A virtual appliance is produced by software installation on a virtual machine and is shown in Figure 1. Our auto-scaling approach for computing distributed jobs is shown in Figure 3.

4. Analysis of Simulation and Result Data

For simulation, we are employing the free Cloud Sim framework, which may be used to imitate cloud services and infrastructure. Every time an image is taken in a computer environment, VM numbers are computed and updated.

The anticipated and observed CPU load periods for a virtualisation system for the server are shown in Figure 3. Additionally, it explains how much CPU power is used, how quickly decisions are made, and how virtual machines are used for task distribution in cloud computing. Use this number to identify the bytes-based type of back-end disk I/O when a disc consumption or latency problem is noticed. This figure represents, in terms of packets, the amount of traffic that arrives on a single instance. This statistic measures the amount of outbound traffic by counting the number of packets delivered for a single case.

5. Conclusion

This study investigates two scaling scenarios for web applications along with distributed computing activities using virtual clusters. The suggested method can manage demand peaks while maintaining higher resource utilisation and using less energy, according to our research. With the help of autoscaling, apps can handle spikes in traffic more easily and at a cheaper cost when fewer assets are required.

References

Liu, J., Yang, Y., Li, H., and Geng, Y. (2021). Event-triggered output-feedback control for networked switched positive systems with asynchronous switching. *International Journal of Control, Automation and Systems*, 19: 3101-3110.

Dhar, N. K., Verma, N. K., and Behera, L. (2017). Adaptive critic-based event-triggered control for HVAC system. *IEEE Transactions on Industrial Informatics*, 14(1): 178-188.

Mishra, T. K., and Tripathi, S. (2018). Congestion control and fairness with dynamic priority for ad hoc networks. *International Journal of Ad Hoc and Ubiquitous Computing*, 29(3): 208-220.

Singh, R., and Agarwal, B. B. (2023). An automated brain tumor classification in MR images using an enhanced convolutional neural network. *International Journal of Information Technology*, 15(2): 665-674.

Singh, R., and Agarwal, B. B. (2022). A hybrid approach for detection of brain tumor with levy flight cuckoo search. *Webology*, 19(1): 5388-5401.

Saxena, D., Singh, A. K., and Buyya, R. (2021). OP-MLB: an online VM prediction-based multi-objective load balancing framework for resource management at cloud data center. *IEEE Transactions on Cloud Computing*, 10(4): 2804-2816.

Shafiq, D. A., Jhanjhi, N. Z., and Abdullah, A. (2022). Load balancing techniques in cloud computing environment: A review. *Journal of King Saud University-Computer and Information Sciences*, 34(7): 3910-3933.

Singh, R., and Agarwal, B. B. (2022). Automatic image classification and abnormality identification using machine learning. In *Proceedings of Trends in Electronics and Health Informatics: TEHI 2021* (pp. 13-20). Singapore: Springer Nature Singapore.

Tian, W., Xu, M., Zhou, G., Wu, K., Xu, C., and Buyya, R. (2021). Prepartition: load balancing approach for virtual machine reservations in a cloud data center. *Arxiv preprint arXiv:2110.09913*.

Chirammal, H. D., Mukhedkar, P., and Vettathu, A. (2016). *Mastering KVM virtualization*. Packt Publishing Ltd.

Study and Comparison of Malware Detection Using Deep Learning Techniques

Geeta Gayatri Behera, Alekha Kumar Mishra

Department of Computer Science and Engineering, NIT Jamshedpur, Jamshedpur, India
E-mail: 2022pgcsis03@nitjsr.ac.in, alekha.cse@nitjsr.ac.in

Abstract

Improvements in computer technology have resulted in increasing virtual living, shifting the focus of cybercriminals to the online realm. Committing crimes online is easier, leading to the proliferation of malicious software (malware) for cyberattacks. Deep Learning (DL) is employed to enhance malware detection, offering efficient and precise detection without manual intervention. This study explores various DL approaches, comparing them with other methods and their own variations.

Keywords: Malware detection, deep learning, DNN, CNN, CNN+LSTM

1. Introduction

For decades, antivirus software has safeguarded devices (Sun *et al.*, 2021). Malware, such as Trojan Horses, rootkits, worms, and viruses, poses threats to users (Malani *et al.*, 2022). Reports indicate the daily creation of 10 lakh malicious software and a projected cybercrime cost of $6 trillion by 2021. The McAfee Covid-19 threat report notes 375 new cybercrimes reported globally every minute, with malware attacks rising by 19.02% in the last four quarters. Market growth for malware analysis is anticipated to reach $11.7 billion USD by 2024, with a CAGR of 31% from 2019 to 2024 (Gopinath and Sethuraman, 2023).

In recent years, Machine Learning (ML) has emerged highly sought after for its capacity to assimilate diverse data and predict outcomes. Deep Learning (DL) techniques, known for their superior accuracy, are now extensively employed in malware detection research.

The main objective of this analysis is to improve accuracy in malware detection through comprehensive analysis of deep learning approaches.

The rest of the paper is divided into 5 sections: Static and Dynamic Malware Analysis in Section 2, DL methods in Section 3, literature survey in Section 4, thorough examination in Section 5, and final observation in Section 6.

2. Static and Dynamic Malware Analysis

Static analysis entails examining executable files without execution, while dynamic analysis observes malware behaviour during execution (Bagane *et al.*, 2021). DL offers a novel idea to address limitations in malware detection and recognition. DL, with its multiple hidden layers, learns from examples. Various DL architectures, such as recurrent neural networks (RNN), deep belief networks (DBN), convolutional neural networks (CNN), and deep neural networks (DNN), have been employed to enhance the model performance (Aslan and Yilmaz, 2021).

Chaper 8 DOI: 10.1201/9781003581215

3. Deep Learning Techniques for Malware Detection

DL, also known as DNNs is a subclass of artificial intelligence (AI) that draws inspiration from the functioning of the brain. DL architectures' primary strength lies in their capacity to interpret big data sets and automatically adjust the concluded meaning based on fresh information, all without requiring domain expertise (Vinayakumar et al., 2019).

3.1. Deep Neural Network

A feedforward neural network (FFN) is depicted as a directed graph without cycles. Its variant, the multi-layer perceptron (MLP), comprises an input layer, multiple hidden layers, and an output layer. Each layer neurons are interconnected, with the optimal number determined by hyperparameter tuning.

3.2. Convolutional Neural Network

The expanded form of ANN, known as CNN, finds widespread use in image processing (Vinayakumar et al., 2019). CNN encompasses various deep learning models, including multi-layered neural networks, featuring a convolutional layer structure (Bagane et al., 2021). The CNN network comprises an Input Layer, multiple convolutional layers, pooling layers, and completely linked layers. The completely linked layer makes the final outcome by processing filtered input images, while the pooling layer shrinks image size for better processing.

3.3. Recurrent Neural Network

RNNs, notable for their bi-directional information flow between layers, are pivotal in NLP and speech recognition. The Long-Short Term Memory Module (LSTM), a kind of RNN, effectively stores data and prevent the gradient problems. RNNs utilise past inputs for consecutive predictions, vital in machine translation, robotic control, speech recognition, music composition, and more (Gopinath and Sethuraman, 2023).

4. Literature Survey

Bokolo et al. (2023) sourcing data from Kaggle. They found that Support Vector Machine (Classifier) (SVC) achieved 92% accuracy and precision, while DL yielded 96% precision and 95% accuracy.

Aslan and Yilmaz (2021) proposed a DNN hybrid for malware categorisation, utilising data from Microsoft BIG 2015, Malevis, and Malimg datasets. Achieving 94.88% and 97.78% accuracy on respective datasets.

Malani et al. (2022) employed ANN, CNN, RNN, and LSTM for malware assessment, achieving 65% accuracy. CNN outperformed others with nearly 98% accuracy, surpassing the hybrid model's 97.6%.

Bagane et al. (2021) utilised CNN, CNN+LSTM and CNN+BiLSTM for malware detection. Malware binaries were transformed into images for CNN processing, with the final layer flattened due to its n-dimensional nature. The dataset was randomly split, and the three deep learning approaches were employed, yielding accuracies of 95.4%, 96.2%, and 97.4%.

Patil et al. (2023) applied the CNN algorithm to both the traditional Microsoft dataset and a balanced dataset generated using SMOTE. Accuracy attained CNN with SMOTE and traditional dataset are 92.09%, 89.51%.

Sun et al. (2021) used ANN to detect malware with a 92.4% F1-Score.

In the following section we have provided key-point observations on the above reported techniques.

5. Analysis

Table 1 shows different DL and their associated data sources, proving effective in addressing gaps in malware detection and categorisation. Standard measures like accuracy, F1-score, and precision based on True Positive (TP), False Positive (FP), True Negative (TN), and False Negative (FN) classifications, assess classifier performance (Vinayakumar et al., 2019).

Figure 1 illustrates various DL methods are shown for their best accuracy and precision: DNN, CNN (used by five researchers), CNN + LSTM, ANN, and GAN. CNN is selected by the majority for its greater accuracy.

6. Conclusion

Identifying the malware detection through normal methods is quite challenging. DL algorithms, however, lessen the drawbacks of both traditional and conventional methods for malware identification and

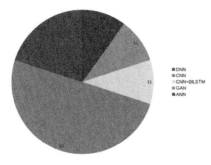

Figure 1: Various deep learning techniques

classification. In this paper, we have discussed several DL strategies, DL generally yields positive results. The acquired data, in specifically, demonstrate the utmost accuracy, F1-score and precision metrics.

References

Aslan, Ö., and Yilmaz, A. A. (2021). "A new malware classification framework based on deep learning algorithms." *IEEE Access*, 9: 87936-87951.

Bagane, P., Joseph, S. G., Singh, A., Shrivastava, A., Prabha, B., and Shrivastava, A. (2021). "Classification of malware using Deep Learning Techniques." In *2021 9th International Conference on Cyber and IT Service Management (CITSM)* (pp. 1-7). IEEE.

Bokolo, B., Jinad, R., and Liu, Q. (2023). "A comparison study to detect malware using deep learning and machine learning techniques." In *2023 IEEE 6th International Conference on Big Data and Artificial Intelligence (BDAI)* (pp. 1-6). IEEE.

Gopinath, M., and Sethuraman, S. C. (2023). "A comprehensive survey on deep learning based malware detection

Table 1: Pre-crisis summary statistics

Author	Dataset	Techniques Used	Result
Bokolo *et al.* (2023)	Microsoft malware classification	DNN	Accuracy 95% Precision 96%
Aslan and Yilmaz (2021)	Microsoft BIG 2015, Malimg, and Malevis	Deep Neural Network (Hybrid deep learning)	Accuracy 96.5%
Malani *et al.* (2022)	IIT Bhilai, Meraz 18	CNN	Accuracy 98%
Bagane *et al.* (2021)	Malimg	CNN+BiLSTM,	Accuracy 97.4%
Patil *et al.* (2023)	9 Microsoft dataset and 26 from malware families.	SMOTE algorithm, CNN	Accuracy 92.09%
Sun *et al.* (2021)	Canadian Institute of Cyber Security	ANN	F1 score 0.924

techniques." *Computer Science Review*, 47: 100529.

Lu, Y., and Li, J. (2019). "Generative adversarial network for improving deep learning based malware classification." In *2019 Winter Simulation Conference (WSC)* (pp. 584-593). IEEE.

Malani, H., Bhat, A., Palriwala, S., Aditya, J., and Chaturvedi, A. (2022). "A unique approach to malware detection using deep convolutional neural networks." In *2022 4th International Conference on Electrical, Control and Instrumentation Engineering (ICE-CIE)* (pp. 1-6). IEEE.

Newaz, S., Imran, H. M., and Liu, X. (2021). "Detection of malware using deep learning." In *2021 IEEE 4th International Conference on Computing, Power and Communication Technologies (GUCON)* (pp. 1-4). IEEE.

Patil, V., Shetty, S., Tawte, A., and Wathare, S. (2023). "Deep learning and binary representational image approach for malware detection." In *2023 International Conference on Power, Instrumentation, Control and Computing (PICC)* (pp. 1-7). IEEE.

Sun, Y., Chong, N. S., and Ochiai, H. (2021). "Network flows-based malware detection using a combined approach of crawling and deep learning." In *ICC 2021-IEEE International Conference on Communications* (pp. 1-6). IEEE.

Vinayakumar, R., Alazab, M., Soman, K. P., Poornachandran, P., and Venkatraman, S. (2019). "Robust intelligent malware detection using deep learning." *IEEE Access*, 7: 46717-46738.

Vo, H. V., Nguyen, H. N., Nguyen, T. N., and Du, H. P. (2022). "SDAID: towards a hybrid signature and deep analysis-based intrusion detection method." In *GLOBECOM 2022-2022 IEEE Global Communications Conference* (pp. 2615-2620). IEEE.

Nature-Inspired Meta-Heuristic Algorithms for Optimizing Neural Network Training

A Focus on Particle Swarm Optimization and Firefly Algorithm

K. P. Swain[1], A. S. Das[2], S. K. Nayak[3], S. K. Mohapatra[1], Parsuram Behera[1], Basudev Das[1]

[1]Department of ETC, TAT, Bhubaneswar, India
[2]Department of CSE, GITA Autonomous College, Bhubaneswar, India
[3]S.K. Nayak, Dept. of CSE, Deputy Curriculum Director, BPUT, Rourkela, Odisha
E-mail: kaleep.swain@gmail.com, ashok.s.das@gmail.com, suryakanta004@gmail.com, sumsusmeera@gmail.com, parshurambehera7735@gmail.com, basudevv.2003@gmail.com

Abstract

The efficiency of Nature-Inspired Meta-heuristic Algorithms in training Simple Neural Networks (SNN), particularly Particle Swarm Optimization (PSO) and Firefly Algorithm (FFA), is explored in this paper. Traditional methods such as Back Propagation (BPN) are renowned for their high resource demands. An alternative approach is presented using PSO and FFA, both derivative-free and potentially more efficient for optimising neural network training. Application of these algorithms to the IRIS dataset demonstrates their effectiveness in multi-level classifications. Comparative analysis with BPN reveals the strengths and limitations of PSO and FFA, emphasising the significance of parameter selection in achieving optimal training results. Highlighting nature-inspired algorithms as viable alternatives for complex problem-solving in neural network training, this study contributes to the field.

Keywords: PSO, FFA, neural network training, meta-heuristic optimization, nature-inspired algorithms

1. Introduction

Within the realm of artificial intelligence and machine learning, neural networks have sparked a revolution, propelling significant strides in tasks like image recognition, natural language processing, and pattern recognition. Among the various types of Neural Networks, Simple Neural Networks (SNN), fully connected Neural Networks (FCNN), and feed-forward neural networks (FNN) have proven to be powerful tools for multi-level classifications. However, one of the key challenges in harnessing their full potential lies in the training process (Mnih *et al.*, 2015, Sutton and Barto, 2005).

Traditionally, supervised learning methods like "Back Propagation" (BPN) have been the de facto approach for training SNNs (Hochreiter, 1998). While BPN is effective in updating the network weights by propagating errors backward, it has notable drawbacks. One of the major limitations is its resource-intensive nature. The iterative process of computing derivatives and adjusting weights can be time-consuming and computationally demanding, especially for large-scale datasets and deep network architectures. Moreover, BPN is susceptible to issues such as vanishing gradients, hindering its ability to effectively learn complex patterns, and occasionally converging to local minima, potentially compromising the network's performance (Hochreiter and Schmidhuber, 1997, Wang *et al.*, 1993).

To overcome these limitations and improve the efficiency of Simple Neural Network training, we turn our attention to "Heuristic Optimization Algorithms." Inspired by natural phenomena, these algorithms provide alternative approaches to optimization without relying on derivatives. Specifically, we investigate the application of Particle Swarm Optimization (PSO) and Firefly Algorithm (FFA) as potential solutions for training SNNs (Rojas-Delgado and Trujillo-Rasúa, 2018, Kang *et al.*, 2008).

2. Literature Review

The study introduces a robust evolutionary algorithm (Yang and Kao, 2001) for neural network training, focusing on efficiency by addressing convergence and local optima issues. Research in (Yu *et al.*, 2007) enhances feedforward neural networks with an advanced PSO algorithm, targeting improved learning and convergence. The book (Yang, 2008) details nature-inspired metaheuristic algorithms, covering Genetic Algorithms, PSO, and Firefly Algorithm, offering insights into their applications in computational intelligence. The article (Mandal *et al.*, 2015) applies the Firefly Algorithm to enhance neural network training, focusing on learning performance and convergence. Further research (Bonyadi and Michalewicz, 2017) examines PSO's role in solving single objective continuous space problems. Articles (Zare *et al.*, 2023), Ghasemi *et al.*, 2022) and Ranjan (2020) discuss the advancements

and Global Best-guided modifications in FFA for engineering optimization.

3. Implementation of Meta-Heuristic Algorithm

In this study, we focused on the IRIS dataset for implementing our work. The IRIS dataset is a commonly used benchmark in machine learning, consisting of four input features and one target output, categorised into three classes of IRIS flowers. For our Artificial Neural Network (ANN), we devised a structure consisting of 4 input nodes, seven hidden nodes, and 3 output nodes, totaling 49 weights for optimisation.

3.1. Particle Swarm Optimization (PSO)

Particle Swarm Optimization (PSO) is an optimisation method inspired by natural phenomena such as bird flocking and fish schooling, offering an innovative approach to problem-solving. As with other evolutionary algorithms, PSO begins with an initial set of candidate solutions and aims to find the best global optimum.

3.1. Firefly Algorithm (FFA)

The Firefly Algorithm, inspired by the captivating behaviour of fireflies, presents a nature-inspired optimisation technique. In this algorithm, each firefly emits light based on its current location, with the intensity of light representing the quality of its privileged position. Higher intensities correspond to more favourable positions within the optimisation landscape. As fireflies move, the relative intensity of their emitted light varies with the changing distances between them. This paper delves into the intricacies of the Firefly Algorithm, exploring how it emulates the collective behaviour of fireflies to efficiently optimise complex problems. By illuminating the core principles of light emission and intensity variations, this research sheds light on the algorithm's convergence properties and effectiveness in solving optimisation challenges.

4. Results and Discussion

This study initially implemented the Back Propagation Neural Network (BPN) on the IRIS dataset, followed by the application of Particle Swarm Optimization (PSO) and Firefly Algorithm (FFA). Our findings revealed that all three methods achieved comparable results in terms of accuracy as shown in Table 1. This consistency across different algorithms underscores their effectiveness in neural network training for classification tasks. The proof of the same is illustrated by the corresponding screenshots given in Figure 1 which are screenshots depicting the outputs of different algorithms applied to the IRIS dataset. Similarly, the other screenshot can be drawn.

Our investigation into the IRIS dataset using Back Propagation Neural Network (BPN), Particle Swarm Optimization (PSO), and Firefly Algorithm (FFA) revealed that all three methods are capable of achieving comparable levels of accuracy. BPN, with its gradient-based approach, showed faster convergence but is susceptible to local minima. In contrast, PSO and FFA, being meta-heuristic algorithms, demonstrated a broader exploration of the solution space, suggesting their potential in handling more complex scenarios where gradient-based methods might falter.

The study also highlighted the importance of time efficiency and parameter selection as discussed in Table 1. BPN was notably faster in training, an advantage in time-sensitive applications. However, PSO and FFA's slower but thorough exploration can be pivotal in applications where finding global optima is crucial, albeit with the caveat of requiring meticulous parameter tuning. These observations highlight the intricate balance between speed, accuracy, and resilience in the training of neural networks, indicating that the selection of the algorithm should be fine-tuned to match the precise demands of the given task.

Table 1: Timing parameters

	BP	PSO	FFA
Real	1.034s	14.644s	16.299s
User	1.01s	14.358s	15.992s
Sys	0.010s	0.209s	0.227s

Figure 1: BPN classification performance on IRIS dataset

5. Conclusion

In conclusion, while the Back Propagation (BPN) algorithm remains the standard method for training Artificial Neural Networks (ANN) through supervised learning, its limitations in dealing with vanishing gradients and convergence to local minima due to derivative reliance necessitate exploration of alternative approaches. Nature-inspired meta-heuristic algorithms like Particle Swarm Optimization (PSO) and Firefly Algorithm (FFA) offer promising derivative-free solutions for optimising complex problems during ANN training. Successfully applied to classify the IRIS dataset, PSO and FFA demonstrate their potential as effective substitutes for BPN in maximising the objective function.

References

Bonyadi, M. R., and Michalewicz, Z. (2017). Particle swarm optimization for single objective continuous space problems: A review. *Evolutionary Computation*, 25(1): 1–54.

Hochreiter, S. (1998). The vanishing gradient problem during learning recurrent neural nets and problem solutions. *International Journal of Uncertainty, Fuzziness and Knowledge-Based Systems*, 6: 107–116.

Hochreiter, S., and Schmidhuber, J. (1997). Long short-term memory. *Neural Computation*, 9(8): 1735–1780.

Kang, Q., An, J., Yang, D., Wang, L., and Wu, Q. (2008). Particle swarm optimization based RBF neural networks learning algorithm. *2008 7th World Congress on Intelligent Control and Automation*, Chongqing, pp. 605–610.

Mandal, S., Saha, G., and Pal, R. (2015). Neural network training using firefly algorithm. *Global Journal on Advancement in Engineering and Science*, 1: 7–11.

Mnih, V., Kavukcuoglu, K., Silver, D., Rusu, A., Veness, J., Bellemare, M. G., Graves, A., Riedmiller, M., Fidjeland, A. K., Ostrovski, G., Petersen, S., Beattie, C., Sadik, A., Antonoglou, I., King, H., Kumaran, D., Wierstra, D., Legg, S., and Hassabis, D. (2015). Human-level control through deep reinforcement learning. *Nature*, 518: 529–533.

Ranjan Senapati, B., Mohan Khilar, P. (2020). Optimization of Performance Parameter for Vehicular Ad-hoc NETwork (VANET) Using Swarm Intelligence. In: Rout, M., Rout, J., Das, H. (eds) Nature Inspired Computing for Data Science. Studies in Computational Intelligence, vol 871. Springer, Cham.

Rojas-Delgado, J. and Trujillo-Rasúa, R. A. (2018). Training neural networks by continuation particle swarm optimization. *Progress in Artificial Intelligence and Pattern Recognition*, 11047. ISBN: 978-3-030-01131-4, DOI:10.1007/978-3-030-01132-1_7

Sutton, R. S., and Barto, A. G. (2005). Reinforcement learning: An introduction. *IEEE Transactions on Neural Networks*, 16(1): 285–286.

Wang, Y.-F., Cruz Jr., J. B., and Mulligan Jr., J. H. (1993). Multiple training concept for back-propagation neural networks for use in associative memories. *Neural Networks*, 6(8): 1169–1175.

Yang, J. M. and Kao, C. Y. (2001). A robust evolutionary algorithm for training neural networks. *NCA*, 10: 214–230.

Yang, X. S. (2008). *Nature-Inspired Meta-heuristic Algorithms*. Luniver Press.

Yu, J., Xi, L. and Wang, S. (2007). An improved particle swarm optimization for evolving feedforward artificial neural networks. *Neural Processing Letters*, 26: 217–231.

Zare, M., Ghasemi, M., Zahedi, A. *et al.* (2023). A global best-guided firefly algorithm for engineering problems. *Journal of Bionic Engineering*, 20: 2359–2388.

PCAFeEx

A Machine Learning-Based Concept for Predicting Heart Disease Using Principal Component Analysis Feature Extraction

Shreeharsha Dash, Subhalaxmi Das

Department of CSE, Odisha University of Technology and Research,
Bhubaneswar, India
E-mail: shreeharshadash1@gmail.com, sdascse@outr.ac.in

Abstract

Early identification is crucial for successful and prompt medical intervention in the treatment of heart disease, which is a significant cause of death world-wide. Important healthcare service data is generated by this model by using PCA (Principal Component Analysis) as a feature extraction approach after pre-processing. Each disease dataset contains a large number of characteristics. The following machine learning (ML) algorithms are used to create a comparison model: LR (Logistic Regression), KNN (K-Nearest Neighbor), GBoost (Gradient Boosting), RF (Random Forest), and DT (Decision Tree). The dataset being compared includes features with and without the hyperparameter GridSearchCV as well as a significant number of features with sequential feature selection (SFS). An improved accuracy of 99.21% and 99.04% achieved with RF and KNN classifiers respectively accompanied by GridSearchCV utilizing PCA demonstrates encouraging performance results for the suggested approach. The findings show that early cardio issue prediction, using the strengths of many base classifiers works well. We outperform the other authors' suggested methods in terms of prediction accuracy. Our model was also compared to models developed by other authors as well.

Keywords: Comparison, feature extraction, hyperparameter, heart disease, preprocessing

1. Introduction

Coronary disease kills many people globally. Identification of this basic illness is crucial to preventing deaths. However, the lack of medical help is not restricted to develop nations and sometimes delays heart disease treatment. This issues also affect undeserved networks in wealthy areas, highlighting the complexity of mandatory medical care access. ML models can examine big datasets and forecast most ailments, including cardiac difficulties, making them helpful tools for such situations (Subhalaxmi et al., 2022).

This research presents a ML based PCE model for early heart illness identification. We standardise data with StandardScalar to ensure model training reliability. Pre-processing is possible (Majumder et al., 2023). After pre-processing, the classification model receives GridSearchCV and SFS features. The classification model uses LR, KNN, GBoost, RF and DT. Finally,

Chapter 10 DOI: 10.1201/9781003581215

compare the two assessment methodologies using the performance matrices.

2. Literature Study

For examining less sophisticated, cheaper and more reproducible approaches may evaluate individuals utilising clinical data in healthcare facilities to predict heart illness (Abuhaija et al., 2023).

For Cardiovascular Disease (CVD) diagnosis, Swati Shilaskar et al. (2013) developed hybrid forward selection by discovering smaller subgroups and enhances diagnostic accuracy over forward inclusion and back-elimination. Aqsa Rahim et al. (2021) created MaLCaDD. It starts with mean replenishment and SMOTE. An ensemble of LR and KNN is recommended for improved cardiac issues prediction. The XGBH machine learning model was proposed by Mengxiao Peng et al. (2023). A unique heart disease detection model, known as QPSO-SVM, was presented by

Fahad Algarni et al. (2023). Subhalaxmi et al. (2021) utilises Rough Set to identify the primary causes of mortality outside of other illness. Using time series data, they forecast these illnesses' exponential rise.

Furthermore, several studies exhibit a deficiency in the comprehensiveness of their feature selection methodologies, which may result in less-than-ideal prediction performance. In order to overcome these constraints, the present study offers a methodology that integrates a diverse set of fundamental classifiers and utilises sophisticated approaches, thereby improving the precision and resilience of the proposed model for predicting heart disease.

3. Proposed Model

Following next Jupyter Notebook and Python are used to implement the algorithms. Find the process schematic in Figure 1 below.

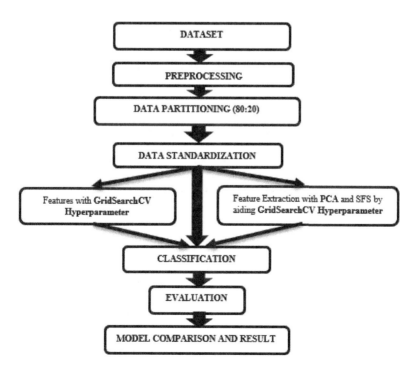

Figure 1: Workflow of the proposed system

3.1. Dataset

An available Kaggle repository provided the study's analysis and comparison dataset (Ulianova, 2019). This dataset contains 12 characteristics derived from 70,000 patient records from clinical trials (Jinjri et al., 2021). This study's heart disease dataset is shown in Table 1.

3.2. Preprocessing

Building a classification model is essential. Analyses are performed on samples. Data cleaning generally involves rectifying mistakes, outliers, and missing real-world data (Majumder et al., 2023). Data was partitioned 80:20 (Jinjri et al., 2021) and standardised with StandardScaler to make them uniform with a 0 mean and 1 standard deviation of (Majumder et al., 2023).

3.3. Feature Extraction

Selecting a comprehensive and accurate collection of features for the forecasting job from a given arrangement of parts is quite challenging (Vikas and Aparna, 2023). The feature extraction method suggested in this study is principal component analysis (PCA) (Anna et al., 2020). We use principal component analysis to choose 4, 6, 8 and 10 elements through SFS, which are quite noticeable. SFS signifies sequential element determination, and it's a passionate computation that works on the display of a predictive model.

3.4. Classification

Classification can be don after completion of feature extraction with ML classifiers such as LR, KNN, GBoost, RF and DT.

LR uses continuous independent factor to predict the influence of the dependent variable to access and predict illness (Subhalaxmi et al., 2022). KNN are delivered using the 'k' closest train data points of relevance. Since KNN does not assume anything about the transmission of information, it finds use in many contexts (Majumder et al., 2023).

Table 1: Fundamental characteristics of patients

Features	Types of features	Description
Age	Objective Features	Days
Height		Centimetres
Weight		Kilograms
Gender		1: Women, 2: Men
Systolic Blood Pressure	Examination Features	ap_hi
Diastolic Blood Pressure		ap_lo
Cholesterol		1: Normal, 2: Above Normal, 3: Well Above Normal
Glucose		1: Normal, 2: Above Normal, 3: Well Above Normal
Smoking	Subjective Features	Binary (Whether patient smokes or not)
Alcohol Intake		Binary (Whether patient takes alcohol or not)
Physical Activity		Binary (Whether patient does physical activities or not)
Presence or absence of cardiovascular disease	Target Feature	Binary (Cardio)

ML techniques like GBoost find utility in many different areas, including identification and regression. RF is a collective learning approach for a variety of tasks, including requested, recursive, and others. Classification and regression problems make heavy use of the technique. DTs are very useful in ML since they are often used in complex ensemble approaches (Majumder *et al.*, 2023).

4. Experimental Results

The results of trials have been carefully examined and recorded at this point with the performance matrices such as accuracy, precision, recall and F1-score. After evaluation of classifiers in relation to all features, the result found out as 97.71% of accuracy by using GridSearchCV hyperparameter along with PCA. Figure 2 details the accuracy of LR, KNN, GBoost, RF, and DT with regard to a significant number of elements.

Figure 3 shows AUC-ROC graph, which compares the performance of the suggested model with and without PCA and also used to thoroughly evaluate our model's performance. Table 2 represents the comparison of our model with previously designed models and check the rate of efficacy.

5. Conclusion

The model includes preprocessing, GridSearchCV hyperparameter feature extraction, and classification with and without PCA. Using the confusion matrix and AUC-ROC curve, we got the best scores of 99.21% and 99.04% with RF and KNN respectively. The methodology and accuracy of our model may be compared to other probes in Table 3 (Majumder *et al.*, 2023). This opens up interesting opportunities for ML and AI research in cardiovascular health.

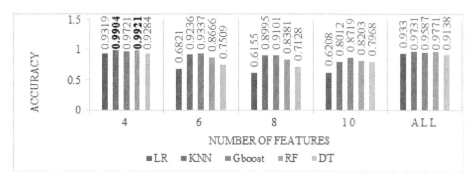

Figure 2: Comparative analysis of classifiers' accuracy in relation to the number of features

Table 2: Evaluation of classifiers in relation to feature

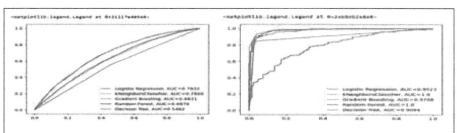

Figure 3: AUC-ROC curve exemplifies comparing model without and with PCA comparably

Table 3: Comparison with prior works

Existing works	Methods	Accuracy
Jinjri et al. (2021)	Support Vector Machine (SVM)	72.66%
Asif et al. (2021)	Hard and Soft Voting Ensemble Classifiers	92%
Polipireddy and Rahul (2022)	OPTUNA Hyperparameter Optimisation Technique with XGBoost	89.3%
Abuhaija et al. (2023)	Weka and Statistical Package for Social Sciences (SPSS) with Decision Tree	95.76%
Mahmud et al. (2023)	XGBoost (XGB)	84.03%
Baghdadi et al. (2023)	CatBoost Model	90.94%
Proposed Model	PCA Feature Extraction Technique with RF and KNN	99.21%(RF) and 99.04%(KNN)

Reference

Abuhaija, Belal, Aladeen Alloubani, Mohammad Almatari, Ghaith M. Jaradat, Barzan Abdallah Hemn, Abdallah Mohd Abualkishik, and Mutasem Khalil Alsmadi. (2023). "A comprehensive study of machine learning for predicting cardiovascular disease using Weka and SPSS tools." *International Journal of Electrical and Computer Engineering*, 13 (2): 1891.

Asif, Md Asfi-Ar-Raihan, Mirza Muntasir Nishat, Fahim Faisal, Rezuanur Rahman Dip, Mahmudul Hasan Udoy, Md Fahim Shikder, and Ragib Ahsan. (2021). "Performance Evaluation and Comparative Analysis of Different Machine Learning Algorithms in Predicting Cardiovascular Disease." *Engineering Letters*, 29 (2).

Baghdadi, Nadiah A., Sally Mohammed Farghaly Abdelaliem, Amer Malki, Ibrahim Gad, Ashraf Ewis, and Elsayed Atlam. (2023). "Advanced machine learning techniques for cardiovascular disease early detection and diagnosis." *Journal of Big Data*, 10 (1): 144.

Chaurasia, Vikas, and Aparna Chaurasia. (2023). "Novel method of characterization of heart disease prediction using sequential feature selection-based ensemble technique." *Biomedical Materials & Devices*, 1-10.

Das, Subhalaxmi, Sateesh Kumar Pradhan, Sujogya Mishra, Sipali Pradhan, and P. K. Pattnaik. (2022). "Diagnosis of cardiac problem using rough set theory and machine learning." *Indian Journal of Computer Science and Engineering*, 13 (4): 1112-1131.

Das, Subhalaxmi, Sateesh Kumar Pradhan, Sujogya Mishra, Sipali Pradhan, and P. K. Pattnaik. "Analysis of heart diseases using soft computing technique." In *2021 19th OITS International Conference on Information Technology (OCIT)*, pp. 178-184. IEEE, 2021.

Elsedimy, E. I., Sara MM AboHashish, and Fahad Algarni. (2023). "New cardiovascular disease prediction approach using support vector machine and quantum-behaved particle swarm optimization." *Multimedia Tools and Applications*, 1-28.

Gárate-Escamila, Anna Karen, Amir Hajjam El Hassani, and Emmanuel Andrès. (2020). "Classification models for heart disease prediction using feature selection and PCA." *Informatics in Medicine Unlocked*, 19: 100330.

Jinjri, Wada Mohammed, Pantea Keikhosrokiani, and Nasuha Lee Abdullah. "Machine learning algorithms for the classification of cardiovascular disease-A comparative study." In *2021 International Conference on Information Technology (ICIT)*, pp. 132-138. IEEE, 2021.

Mahmud, Tanjim, Anik Barua, Manoara Begum, Eipshita Chakma, Sudhakar Das, and Nahed Sharmen. "An improved framework for reliable cardiovascular disease prediction using hybrid ensemble learning." In *2023 International Conference on Electrical, Computer and Communication Engineering (ECCE)*, pp. 1-6. IEEE, 2023.

Majumder, Annwesha Banerjee, Somsubhra Gupta, Dharmpal Singh, Biswaranjan Acharya, Vassilis C. Gerogiannis, Andreas Kanavos, and Panagiotis Pintelas. (2023). "Heart Disease Prediction Using Concatenated Hybrid Ensemble Classifiers." *Algorithms*, 16 (12): 538.

Peng, Mengxiao, Fan Hou, Zhixiang Cheng, Tongtong Shen, Kaixian Liu, Cai Zhao, and Wen Zheng. (2023).

"Prediction of cardiovascular disease risk based on major contributing features." *Scientific Reports*, 13 (1): 4778.

Rahim, Aqsa, Yawar Rasheed, Farooque Azam, Muhammad Waseem Anwar, Muhammad Abdul Rahim, and Abdul Wahab Muzaffar. (2021). "An integrated machine learning framework for effective prediction of cardiovascular diseases." *IEEE Access*, 9: 106575-106588.

Srinivas, Polipireddy, and Rahul Katarya. (2022). "hyOPTXg: OPTUNA hyper-parameter optimization framework for predicting cardiovascular disease using XGBoost." *Biomedical Signal Processing and Control*, 73: 103456.

Shilaskar, Swati, and Ashok Ghatol. (2013). "Feature selection for medical diagnosis: Evaluation for cardiovascular diseases." *Expert Systems with Applications*, 40 (10): 4146-4153.

Ulianova, Svetlana. (2019). "Cardiovascular disease dataset." *Kaggle.com.*

Effectiveness of SOSA, FFA and GOA Algorithms for Epileptic Seizure Detection Using DWT and SVM

Puspanjali Mallik[1], Ajit Kumar Nayak[2], Sumanta Kumar Mohapatra[3], K. P. Swain[3]

[1]Department of Computer Science and Engineering, Institute of Technical Education and Research, Siksha 'O' Anusandhan (Deemed to be) University, Odisha, India

[2]Department of CS and IT, Institute of Technical Education and Research, Siksha 'O' Anusandhan (Deemed to be) University, Odisha, India

[3]Department of Electronics and Telecommunication Engineering, Trident Academy of Technology, Odisha, India

E-mail: mallickpuspa@gmail.com, ajitnayak@soa.ac.in, sumsusmeera@gmail.com, Kaleep.swain@gmail.com

Abstract

Epilepsy is the unusual electrical signal flow in the brain that causes different varieties of seizures and affects the normal activities of a person. A lot of emphasis is put in the field of accurate epilepsy detection so that preventive steps can be taken to recover from the epilepsy. In this research, we work with three meta-heuristic search-based optimisation techniques for epilepsy detection: symbiotic-organism search algorithm (SOSA), grasshopper optimisation algorithm (GOA), and farmland fertility algorithm (FFA). We obtained accuracy, sensitivity, specificity, positive predictive value and area-under curve value for all these three algorithms and the obtained values of these parameters for FFA are: 99.8%, 99.35%, 99.82%, 96.85% and 1 which proves that FFA is performing efficiently than SOSA and GOA. All the methods can be integrated with hybrid optimisation techniques in future.

Keywords: SOSA, FFA, GOA and SVM\

1. Introduction

Epilepsy is a brain irregularity generated by the unusual electrical signal flow which is the vital means of seizure symptoms. These disorder activities divert the natural harmony of the human and affects the consciousness of a healthy person (Ahmad (2022)). As per the data given by World Health Organization (WHO) nearly 52 million people are affected by epilepsy which claims that the detection of epilepsy must be conducted in a fast, accurate and secure manner. According to the medical practitioner's statement, epileptic seizure can be classified into partial seizure and generalized seizure (Tang (2020)). Partial seizures are caused due to the affected portion of only a section of the cerebral hemisphere which are categorized

Chapter 11 DOI: 10.1201/9781003581215

as simple partial and complex partial. As per the observation, it is known that a simple partial is not harmful, it only affects the proper communication of the patient. In case of complex-partial, the patient behaves abnormally and lost consciousness. In case of generalized seizures, the toral portion of the brain is affected quickly. By electroencephalogram (EEG) early and accurate detection of the brain waves can be possible to ensure the recovery of the patient from the affected seizure. For the diagnosis, medical practitioners use intra-cranial EEG signals (Turang (2018)). Presently advanced meta-heuristic optimization techniques are used to help in real-time detection of seizure symptoms . The above stated algorithms ensure in experimental basis the accurate and most secure seizure detection.

2. Literature Review

In earlier times, the epilepsy symptom analysis is carried out by electroencephalogram (EEG) using metal discs attached to the scalp part of the brain cells which is time consuming and will not serve the need in real-time incidents . In order to overcome this barrier, fast and accurate seizure defection techniques are devised for automatic seizure detection. Presently this automatic seizure detection follows the various metaheuristic optimization and deep learning techniques (Goldanloo and Gharehchopogh (2022)). In paper (Shayanfar and Gharehchopogh (2018)) authors introduced a meta-heuristic optimization algorithm which is called as farmland fertility algorithm and in paper (Truong and *et. al.* (2020)) authors introduced nature inspired quasi-oppositional chaotic symbiotic organism search algorithm. Authors S. K. Mohapatra and S. Patnaik in year 2022 presents an improved atom search optimization algorithm by introducing levy flight as a stochastic process to enhance the exploration and exploitation in the search space of the object (S. K. Mohapatra and S. Patnaik (2022)).

3. Materials and Methods Used

3.1. Dataset Used

The Child Health Boston (CHB) is a dataset for epileptic intractable seizure recordings of paediatric subjects. The dataset link is: https://archive.physionet.org/pn6/chbmit/.

3.2. Feature Extraction

The statistical features extracted after wavelet transformation are expressed by: Mean, variance, standard deviation and entropy. In order to obtain the time-frequency domain discrete wavelet transformation is used.

3.3. Optimisation Algorithms

3.3.1. *Grasshopper optimisation algorithm*

Grasshopper optimisation algorithm is used in most of the heuristic search cases which are multi-constraint in nature. The foragingness of the swarm is mathematically defined as per the reference of paper (Arora and Anand, 2019).

$$Y_i = S_i + G_i + W_i \tag{1}$$

Yi is the current position of the ith grasshopper, Si is the social gatherings, Gi is the gravitational force and Wi is the wind flow. The updated position is:

$$Y_i^d = c_1 \left(\sum_{i=1, j \neq i}^{N} c_2 \frac{ub_d - lb_d}{2} s\left(\left|x_j^d - x_i^d\right|\right) \frac{x_j - x_i}{d_{ij}} \right) + \widehat{T_d} \tag{2}$$

Algorithm 1: Pseudocode of Grasshopper optimization algorithm
1 Initialize parameters of the eco-system
2 Initialize grasshopper position,
3 Compute fitness value of each grasshopper
4 while (target not met)
5   Normalize d_{ij} and update Y_i^d
6   Update
7 end
8 Return T

3.3.3. Farmland Fertility Algorithm

It is inspired by the nature to increase the soil productivity. The algorithm follows mainly 6 steps.

Algorithm 2: Pseudocode of Farmland Fertility algorithm

---

1   Initial solution

2   Test soil quality

3   Update memory

4   Change soil quality

5   Soil combination

6   *if result matches*

7   *go to step 3*

8   else final solution

3.3.4. Symbiotic Organisms Search

It is based on the mutualism, commensalism and parasitism nature of the same or different organism in an ecosystem where each of these organisms has a fitness value.

Algorithm 3: Pseudocode of Symbiotic Organism Search Algorithm (Truong *et al.*, 2019)

---

1   Initialise ecosystem

2   *while stopping condition not satisfied*

3   *for i=ecosize*

4   *Update mutualism phase*

5   *Update commensalism phase*

6   *Update parasitism phase*

7   *return* X_{best}

4. Evaluation and Experimental Results

Proposed model contains the extraction method and classifier and it uses 10 fold cross validation

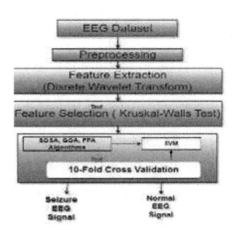

Figure 1: Proposed Model

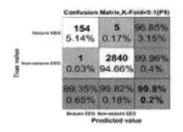

Figure 2: Confusion Matrix

5. Future Scope and Conclusion

Our experimental results show that the farmland fertility algorithm (FFA), when combined with Discrete Wavelet Transform (DWT) and Support Vector Machine (SVM), outperforms the other two algorithms tested—symbiotic-organism search algorithm (SOSA) and grasshopper optimisation algorithm (GOA)—in the form of accuracy, sensitivity, specificity, positive predictive value, and area under the curve. FFA's higher performance metrics suggest its potential as a more reliable and accurate tool in medical diagnostic applications, particularly in epilepsy detection. In future, we plan to integrate with Internet of Medical Things (IoMT)

Table 1: The experimental cases

Experimental Cases	ACC (%)	SEN (%)	SPE (%)	PPV (%)	MCC (%)	AUC
DWT-SOSA-SVM	99.4	96.84	99.29	94.84	95.83	0.995
DWT-FFA-SVM	99.8	99.35	99.82	96.85	98	1
DWT-GOA-SVM	99.2	95.65	99.47	91.66	93.61	0.998

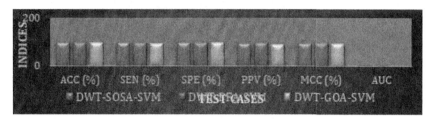

Figure 3: Measurement paramerters

Table 2. Comparison with Existing methods

Sl. No.	Authors and Reference. Numbers	Methods Used	Performance Measure Parameters		
			SPE (%)	SEN (%)	ACC(%)
1	L. Tang, and et al. [2]	LFDA+WPD	95.32	91.35	98.0
2	N. D. Truong and et al. [3]	STFT+CNN	87.23	86.46	97.80
3	Mohapatra S. K and et al. [8]	ESA-ASO + TQWT + LSSVM	91.46	91.11	98.37
4	Proposed Method	(Best Result Finding: DWT-FFA-SVM	99.82	99.35	99.8

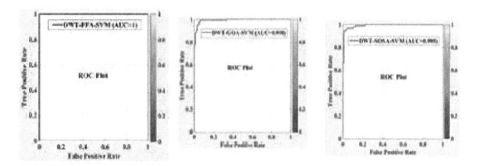

Figure 4: ROC Curves of DWT-FFA-SVM, DWT-GOA-SVM and DWT-SOSA-SVM

References

Ahmad J, Wang X, and *et. al.*, EEG-Based Epileptic Seizure Detection via Machine/Deep Learning Approaches: A Systematic Review", Hindawi, *Computational Intelligence and Neuroscience* Volume 2022, Article ID 6486570, 20pages, https://doi.org/10.1155/2022/6486570. pp. 1-20.

Arora, S. and Anand, P. (2019). "Chaotic grasshopper optimization algorithm for global optimization." *Neural Computing and Applications*, 31, 4385-4405.

Goldanloo M. N. and Gharehchopogh F. S. (2022). "A hybrid OBL-based firefly algorithm with symbiotic organisms search algorithm for solving continuous optimization problems." *The Journal of Supercomputing*, 78, 3998-4031, Springer.

Mohapatra S. K, and Patnaik S, "ESA-ASO: An enhanced search ability based atom search optimization algorithm for epileptic seizure detection", *Measurement Sensors*, Volume 24, Elsevier. December 2022, 100519.

Selesnick, I. W. (2011). "Wavelet transform with tunable Q-factor IEEE transactions on signal processing." 8 (59): 3560-75. doi: 10.1109. *TSP*.

Seng C. H., Demirli, R., *et al.* (2012). "Seizure detection in EEG signals using support vector machines." *38th Annual Northeast Bioengineering Conference (NEBEC)*, IEEE, 2012.

Shayanfar H, Gharehchopogh S F, "Farmland Fertility: A New Metaheuristic Algorithm for Solving Continuous Optimization Problems", *Applied Soft Computing Journal* xxx (2018) xxx-xxx, Elsevier, 1568-4946/ © 2018. pp. 728-746

Tang, L., Zhao, M., and Wu, X. (2020). "Accurate classification of epilepsy seizure types using wavelet packet decomposition and local detrended fluctuation analysis." *Electronics Letters*, 56 (17): 861-863.

Truong, K. H., Nallagownden, P., *et al.*" (2019). A quasi-oppositional-chaotic symbiotic organisms search algorithm for global optimization problems." *Applied Soft Computing Journal*, Elsevier.

Truong, N. D., et al. " Convolutional neural networks for seizure prediction using intracranial and scalp electroencephalogram", *Neural Network.*, vol. 105, pp. 104–111, 2018

Zhang, X., Yao, L., Dong, M., Liu, Z., Zhang, Y., and Li, Y. (2020). "Adversarial representation learning for robust patient-independent epileptic seizure detection." *IEEE Journal of Biomedical and Health Informatics*, 24 (10), 2852-2859.

A Novel Approach to Detect Epileptic Seizure from EEG Signal Using WPT and LSSVM Classifier Optimised by Tunicate Swarm Algorithm

Sumanta Kumar Mohapatra[1], K. P. Swain[1], S. K. Nayak[2],
Parshuram Behera[1], Basudev Das[1]

[1]Department of ETC, Trident Academy of Technology, Bhubaneswar, India
[2]BPUT, Odisha, India
E-mail: sumsusmeera@gmail.com, Kaleep.swain@gmail.com

Abstract

In neuroscience, epilepsy is treated as a critical brain disorder that generates abnormal symptoms called a seizure. The importance of this experimental study is to implement a new patient specific real time machine intelligence learning model to automate the complexity of epileptic seizure detection. Here we propose, a novel machine learning method that uses Tunicate Swarm Algorithm (TSA) to improve the global search capabilities. For decomposing signals and extracting the features, the integrated Wavelet Packet Transform (WPT) method and Kruskal-Wallis (KW) method is applied to choose the appropriate features. To validate the accuracy of our experiment, we have taken publicly available Bonn-University EEG datasets. The proposed WPT-TSA-LSSVM method achieved the highest accuracy (ACC) of 99.8% per C-E set and 99.67% for AB-CD-E set with AUC=1. The comparison results of proposed method with existing methods, confirm that the works of our proposed method is effective for proper medical diagnosis of epileptic seizure detection.

Keywords: WPT, LSSVM, Tunicate Swarm Algorithm, EEG signal, epileptic seizure signals

1. Introduction

Recently, the stroke is a very general symptom of epilepsy Khan (2020). The analysis of epilepsy shows that the seizures generated due to abnormality of EEG signals in neurons Raghu (2018). From the statistical data of World Health Organization, the young mass who are continuing between the ages of 10 and 20 are most affected Mohapatra (2022), Bandarabadi (2015). The sudden unexplained death rate rapidly increases due to uncontrolled seizure. Therefore, many researchers are willing to develop automated epileptic seizure detection models. In the last 10 years, the researchers are developed very rapid detection methods for proper and accurate clinical operations. In this context, there are two phases: seizure prediction and seizure detection Zabini (2016), Gill (2015), Cheng (2021). The symptom of seizure prediction indicates the formation of continuous deviations in normal functioning of the brain before the seizure on set and then analysis of the obtained signal by classifying to seizure EEG and non-seizure EEG signals Jiang (2021), Peng (2021).

2. Literature Review

LDA and Navie Bayesian for EEG classification was presented by Zabihi. In this case, 24-subjects are used. It presented with the classification accuracy of EEG 93.11%, specificity93.21% and sensitivity 88.27% Chandel (2019). A GMM classifier for classification was presented by Gill Gill (2015). In this paper, 12- subjects are used to complete the whole experimental analysis. It presented with the classification accuracy of EEG 86.93%, specificity 87.58% and sensitivity 86.26%. Cheng et al. proposed a novel classification method of accuracy 99.34%, and sensitivity of 99.60% Cheng (2021). Jiang et al. presented a novel method in the context of synchro extracting chirp let transform (SECT) Satnam (2020). In this paper, SVM classifier and CHB-MIT EEG signal is used to perform the experimental analysis. It performed accuracy rate of 99.17, specificity of 99.33% and sensitivity of 99.28%.

3. Materials and Methods Used

3.1. EEG Dataset

In this work, University of Bonn scalp EEG dataset (Satnam Kaur *et al.*, 2020) is used and is presented in Table-1. In this work

five cases of date sets are taken as A-E, C-E, A-D-E, AB-CD-E and ABCD-E.

3.2. Feature Extraction

All the EEG datasets including scalp EEG data are decomposed by WPT and feature extraction is done by statistical methods viz. mean, standard deviation, variance and Shanonn entropy.

Tunicate Swarm Algorithm (Chandel *et al.*, 2019)

3.3. Performance Evaluation

All the EEG datasets including scalp EEG data are decomposed by WPT and feature extraction is done by statistical methods viz. mean, standard deviation, variance and Shanonn entropy.

4. Experimental Results and Discussion

We prepared five subsets as A-E, C-E, A-D-E, AB-CD-E and ABCD-E. The correctness of the stated optimization technique has been performed by LSSVM. Table 2 represents the results of test cases whereas Table 3 showcases the comparison matrix with existing model. Fig. 3 indicates the performance curve for C-E EEG dataset and ROC curve for different EEG datasets.

Tunicate Swarm Algorithm [11]
The steps of tunicate swarm algorithm are represented in Fig.1.

Table 1: EEG Dataset

Setup	Data Set A	Data Set B	Data Set C	Data Set D	Data Set E
Subjects and Signal types	Healthy EEG signal	Healthy EEG signal	Non-Seizure EEG signal	Non-Seizure EEG signal	Seizure EEG signal
Electrode types	surface	surface	Intra-cranial	Intra-cranial	Intra-cranial
channels	100	100	100	100	100
Segment duration in second	23.6	23.6	23.6	23.6	23.6

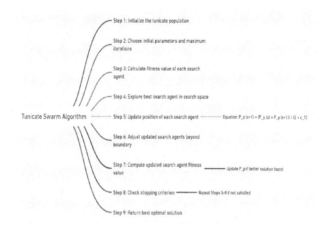

Figure 1: Tunicate Swarm Algorithm

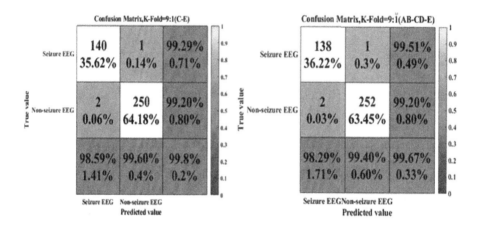

Fig 2: Confusion matrix for C-E and AB-CD-E dataset.

Table 2: Experimental Cases

Test Cases	ACC(%)	SEN(%)	SPE(%)	PPV(%)	MCC (%)	AUC
A-E	99	96.84	99.29	94.84	95.83	0.995
C-E	99.8	99.35	99.82	96.85	98	1
A-D-E	99.25	95.65	99.47	91.66	93.61	0.998
AB-CD-E	99.67	99.15	99.73	98	98.59	1
ABCD-E	97.5	93.4	98.23	90.42	91.89	0.981

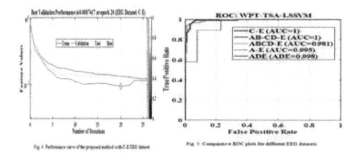

Figure 3: Performance curve for C-E EEG dataset and ROC plot for different EEG

Table 3: omparative analysis of the proposed method with existing methods

Sl. No.	Authors and Reference. Numbers	Methods Used	Performance Measure Parameters		
			SPE (%)	SEN (%)	ACC(%)
1	Zabihi *et al.* (2016)	LDA+Navie Bayesian	93.21	88.27	93.11
2	Gill *et al.* (2015)	GMM Classifier	87.58	86.26	86.93
3	Cheng *et al.* (2021)	Wavelet Energy+Bi-LSTM	99.60	99.34	99.47
4	Jiang *et al.* (2021)	Synchroextracting Chirpet Transform (SECT)	99.33	99.28	99.17
5	Proposed Method	(Best Result Finding: WPT-TSA-SVM) for C-E Dataset	99.82	99.35	99.8

5. Conclusion and Future Scope

This work represents an effective algorithm named as WPT-TSA-LSSVM for epileptic seizure detection with Bonn University EEG signals. Here the LSSVM is employed for classification. When compared with other methods, the proposed method achieved high accuracy of 99.8%, sensitivity of 99.35% and specificity of 99.82%. In future, to improve the detection ability different classifiers with existing hybrid optimization algorithms can be used with Internet of medical Things (IoMT).

References

Khan, K. A., Shanir, P. P., Khan, Y. U., and Farooq, O. (2020). "A hybrid Local Binary Pattern and wavelets-based approach for EEG classification for diagnosing epilepsy." *Expert Systems with Applications,* 140: 112895.

Raghu, S. and Sriraam. (2018). "Classification of focal and non-focal EEG signals using neighbourhood component analysis and machine learning algorithms". *Expert Systems with Applications,* 113: 18-32.

Mohapatra, S. K. and Patnaik S. "ESA-ASO: An enhanced search ability based atom search optimization algorithm for epileptic seizure detection." *Measurement Sensors,* Elsevier.

Bandarabadi, M., Rasekhi, J., Teixeira, C. A., Karami, M. R., and Dourado, A. (2015). "On the proper selection of preictal period for seizure prediction." *Epilepsy and Behavior,* 46: 158-166.

Zabini, M., *et al.* (2016). "Analysis of high-dimensional phase space via Poincare section for patient- specific seizure detection." *IEEE Transaction Neural Systems Rehabiliation Engineering*, 24 (3): 386-398.

A. F. Gill *et al.*, Analysis of EEG signals for detection of epileptic seizure using hybrid feature set, In Theory and Applications of Applied Electromagnetics (2015) 49-57.

Cheng *et al.* , Patient- specific method of sleep electroencephalography using wavelet packet transform and Bi-LSTM for epileptic seizure prediction, Biomedical signal processing and control 70 (2021).

Jiang Yun *et al.*, Synchroextracting chirplet transform- based epileptic seizure detection using EEG, Biomedical signal processing and control, 68 (2021).

Peng *et al.* ,Automatic epileptic seizure detection via Stein Kernel-based sparse representation, Computers in Biology and Medicine, 132 (2021).

G. Chandel, P. Upadhyaya, O. Farooq, Y.U. Khan, Detection of seizure event and its onset/offset using orthonormal triadic wavelet based features, IRBM 40 (2) (2019) 103–112.

Satnam Kaur, Lalit K. Awasthi, A.L. Sangal, Gaurav Dhiman Tunicate Swarm Algorithm: A new bio-inspired based metaheuristic paradigm for global optimization, Engineering Applications of Artificial Intelligence, Volume 90, April 2020, 103541.

Antral Gastritis Preliminary Diagnosis by Endoscopic Image Analysis Using Deep Learning Approach

Lalit Kumar Behera, Satya Narayan Tripathy

Department of ComputerScience, Berhampur University,Berhampur, India,
E-mail: lkb.rs.cs@buodisha.edu.in, snt.cs@buodisha.edu.in

Abstract

Diagnosis methods of diseases automatically are the need of the hour in the medical healthcare sector. For this purpose, artificial intelligence, machine learning and subfields like deep learning are profusely used to analyze different images produced in mammography, x-ray, radiography, ultrasound, angiography, echo-cardiograph endoscopic imaging etc. Along with this, the amalgamation of explainable artificial intelligence is playing a pivotal role in the trustworthy prediction of various concerns of diseases. Nowadays antral gastritis has become a common gastrointestinal problem faced by a significant amount of the population in the world. For the effective treatment of the patients, accurate diagnosis at the proper time can lead to solving this type of gastrointestinal problem. In this paper, we have approached to analyze preliminarily the endoscopic images of antral gastritis with the help of a convolutional neural network (CNN). This is the primary and most studied method among many deep learning methods. Such type of visual explanation methods can assist medical endoscopists in their diagnosis of various aspects of antral gastritis providing a direction from the very beginning of the process of treatment.

Keywords: Antral gastritis, deep learning, endoscopic image

1. Introduction

Physically and mentally wellness without any disease infers good health. However, due to various factors, many people are suffering from gastrointestinal (GI) problems in the world. The increasing number of cases is creating a challenge in the healthcare service domain. If not treated at the proper time antral gastritis may develop into gastric cancer which is a chronic condition of the gastrointestinal path. The mucus layer on the mucosa of the stomach guards it from acid that absorbs nutrients from diet. Due to different reasons, this lining of protection of the stomach gets disturbed. This results in the inflammation of mucosa causing gastritis. Antral gastritis is related most of the time to Helicobacter pylori infection, different factors of the environment mostly unidentified, and also the autoimmunity of glandular cells. Gastritis is proliferating as a silent killer of a huge population with severe implications due to peptic ulcer and gastric cancer. More than 50% of the population worldwide has chronic gastritis at the present time (Azer et al., 2023). Gastritis proliferates from childhood by *H. pylori* having a microbial origin. From this point, the epidemiology and the course of the disease can be traced out. The structural abnormality (atrophic gastritis) of stomach mucosa is the final consequence of an aggressive inflammation persisting throughout life. This dysfunction of stomach mucosa leads to a stomach having a lack of required acid. The maximum causative factor for gastric problems is the lack of essential acids in the stomach. These malignant conditions result in the potential reduction, even failure in the absorption and assimilation of essential micronutrients from food, vitamins and medicines. In Chronic cases, atrophic gastritis is generally considered a precondition of cancer (Zhao *et al.*, 2022).

2. CNN in the Domain of Deep Learning

Growth of deep learning is inevitable due to its satisfactory throughput. This scenario also involves the increment of the capability of computer devices with respect to computing. . It will be assisted by the improvement of graphics processing units (GPU). In particular convolutional neural networks can categorise pathological tissue images. Now a day it is used in the case of brain, breast, lung, prostate, and skin cancer-related images (Ibrahim *et al.*, 2022). However, the complex structure with many perceptron layers, filter designing for feature extraction, designing the pooling layer and sharing of obtained information among different layers leads to non-linearity in the operations of convolutional neural network models. Again non-standard images for the input are posing another challenge. This happens as they are frequently influenced by camera variations, image distortions, and the skill of endoscopist. In the light of above points now it is a fact that convolutional neural networks extract local features with their well designed filters and the max pooling operations working iteratively in a loop that preserves the information but reduces the error (Abdelhafiz *et al.*, 2019). Having the foundation of multilayer perceptron (MLP), a convolutional neural network implements its convolution dynamically. The representative image is organised in many dimensions like height, width and depth in each layer of a neural network model. Filters use bias and weight to prepare feature maps. The product between the input and the weights added with bias is passed as a parameter for the activation function which at last gives the throughput. This is given in eq. 1 below.

$$h_k = f(W_k * x + b_k) \qquad (1)$$

Here bk denotes the bias , wk denotes weight and x is the input to the input layer of the neural network. hk represents the output and k represents the number of filters. The pooling function is applied on all feature maps for the area of filter size. In the end, the feature collection layers obtain the pooled features, culminating with proper inferences but maintaining a high abstraction level (Alzubaidi *et al.*, 2021). The basic framework is given in Figure 1.

3. Data Preparation

Dealing with an excessive number of endoscopic images is becoming a tedious task day by day. However, medical image datasets commonly available are scanty in the public domain. Because data acquisition, preparation, and curation add complexities to their appearance in the public domain (Canales Fiscal *et al.*, 2023). For our proposed work we took images from GitHub site. The URL is https://github. com/DebeshJha/GastroVision. Most of the images are JPEG type with pixel size 1280x1024.

4. Model Development

To verify the performance of the CNN design proposed here for preliminary disease detection, we have proposed a binary classification operation. For our model, we used the google colab platform to run the programme. First, we imported required library packages like TensorFlow,

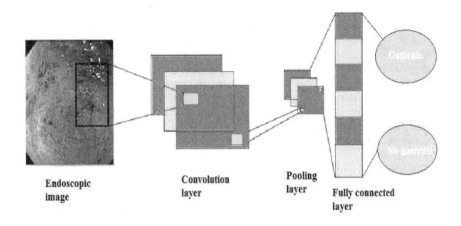

Endoscopic image   Convolution layer   Pooling layer   Fully connected layer

Figure 1: Basic framework of deep learning using convolution neural network for our research work

Keras , Sequential, Dense, Conv2D, MaxPooling2D, Flatten etc. Then using keras.utils.image_dataset_from_directory( ) function we prepared the training dataset. In a similar way validation dataset was prepared. We normalised the size of image. At this point, our data is ready to be processed by the neural network. Here the image size was maintained to 256x256 with RGB value 3. Thus basic image processing steps were done for medical images before diagnosis. This includes resizing images, sometimes grayscale conversion as well as normalisation. In a similar manner other layers like pooling and feature collection are implemented with the conv2d( ) and dense( ) functions with proper signatures. In the last output layer sigmoid activation function ( ) is implemented to deal with the nonlinearity.

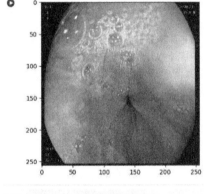

```
[23]  1 val=model.predict(test_input)
      2 if val==0:
      3   print(" it is a case of gastric")
      4 else:
      5   print(" it is not a case of gastric")
```

Figure 2: Sample output on running the CNN model

5. Deployment of the Model

For the deployment of the model, we compiled the designed model with the help of model.compile() function with a signature of binary cross entropy. Here we used adam optimiser as our approach is binary type. We gave accuracy as a parameter for matrices of compile function. We ran model.fit () with a train dataset with different epoch values. The validation accuracy is 81.82% with an epoch value of 10. With varying epoch size, the accuracy also shows variation. Finally, to get the inference we ran the code which is given in Figure 2.

```
1 history=model.fit(train_ds,epochs=12,validation_data=validate_ds)
```

```
Epoch 1/12
3/3 [==============================] - 3s 642ms/step - loss: 5.4219e-04 - accuracy: 1.0000 - val_loss: 0.7417 - val_accuracy: 0.8182
Epoch 2/12
3/3 [==============================] - 2s 421ms/step - loss: 4.5198e-04 - accuracy: 1.0000 - val_loss: 0.7146 - val_accuracy: 0.8182
Epoch 3/12
3/3 [==============================] - 2s 465ms/step - loss: 2.9350e-04 - accuracy: 1.0000 - val_loss: 0.6817 - val_accuracy: 0.8636
Epoch 4/12
3/3 [==============================] - 2s 637ms/step - loss: 2.0721e-04 - accuracy: 1.0000 - val_loss: 0.6504 - val_accuracy: 0.8636
Epoch 5/12
3/3 [==============================] - 3s 590ms/step - loss: 1.6593e-04 - accuracy: 1.0000 - val_loss: 0.6257 - val_accuracy: 0.8636
Epoch 6/12
3/3 [==============================] - 2s 416ms/step - loss: 1.5427e-04 - accuracy: 1.0000 - val_loss: 0.6153 - val_accuracy: 0.8636
Epoch 7/12
3/3 [==============================] - 2s 412ms/step - loss: 1.6377e-04 - accuracy: 1.0000 - val_loss: 0.6130 - val_accuracy: 0.8636
Epoch 8/12
3/3 [==============================] - 2s 425ms/step - loss: 1.6259e-04 - accuracy: 1.0000 - val_loss: 0.6171 - val_accuracy: 0.8636
Epoch 9/12
3/3 [==============================] - 2s 412ms/step - loss: 1.4827e-04 - accuracy: 1.0000 - val_loss: 0.6267 - val_accuracy: 0.8636
Epoch 10/12
3/3 [==============================] - 2s 416ms/step - loss: 1.2813e-04 - accuracy: 1.0000 - val_loss: 0.6439 - val_accuracy: 0.8636
Epoch 11/12
3/3 [==============================] - 3s 679ms/step - loss: 1.0085e-04 - accuracy: 1.0000 - val_loss: 0.6679 - val_accuracy: 0.8636
Epoch 12/12
3/3 [==============================] - 3s 612ms/step - loss: 8.3479e-05 - accuracy: 1.0000 - val_loss: 0.7001 - val_accuracy: 0.8636
```

Figure 3: Sample output of model.fit() function showing validation accuracy of 86.36%

6. Discussion

There is a high risk of gastric adenocarcinoma in case of patients with antral gastritis (Lin *et al.*, 2021). This work develops the convolution neural network with certain layers with specific characteristics. In our work the convolutional neural network detects certain portions of the gastric antrum, producing binary throughput. For an epoch value of 12, the validation accuracy is 86.36% as shown in Figure 3 below justifying the performance of this model. This novel model can help clinicians in preliminary disease analysis. This type of artificial intelligence system which is accompanied by real-time endoscopic images has the potential to replace and upgrade the existing diagnostic modalities.

7. Conclusion

Antral gastritis is becoming a common health issue day by day. Millions of new cases are observed each year. Such type of health problems generally puts the patient in mental distress and continuous financial loss. In this research work, a novel deep learning approach like an artificial convolutional neural network for image analysis will perceive a way to further develop the models. This model offers a promising base to expand the field of medical image utilisation properly for preliminary-level diagnosis of the disease.

References

Abdelhafiz, D., Yang, C., Ammar, R., *et al.* (2019). "Deep convolutional neural networks for mammography: advances, challenges and applications." *BMC Bioinformatics*, https://doi.org/10.1186/s12859-019-2823-4.

Alzubaidi, L., Zhang, J., Humaidi, A. J., *et al.* (2021). "Review of deep learning: concepts, CNN architectures, challenges, applications, future directions." *Journal of Big Data*, https://doi.org/10.1186/s40537-021-00444-8.

Azer, S. A., Awosika, A. O., Akhondi, H. (2023). "Gastritis. [Updated 2023 Oct 30]." In: StatPearls [Internet]. Treasure Island (FL): StatPearls Publishing; 2023 Jan.

Canales-Fiscal, M. R. and Tamez-Peña, J. G. (2023). "Hybrid morphological-convolutional neural networks for computer-aided diagnosis." *Frontiers in Artificial Intelligence* 6: 1253183. doi: 10.3389/frai.2023.1253183.

Ibrahim, A., Mohamed, H. K., Maher A., and Zhang, B. (2022). "A survey on uman cancer categorization based on deep learning." *Frontiers in Artificial Intelligence*, 5: 884749. doi: 10.3389/frai.2022.884749.

Jin, Z., Gan, T., Wang, P., *et al.* (2022). "Deep learning for gastroscopic images: computer-aided techniques for clinicians." *Biomedical Engineering Online*, 21(1): 12. https://doi.org/10.1186/s12938-022-00979-8.

Lin, N., Yu, T., Zheng, W., *et al.* (2021). "Simultaneous recognition of atrophic gastritis and intestinal metaplasia on white light endoscopic images based on convolutional neural networks: a multicenter study." *Clinical and Translational Gastroenterology*, 12(8): e00385. https://doi.org/10.14309/ctg.0000000000000385.

Zhao, Q., Jia, Q., and Chi, T. (2022). "Deep learning as a novel method for endoscopic diagnosis of chronic atrophic gastritis: a prospective nested case-control study." *BMC Gastroenterology*, 22 (1): 352. https://doi.org/10.1186/s12876-022-02427-2.

Design and Development of Privacy Preservation Approach in Data Mining Using Multivariate Framework in Continuous and Multi-Dimensional Data

Shailesh Kumar Vyas, Swapnili Karmore

Department of CSE, G H Raisoni University, Saikheda, India
E-mail: shailesh.pk.29@gmail.com,swapnilikarmore@gmail.com

Abstract

Data mining raises serious privacy issues. For the purpose of protecting privacy in data mining, techniques have been developed. Privacy-preserving data mining for multi-dimensional data sets, however, results in significant data loss, information leakage. An innovative method called Multidimensional Noise Additive Model (MNAM), which improves cluster identification and privacy while reducing information leakage, is suggested in this study. Anonymization technique that incorporates aggregation is applied in multi-dimensional data sets which leads to the reduction of information leakage and enhances the privacy. The domain-specific range of the subspaces are taken into consideration when performing noise addition. The MNAM method considers Euclidian distances among the neighbouring clusters. Then, anonymized subspaces are subjected to random noise inside the subspace domain to improve cluster detection and minimize data loss.

Keywords: Data mining, MNAM, data loss, privacy, noise

1. Introduction

In order to achieve privacy, PPDM algorithms alter the data using either noise generation techniques (Liew *et al.*, 1985) or anonymization (Sweeney, 2002) obtaining privacy. The following difficulties face privacy preservation when working with high dimensional continuous datasets. Data is contained in subspaces, and there are a lot of them (Agrawal, 2005). The following are the contributions made by this paper: it suggests a brand-new method named Multidimensional Noise Additive Model (MNAM). 1) MNAM locates data in diversified dimensions of data sets. The clusters of diversified dimensions are aggregated to reduce the leakage of

information (Matthias et al. 2015). 2) Additive noise model identifies the clusters in diversified data sets that leads to the reduction of information leakage and preserves the privacy. 3) The anonymized subspaces are given more noise, lowering the chance of disclosure and enhancing privacy (Taipale, Kim A 2003).

2. Literature Review

Generalization and suppression rules for maintaining privacy were proposed by the K-anonymity technique (Defays and Nanopoulos, 1992). It made sure that groups of k comparable records were present in the released data. Micro aggregation has also been used to produce anonymization (Defays and Nanopoulos, 1992). There are numerous methods for achieving micro aggregation. Micro aggregation was accomplished using the individual ranking approach (Defays and MN., 1998). Each dimension was used to order the records, after which associations of k records have been established and the group aggregate was used to replace them (Panagopoulos et al. 2015). For both

3. Multidimensional Noise Additive Model (MNAM)

The noise introduced by existing privacy preservation method affects the cluster identification.

$$\text{upper}(S_{sd}) = ud_{sd} \tag{1}$$

$$\text{lower}(S_{sd}) = ld_{sd} \tag{2}$$

$$(X'[i][j] < ud_{sd}) \&\&((X'[i][j] > ld_{sd})) \tag{3}$$

$$X[i][j] = X_{sn}[i][j] + \text{random}(P) \tag{4}$$

In Eq. (4), represents the integration of random noise in the diversified multidimensional data sets (Table 1).

3.1. Proposed Algorithm MNAM

Input: High Dimensional DataSet H

   Input: d (Anonymization degree), S(restraint of noise), g' and l'

1 Finding compact subspaces S_{rd} and non-compact subspaces S_m
2 while$(i \le n)$):
3 while$(j \le n)$:
4 Computing S_{sd} (anonymized compact subspace)
5 end while loop
6 end while loop
7 while$(i \le n)$:
8 While$(j \le n)$:
9 Computing S_{rd}
10 end while loop
11 end while loop
12 for each $S_{rd:}$
13 $Ud_{sd} = \max(S_{rd:})$
14 $Ld_{sd} = \min(S_{rd})$
15 while$(i \le n)$:
16 $X'[i][j] = [i][j] + \text{rand}(p)$

Table 1: Actual high dimensional data set X

	X_1	X_2	X_3	X_4	X_5	X_6	X_7	X_8
Y_1	28	70	28	39	70	93	28	71
Y_2	22	65	24	51	64	84	23	65
Y_3	28	65	29	54	65	78	30	65
Y_4	28	64	27	48	64	95	28	63
Y_5	24	69	23	43	71	88	22	69
Y_6	23	70	22	50	71	90	22	71
Y_7	29	64	27	56	64	78	28	65
Y_8	23	70	23	37	70	84	24	70

17 $((X'[i][j] < Ud_{sd}) \&\& (X'[i][j]\ Ld_{sd})$
18 end while loop
19 end while loop
20 for each $S_{m:}$
21 $Un_{sn} = \text{maximum}(S_{sn})$
22 $Ld_{sn} = \text{minimum}(S_{sn})$
23 while$(i \le n)$:
24 $X'[i][j] = X_{sn}[i][j] + \text{rand}(p)$
25 $(\ (X'.[i][j] < Ud_{sn}) \&\& (X'[i][j] > Ld_{sn}.)\ )$
26 end while loop
27 end for loop

3.2. Explanation of MNAM Algorithm

Subspaces consist of data with high dimensions. As an outcome, Line 1 applies subspace clustering to determine compact subspaces S_{rd} and non-compact subspaces (S_m). The compact and non-compact subspaces are separately transformed to optimize the value of data and reduce information loss. To obtain anonymized subspaces $(S_{rd.})$, adulteration of noise is given by lines 2 to 6 along each dimension of data set. Lines 7 to 11 now anonymize non-compact subspaces to generate anonymized non-compact subspaces (S_{sn}). To improve cluster identification the anonymized subspaces are then given to noise addition. Two important parameters Ud_{sn} and Ld_{sn}. are evaluated for the upper limit and lower limit in each cluster of diversified data set that is given by line number 13 and 14 in the algorithm. As stated in Eq. (6) corresponds to the line number 16 to 19 that evaluates a parameter random (in the compact cluster of

Table 2: Anonymized subspaces with noise addition

	X_1	X_2	X_3	X_4	X_5	X_6	X_7	X_8
Y_1	29	71	29	43	71	92	29	71
Y_2	24	66	24	53	65	80	24	65
Y_3	28	65	29	52	66	76	28	65
Y_4	28	65	27	43	65	92	29	65
Y_5	24	70	23	42	71	91	24	70
Y_6	24	71	23	52	70	91	23	70
Y_7	29	66	29	53	65	80	29	65
Y_8	24	70	23	42	71	80	24	70

the diversified data set. Equations (3) and (4) corresponds to the line number 21 and 22 of the algorithm that evaluates the parameters Ud_{sn} and Ld_{sn} for the upper and lower limit of non-compact cluster in the diversified data set. Line number 24 to 27 evaluatethe parameter random (q) that integrates the random noise between lower and upper limit of non-compact cluster (Table 2).

Eq. (4) specifies the sum of square error (SSE).

$$\text{Mean} = \sum_{i=0}^{n} X[i][j]/n \qquad (5)$$

$$\text{Medean} = X\left[\frac{n}{2}\right] \qquad (6)$$

$$\text{Standard Deviation} = \sqrt{\frac{\Sigma(X_i - \mu)2}{N}} \qquad (7)$$

$$\text{Variance} = \Sigma\left(X_i - \bar{X}\right)/n - 1 \qquad (8)$$

Equation (5) specifies mean, Equation (6) specifies median, and Equation (7), Standard deviation is determined by Equation (7), while Variance is determined by Equation (8). The Eqs. (5)–(8) represents the reduction of data loss.

4. Result and Discussion

Associated works SNA, NALT, and NANLT techniques are compared with the suggested work MNAM. Below is the explanation of these techniques. Using additive noise, SNA protects privacy (Brand R, 2021).It raises the unbiased covariance value by adding a random value drawn from an error vector with a normal distribution. The distribution function is calculated using NANLT taking into account the range for each dataset dimension (Templ M., 2017). The distribution of the original data and the covariance are preserved by NANLT, but information loss is substantial and clusters are not recognized (Dittcompact D, Kenneally E, 2011).

δ is used as a metric for privacy measurement (Parsons et al. 2004). The best scenario is taken into consideration when comparing privacy to demonstrate that MNAM always offers significantly more privacy than the available techniques. In the best-case scenario, the disclosure risk probability is the same for all currently used strategies. However, it is clear that MNAM provides greater privacy than the methods now in use (Beyer K, Goldstein J, 1999). MNAM has a low disclosure chance of 0.301 even when s=2. Fig. 2 shows the Mean Variation Errors (MVE) within the Madelon dataset. Low MVE is the ideal.

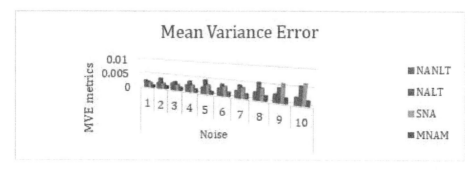

Figure 1: Mean variance error

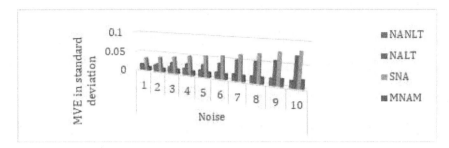

Figure 2: MVE in standard deviation

5. Conclusion

For the Arcene Dataset, MNAM minimizes disclosure risk by 80% to 50%, and for the Madelon Dataset, it reduces it by 90% to 55%. As MNAM incorporates noise to anonymized subspaces, it lowers the chance of disclosure (Gaby et al. 2015). The correlations among these dimensions as well as the data properties found in the original dataset are not taken into account by the current noise addition algorithms, which introduce noise to every dimension separately.

References

Agrawal R, Gehrke J,Gunopulos D, Raghavan R (2005) Automatic subspace clustering of high dimensional data for data mining applications. Data Min Knowl Disc 11(1):5–33

Defays D, Nanopoulos P (1992) Panels of enterprises and confidentiality: the small aggregates method. In: Proceedings of the symposium on design and analysis of longitudinal surveys. Statistics Canada, Ottawa, pp 195–204

Defays DA, MN. (1998) Masking microdata using micro-aggregation. 14(4): 449–461

Taipale, K. A. (2003). Data mining and domestic security: Connecting the dots to make sense of data. *Columbia Science and Technology Law Review*, 5(2): 242-251.

Dittcompact, D. and Kenneally, E. (2011). TheMenlo report: Ethical principles guiding information and communication technology research. US Department of Homeland Security.

Sweeney, L. (2002). k-anonymity: A model for protecting privacy. *International Journal Uncertain Fuzziness and Knowledge-based System*, 10: 557–570.

Gaby, G., Iqbal, M., and Fung, B. (2015). Fusion: Privacy-preserving distributed protocol for high-dimensional data mashup. *IEEE 21st International Conference on Parallel and Distributed Systems*.

Liew, C., Choi, C., and Liew, J. (1985). A data distortion by probability distribution. *ACM Transactions on Database System (TODS)*, 10(3): 395–411.

Brand, R. (2002). Microdata protection through noise addition. *Lecture Notes in Computer Science*. Springer.

Matthias, T., Alexander, K., and Bernhard, M. (2015). Statistical disclosure control for micro-data using the R package sdcMicro. *Journal of Statistical Software*, 67(4): 1–36. doi: 10.18637/jss.v067.i04

Templ, M. (2017). Disclosure risk. In *Statistical Disclosure Control for Microdata* (pp. 49–87). Springer.

Panagopoulos, P., Pappu, V., Xanthopoulos, P., and Pardalos, P. M. (2015). Constrained subspace classifier for high dimensional datasets. *Omega*. doi: 10.1016/j.omega-.2015.05.-009i

Beyer, K., and Goldstein, J. (1999). When is nearest neighbor meaningful?' *Proceedings of the 7th International Conference on Database Theory*. In: Database theory – ICDT'99, vol. 1540, pp. 217–235.

Parsons, L., Haque, E., and Liu, H. (2004). Subspace clustering for high dimensional data: A review. *ACM SIGKDD*, 6(1): 90–105.

Virtual IGA-GRU

An Approach for Load Balancing in Cloud Computing

Nihar Ranjan Sabat[1], Rashmi Ranjan Sahoo[2]

[1]Faculty of Engineering, Biju Patnaik University of Technology, Rourkela, India,
E-mail:n.ranjan9@gmail.com
[2]Department of Computer Science & Engineering, Parala Maharaja Engineering College,
Berhampur, India, E-mail: rashmiranjan.cse@pmec.ac.in

Abstract

In cloud computing, load balancing evenly allocates traffic and tasks to improve performance. We propose an intelligent genetic load balancing strategy for hybrid clouds in this paper. A genetic algorithm (GA) solves hybrid cloud load balancing problems, while a Gated Recurrent Unit (GRU) model classifies and ranks virtual machines. TensorFlow framework analysed the cloud environment and load balancer. The proposed approach achieved an average success index of 94%. The success rate of the suggested strategy has been compared to that of existing methodologies.

Keywords: Load balancing, Gated Recurrent Unit, and virtual machine

1. Introduction

Cloud computing enables clients to hire computational resources online to accelerate resource allocation, and cost savings. Throughput and response time depend on load balancing in dynamic cloud systems. Task allocation to computer resources enhances service quality, performance, and contract compliance. Computational complexity, client data loss, host, and new VM memory utilisation require load balancing modifications to fulfil processing and storage demands. To meet network resource restrictions, network management-compliant load-balancing was used to distribute traffic efficiently (Bhosale *et al.*, 2023). DCCO reduces cloud server workload and chooses the appropriate load-balancing VM using CMBO and DXO (Geeta, 2023).

The remainder of the paper is structured as follows. Section 2 reviews the existing literature. Section 3 explains the proposed methodology. Section 4 discusses the empirical findings. Section 5 summarizes the paper.

2. Literature Review

The Virtualised Intelligent Genetic Load Balancer method allocates resources while maintaining load and lowering span time for hybrid cloud load balancing (Rajkumar and Jeevaa, 2023). A unique efficient genetic load-balancing approach concerns cloud computing models, virtual machine allocation, and load balancing (Manikandan, 2022). The GRU network predicts cloud data centre host machine workload using statistics, spatial attribute extraction, and attention optimisation. (Dogani *et al.*, 2023a).

The CNN-GRU model employs SMOTE to classify Microsoft Azure virtual machine workload as Delay-Insensitive or Delay-Sensitive (Khodaverdian *et al.*, 2023). BiGRU calculates and predicts workload

Chapter 15 DOI: 10.1201/9781003581215

based on temporal correlation (Dogani et al., 2023b). Hybrid VTGAN models utilise technical indicators and transforms to evaluate sliding window size, prediction stages, and training data (Maiyza et al., 2023).

Proposed Methodology

The proposed approach leverages the Genetic Algorithm (GA) for hybrid cloud load balancing and the Gated Recurrent Unit (GRU) model for virtual machine classification and ranking, respectively.

3.1. Genetic Algorithm

Genetic algorithms can develop optimal solutions in complicated applications, and build hybrid cloud load balancing representation, fitness function, and parameters. The fitness f_i can be expressed as the root mean square error (RMSE) between the projected task load values and the actual task load values, determined by equation (1).

$$f_i = \frac{k}{f_i} \qquad (1)$$

The probability of selecting an individual i from the population, represented as Pi, can be computed using equation (2).

$$P_i = \frac{f_i}{\sum_{j=1}^{N} f_j} \qquad (2)$$

Bit-flip mutation was used because binary coding's unique properties make gene values opposite probability P_i. The interval has been partitioned into $(-a+a) \times 10^b$ discrete values for accurate b-bit binary to decimal conversion. The expression x^{bin} was subsequently modified to x^{dec}. Subsequent to that, the process of decrypting has been successfully executed, as indicated by equation (3).

$$x = -a + x^{dec} * \frac{2a}{2a^b - 1} \qquad (3)$$

GA dynamically adapts to conditions and optimises cloud computing performance by finding load distribution patterns.

3.2. Gated Recurrent Unit (GRU)

GRU is a popular deep learning RNN for sequence problems. A detailed analysis of the data, training process, and model integration into the cloud architecture is required before employing a GRU-based model. At each time step, the GRU changes the network's hidden state. Hidden state h_t in the GRU is calculated as follows:

GRU calculates the Update Gate using a sigmoid function and the current input, as illustrated in equation (4), to determine the extent to which to update the previous hidden state.

$$z_t = \sigma \left(W_z \cdot [h_{t-1}, x_t] + b_z \right) \qquad (4)$$

Here Previous hidden state h_{t-1}, gate vector z_t, sigmoid function σ, and updated gate weight matrix W_z. Update gate bias is b_z, current input is x_t, and final concatenation is $[h_{t-1}, x_t]$. The Reset Gate can be computed by applying a sigmoid function to reject previous data, as specified in equation (5).

$$r_t = \sigma \left(W_r \cdot [h_{t-1}, x_t] + b_r \right) \qquad (5)$$

Gate vector r_t, the weight matrix is W_r, and the bias term is b_r. The reset gate uses a hyperbolic tangent function in equation (6) to find the Candidate hidden state.

$$h_t' = tanh \left(W \cdot [r_t * h_{t-1}, x_t] + b \right) \qquad (6)$$

The weight matrix in this case is W, the candidate hidden state is h_t', element-wise multiplication is *, and bias is b. The update gate uses equation (7) to assess its influence, but linear interpolation finds the final hidden state by evaluating previous and potential hidden states.

$$h_t = (1 - z_t) * h_{t-1} + z_t * h_t' \qquad (7)$$

Here h_t is the Final hidden state. GRU's context-aware gating selectively updates its hidden state, making it ideal for cloud load balancing like classifying and ranking virtual machines.

4. Result and Discussion

All VM execution times were recorded. The host, guest, and virtualisers have been chosen. Guests picked components and communication, hosts handled VMs. With time data and the same load and weight variables, each virtual machine's accuracy factor was computed. An accuracy, precision, recall, measure, and success index were calculated. Iterations of trained and test data using index calculation methods are shown in Table 1.

The assessment of the Success Index and other associated outcomes of deep learning have been carried out using Table 1. Several iterations are conducted throughout the simulation of the cloud environment. An average Success Index of 94% has been obtained by our proposed virtual IGA-GRU load balancing approach.

Figure 1 illustrates a comparative analysis of the success index between our proposed IGA-GRU approach and other contemporary deep DBN, LSTM, MLP, etc.

5. Conclusion

Cloud computing uses hybrid load balancing to handle changing workloads and applications. Deep learning is required for cloud computing load balancing because it can analyse complex patterns, forecast, detect abnormalities, and respond to changing demand based resource allocation and deallocation. An intelligent genetic load-balancing technique in this research, a virtual IGA-GRU for hybrid clouds is developed, which uses a Gated Recurrent Unit (GRU) model for virtual machine classification and ranking, and a genetic algorithm (GA) to handle hybrid cloud load balancing challenges. The proposed hybrid approach has been simulated and tested using the TensorFlow framework. The proposed approach's Success Index has been compared to various important deep learning algorithms, including CNN, DBN, LSTM, MLP, etc. It was observed that the proposed method has the highest Success Index of 94% among all approaches. This approach has the potential to be implemented in the future for automated and dynamic datasets.

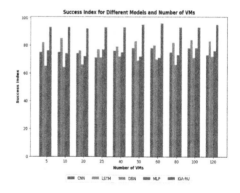

Figure 1. Success index analysis

Algorithm: *Proposed Virtual IGA–GRU Load Balancing Algorithm*

Input: Tasks T_k , $1 \leq k \leq n$
Output: Optimised Task Load Distribution

1:	Create a VMs and Data centres.	11:	Set up a GRU RNN model to rank VMs.
2:	Monitor load and virtual machine load status.	12:	Prioritise lower GRU for VMs.
3:	Cluster VMs by load or availability.	13:	Compute the max execution time and allocate VM.

4:    Assign requests to the lowest-priority VM cluster.

$$\text{Threshold}_{\text{High}} \rightarrow VM(D)\,|$$
$$\text{Min}\left(\sum \text{Threshold}(\text{High})\right) -$$
Minimum Task Execution Time

5:    Set dynamic quantum time and assess request VM load.

$$\text{Threshold}_{\text{Medium}} \rightarrow VM(D)\,|$$
$$\text{Min}\left(\begin{array}{c}\sum \text{Threshold}(\text{High}) + \\ \sum \text{Threshold}(\text{Medium})\end{array}\right) -$$
Minimum Task Execution Time

6:    Select quantum index, begin with lowest, allocate, and repeat until idle.

$$\text{Threshold}_{\text{Low}} \rightarrow VM(D)\,|$$
$$\text{Min}\left(\sum \text{Threshold}(\text{Low})\right) -$$
Minimum Task Execution Time

7:    Obtain each VM's load and capacity.

8:    Check VM overload / underload.

9:    Remove overloaded VM and resume round robin till completed.

10:    Group and update VMs by load, weight, and capacity.

14:    Adjust the process to the result and repeat till it stops.

15:    Find VM's burst, turnaround, and completion time.

16:    Set execution times from lowest to highest.

17:    Allocate VM tasks until threshold is zero or lower.

18:    Repeat step 1 till VMs finish.

Table 1: VM accuracy, precision, recall, measure, and success index

VMs	Accuracy	Precision	Recall	Measure	Success Index
5	0.95	0.15	0.87	0.77	93
10	0.95	0.11	0.88	0.79	93
20	0.96	0.15	0.87	0.78	92
25	0.96	0.12	0.86	0.76	93
40	0.96	0.13	0.89	0.74	93
50	0.95	0.15	0.88	0.78	95
60	0.95	0.12	0.87	0.75	96
80	0.96	0.13	0.87	0.78	93
100	0.95	0.15	0.88	0.75	93
120	0.96	0.12	0.88	0.79	95

References

Bhosale, N., Tejashri, S., and Shantanu, N. (2023). "Load balancing techniques in cloud computing." *Vidhyayan*, 8(7): 707-728.

Dogani, J., Farshad, K., Mohammad, R. M, and Mehdi, S. (2023). "Multivariate workload and resource prediction in cloud computing using CNN and GRU by attention mechanism." *Journal of Supercomputing*, 79(3): 3437-3470.

Dogani, J., Farshad, K., and Mehdi, S. (2023). "Host load prediction in cloud computing with Discrete Wavelet Transformation (DWT) and Bidirectional Gated Recurrent Unit (BiGRU) network." *Competition Commission*, 198: 157-174.

Geeta, K. and Kamakshi Prasad, V. (2023). "Multi-objective cloud load-balancing with hybrid optimization." *International Journal of Computers and Applications*, ˈ45(10): 611-625.

Khodaverdian, Z., Hossein, S., Seyed, A. E., and Mojdeh, N. (2023). "An energy aware resource allocation based on combination of CNN and GRU for virtual machine selection." *Multimedia tools and applications*, 1-28.

Maiyza, A. I., Noha, O. K, Karim, B., Hanan, A. H, and Walaa, M. Sheta. (2023). "VTGAN: hybrid generative adversarial networks for cloud workload prediction." *Journal of Cloud Computing*, 12(1): 97.

Manikandan, S. and M. Chinnadurai. (2022). "Virtualized load balancer for hybrid cloud using genetic algorithm." *Intelligent Automation and Soft Computing*, 32(3): 1459-1466.

Rajkumar, S., and Jeevaa, K. (2023). Virtualized intelligent genetic load balancer for federated hybrid cloud environment using deep belief network classifier. *Journal of Cloud Computing*, 12(1), 138.

Fuzzy Logic Based Usability Sub-CharacteristicAnalysis for M-Learning Application Utilizing GQM & ISO 9241-11

Manish Mishra[1], Reena Dadhich[2]

[1]UOK, Kota, India
[2]Department of Computer Science & Informatics, University of Kota(UOK), Kota, India
E-mail:kota.phd@gmail.com, profrdadhich@uok.ac.in

Abstract

A mobile application has many quality characteristics. Usability is one of the characteristic which is used to and evaluate as per user expectation. Usability has many sub-characteristics, which defines the overall impact of usability. Usability is a qualitative framework specifically designed as per needs of customers, but the question is: are we develop mobile application in right direction as per usability expectation? This will only be fulfilling by numeric quantity of quality perception of sub-characteristics of usability. This research propose a fuzzy logic based evaluation criteria which converts qualitative framework of usability to numeric value by analyzing sub-characteristics of usability so that mobile development companies may take corrective measure by providing feedback to stakeholders, developers, and testers during development of mobile application. The goal of this paper is to provide a design qualitative framework for M-Learning application and convert in to quantitative measurement for sub-characteristics of usability expectation as per ISO 9241-11 quality model and GQM paradigm. The assessment is conducted to ensure user satisfaction and enhance the overall quality of the M-Learning application

Keywords: Usability framework, M-Learning, GQM, ISO 9241-11, Questionnaire.

1. Introduction

Usability refers to the extent to which software is user-friendly, efficient, and able to suit the diverse demands of its users. Various factors may influence the level of learning experienced by end-users. Optimal usability is essential for M-Learning systems. The reason for this is that usability plays a crucial role in determining the success of software. This research presents a method for evaluating the usability of mobile learning applications, which assesses the sub-characteristics of usability. This approach is based on the ISO 9241-11 quality model and GQM paradigm, which establishes usability metrics for measuring the sub-characteristics of usability and efficacy of M-Learning applications. The framework is then used in a case study with an M-Learning application to showcase its applicability by converting qualitative framework to quantitative measurements, which further uses for corrective measures if it is required.

Chapter 16 DOI: 10.1201/9781003581215

2. Literature Review

Software development is defined as continuous state of change. According to Basili (1989), the approach to software development should prioritise exploration rather than theory. There is a need for a quantitative assessment mechanism to assess the performance of software engineering approaches. According to Basili (1985), assessment of overall quality with the help of quantifiable metrics are used to convert qualitative to quantitative measurement. A 1977 technical study conducted by the Rome Air Development Center emphasizes the need of using optimum metrics to develop software that is dependable and free from errors. These elements enhance the quality and expertise of the Air Force system. GQM paradism, devised by Basili (1992), facilitates the modeling and assessment of software. Software metrics are essential for anyfirm as they enhance the quality, cost-effectiveness, and efficiency of software development. According to Grady and Caswell (1987), it is recommended for companies to analyze challenges, establish performance goals, and assess software. According to Bevan (1995), usability should be the main focus in the design of interactive software. A survey paper by Kumar and Mohite (2017), explores all aspect of quality factors of usability for M-Learning application. This research study proposes a qualitative to quantitative method for evaluating the quality factors of usability in M-Learning applications. This methodology employs fuzzy logic, GQM paradism, and the ISO 9241- 11(2015) standard, which provide a framework for usability or user interaction.

3. GQM Paradigm and ISO9241-Methodology

The Goal Question Metric (GQM) paradigm used to define metric framework used for evaluation. This paradigm has three steps. The software company first establishes crystal-clear and precise goals that it intends to accomplish via the creation of a particular mobile application. These goals are collectively agreed upon by all parties concerned. Questions to be asked, providing a framework that is entered on evaluation. Metrics are measurable metrics that offer data based on queries. The answers to these questions provide a method for converting qualitative features into quantitative measurements. ISO 9241-11 is quality framework specially for usability.

4. Fuzzy Logic

Fuzzy logic serves as an alternative to standard crisp logic, which relies on a strict true/false categorisation for statements. Fuzzy logic, in contrast, permits the incorporation of any possible value ranging from true (1) To false (0). Fuzzy logic has many applications in the area of engineering as per Ross (2004). It is crucial that all input variables include a range spanning from 0 to 1. Hence, fuzzy logic is an influential instrument, particularly in situations when establishing definitive judgments is not feasible. The paper adopts fuzzy triangle functions, which serve as fuzzy rating parameters by Singh and Vidyarthi (2008) as illustrated in Table 16.1.

5. Proposed Methodology & Case study

In order to measure the whole quality of M-Learning mobile apps' usability, this article suggested the following method.

Step1: The primary objective and set of questions for the mobile learning app

Table 16.1: Fuzzy wt & rate

Ling. Var.	Fuzzy Wt	Fuzzy Rate
LV1	(0.0,0.0, 0.25)	(0.0,0.1, 0.3)
LV2	(0.0,0.25,0.50)	(0.1,0.3, 0.5)
LV3	(0.25,0.50 ,0.75)	(0.3,0.5, 0.7)
LV4	(0.50 ,0.75, 1.0)	(0.5,0.7, 0.9)
LV5	(0.75 ,1.0 ,1.0)	(0.7,0.9, 1.0)

should inform the first step in developing a usability metric framework.

Step 2: Assign stakeholders or decision maker's perspectives on

Step 3: Determine an overall grade for the six goals: features, time taken, learn-ability, accuracy, and security.

Overall fuzzy rating = $r_1 \times w_1 + r_2 \times w_2 + \ldots + r_n \times w_n = \Sigma\, r_i \times w_i$

Step 4: While evaluating usability, take into account the aspects of efficiency, effectiveness, and satisfaction to get an overall fuzzy assessment. The following steps will illustrate process of converting usability qualitative framework to quantitative mersurements.

Step1: Design a usability framework for an M- Learning Application as illustrated in Table 16.2.

Step 2: Evaluation of overall rating for six goals as illustrated in Table 16.3. where U1,U2 and U3 stands for

Efficiency, Effectiveness and Satisfaction and G1,G2,G3,G4,G5 and G6 stands for feature, time-taken, learnability, accuracy, security and beta-user's feedback as illustrated in Table 16.3.

Step 3: Evaluation of overall fuzzy rating for U1, U2 and U3 calculated as (0.20, 0.68, 1.0), (0.30, 0.9, 1.0), (0.30, 0.9, 1.0).

Step 4: Crisp value evaluation of M-Learning application calculated as 0.6258, 0.7357 and 0.7357 with the help of centroid method of defuzzification.

Step 5: Comparison of three usability factors as illustrated in Table 16.4.

It is thus a feedback to those who are responsible for making decisions or stakeholders that efficiency has to be taken into consideration prior to the final release of the Mobile learning application. Important information about the quality of the M-Learning application is provided by this fuzzy-based mathematical framework.

Table 16.2: Usability framework

Question	Metric
Clarity of content display?	About text, font, colour
	Easy to visible for Longer time.
	About night vision.
About Content	Date-wise weekly schedule available as per date, subject and topic.
	Today's class schedule as per subject and topic-wise.
	Downloaded video and documents as per logged student.
	Online and recorded lectures are available.
What about motivational effort for learning management?	Easy to download video and Documents as per logged student.
	Intuitive learning is easy.
	Multimedia content is available as per requirement of each section.
What about assignment and evaluation?	Scholarship as per student's performance.
	Reward as per their attendance and weekly assignment and weekly test.
Time to load content?	Feasible time to load online content Feasible time to load offline content Feasible time to load multimedia content

(Continued)

Table 16.2: *(Continued)*

Intuitive learning about mobile app?	learning experience about interface
	content management
Customer service?	Demo version
	Text version
What about Mobile app response as per action?	Mobile app respond properly as per action.
What about authorisation for a particular student?	Adequate information about privacy policy
	Secure socket layer
	About payment methods
	Mode for verification
Overall feedback	Feature
	About learning
	About accuracy
	About security

Table 16.3: Overall Rating for Six Goals

Usability factor	Goal	Overall fuzzy rate
U1	G1	(0.26,0.675,1.0)
	G2	(0.26,0.675,1.0)
U2	G3	(0.26,0.675,1.0)
	G4	(0.394,0.9,1.0)
U3	G5	(0.394,0.9,1.0)
	G6	(0.394,0.9,1.0)

Table 16.4: Crisp value (Quality)

Usability factor	Overall quality
Efficiency	62.58%
Effectiveness	73.57%
Satisfaction	73.57%

6. Conclusion

This research paper presents the conversion of qualitative usability framework to quantitative measurement for M- Learning mobile applications. The quantification was conducted using fuzzy logic, ISO 9241-11, and the GQM approach. This method may function as an automated usability testing procedure, which may be implemented during the agile development of mobile applicationsin accordance with budget and schedule constraints. This research article presents a thorough framework for utilizing fuzzy logic to quantify the sub-characteristics of usability expectation of domain specific any mobile application. The proposed methodology exhibits promise for implementation across diverse application domains and presents a bright outlook for the progression of mobile applicationdevelopment as per its quality standard.

References

Basili, V. R. (1989). "Software development: a paradigm for the future." *Proceeding of COMPSAC*, September, 1989.

Basili, V. R. (1985). "Quantitative evaluation of software engineering methodology." *Proceeding of First Pan Pacific Computer Conference*, September 1985, Melbourne, Australia.

Basili, V. R. (1992). "Software modeling and measurement: the goal question metric paradigm." *Computer Science Technical Report Series*, CS-TR-2956.

Bevan, N. (1995). "Usability is quality of use." *Advanced Human Factors/Ergon*, 20: 349-354.

Grady, R. B. and Caswell, D. L. (1987). *Software Metrics –Establishing a Company-wide Program*, Prentice Hall.

ISO9241-11, (2015). *Ergonomics of Human-system*.

Kumar, B. and Mohite, P. (2017). "Usability of mobile learning applications: a systematic literature review." *Journal of Computer Education*, Vol 5 (1), pp. 1-17.

Rome Air Development Center. (1977). Factors in Software Quality, RADC TR-77-369.

Ross, T. J. (2004). *Fuzzy Logic with Engineering Applications*, 2nd Ed, Wiley India Pvt. Ltd, New Delhi, India.

Singh, A. P. and Vidyarthi, A. K. (2008). "Optimal allocation of land fill disposal site: A fuzzy multi criteria approach." *Iranian Journal of Environment Health Science and Engineering*, 5 (1):25-34.

Analytical Studies of Load Variation on Terminal Voltage and Frequency of Three PhaseSynchronous Generator Using Fractional Order PIController

Dheeman Panigrahi

Department of Electrical Engineering, IGIT Sarang, Dhenkanal, India
E-mail: panigrahidheeman@gmail.com

Abstract

Synchronous generator converts mechanical energy to AC electrical energy. The generator is synchronized onto a large interconnected grid thus operating in unison or in parallel with all other generators of the overall system supplying energy to all the loads connected to it. It is essential to study the effect of the load variations on the magnitude and frequency of the voltage. The work in this paper uses proportional-integral (PI) and fractional-order proportional-integral (FOPI) controllers to study the effect of electrical-load-variations on the voltage magnitude and frequency of an isolated three phase synchronous generator. The PI controllers, FOPI controllers and the three-phase synchronous machine have been modelled and simulated in MATLAB Simulink environment. The results of the simulation work here show that the frequency and the magnitude of the terminal voltage of the synchronous generator remain constant at their rated values irrespective of variation in connected load (both real power and reactive power) with less ripple in 'developed-electrical torque' and 'rotor speed' in case of FOPI controllers.

Keywords: Three phase synchronous generator, PI controllers, FOPI controllers, MATLAB Simulink software, real power, reactive power

1. Introduction

Synchronous generator converts mechanical energy to AC electrical energy. Magnetic field is produced in the synchronous generator when a DC current is allowed to flow in the rotor winding. A rotating magnetic field is produced in the machine by rotating the generator rotor by a prime-mover. The rotating magnetic field thus produced generates a set of three-phase balanced voltages in the three-phase balanced distributed stator winding of the generator (Chapman, 2005; Say, 1990; Fitzgerald et al. 2003, Elgerd, 1989). The generator would be synchronized onto a large interconnected grid thus operating in unison or in parallel with all other generators of the overall system supplying energy to all the loads connected to it. During the machine operation, the frequency and the terminal voltage level must be controlled by the means of the prime mover torque and the field current respectively (Sanjenbam and Singh, 2022, 2023a, b, Sarker, 2024).

Chapter 17 DOI: 10.1201/9781003581215

2. Relationship between Real Power Mismatch and Frequency

The real power and reactive power of an isolated generator are determined by the load connected to its terminals. A synchronous machine can be considered initially running at rated speed with no load and a full terminal voltage. Its terminals are connected to a load having three star-connected resistors. As the load is being applied, three-phase balanced stator currents are created thus leading to development of an air-gap torque. This torque tends to decelerate the rotor. In order to maintain constant speed, it is thus necessary to increase the prime-mover torque until torque balance is restored. The air-gap torque thus creates the mechanism by means of which the prime-mover mechanical power is transformed into electrical power to be used by the load impedances. Thus, the real power output depends upon the load and is controlled by regulating the prime-mover torque to achieve torque balance and, as a result, constant speed, and frequency (Chapman, 2005; Say, 1990; Fitzgerald et al., 2003; Elgerd, 1989; Sanjenbam and Singh, 2022; Sanjenbam and Singh, 2023a, b; Sarker, 2024).

3. Reactive Power Mismatch and Terminal Voltage Relationship

A voltage drop occurs across the synchronous-reactance because of the stator current. As a result, the terminal voltage drops below its rated value. To restore constant terminal voltage, the excitation level is increased, thereby boosting the emf magnitude. One can say that the terminal voltage is maintained constant by regulating the excitation level. However, there is a cross-coupling between the two control channels. The real power output increases when the excitation level increases. This increase can be brought back to original value by adjusting the prime mover torque. When a single generator is running in isolation, it will supply the real power (P) and reactive power (Q) as demanded by the load connected to its terminals. It can be inferred from this analysis that for any given real power, the operating frequency of the generator is controlled by the governor set points and for any given reactive power, the terminal voltage of the generator is controlled by the field current (Chapman, 2005; Say, 1990; Fitzgerald et al., 2003; Elgerd, 1989; Sanjenbam and Singh, 2022; Sanjenbam and Singh, 2023a, b).

4. Modelling of the Synchronous Generator using MATLAB Simulink Software

The synchronous generator operating alone has been modelled by using MATLAB Simulink software (version R2023a). The Simulink blocks used to model the synchronous machine are taken from the library of the MATLAB. Figure 1 shows the detailed model of the three-phase synchronous machine with PI speed regulator and PI voltage regulator. Figure 1 represents the MATLAB Simulink model of the isolated synchronous generator and feeding three-phase RLC balanced load

Two regulators - one PI voltage regulator and the other the PI speed regulator are used here. In the PI voltage regulator, the error between the actual pu terminal voltage and 1 pu rated voltage is fed to the PI regulator and the output obtained is fed as an excitation voltage to the synchronous machine. The proportional and integral constants are tuned at values 10 and 100 respectively. The purpose of this regulator is to maintain the terminal voltage constant at its rated value irrespective of variation in the reactive power load on the generator terminals. Similarly, in the PI speed regulator, the error between the actual pu rotor speed and 1 pu rated speed is fed to the PI controller and the output is fed as the mechanical power input to the alternator from the prime- mover. The proportional and integral constants are tuned at values 25 and 550 respectively. The purpose of this controller is to maintain the frequency of the terminal voltage at its rated value irrespective of variation in real power load on the generator terminals.

5. Simulation and Results

The model of the alternator shown in Figure 1 was simulated using MATLAB Simulink software. The machine parameters are assigned the following values. Three-phase synchronous generator: Nominal rated power, line-to-line voltage, frequency [Pn (VA) Vn (Vrms) fn (Hz)]: [2000e3, 600, 50]

Inertia, damping factor, pairs of poles [H(sec) Kd(pu_T/pu_w) p()]: [0.6, 0.0, 2]

Internal impedance [R(pu) X(pu)]: [0.0036 0.16]

Initial conditions [dw(%) th(deg) ia,ib,ic(pu) pha,phb,phc(deg)]: [07.01083 0.824535 0.824535 0.824535 -14.0377 -134.038 105.962]

Loads: Load1 – 600 V, 50 Hz, 850kW, 400 kVAR; Load2 – 600 V, 50Hz, 850 kW, 125 kVAR, -100 kVAR; Load3 – 600 V, 50 Hz, 850 kW, 125 kVAR, -100 kVAR

'Load1' was connected to the terminals of the synchronous generator permanently. 'Load2' was switched off at time t = 10 s. 'Load3' was switched on at time t = 15 s. The simulation was done for a time of 25 seconds. The results and waveforms as obtained from the simulation are shown in Figures 2-4.

Figure 2 shows the variation in stator three-phase currents, rotor speed and

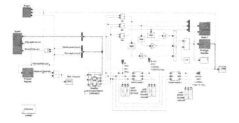

Figure 1: Simulink model of the synchronous generator

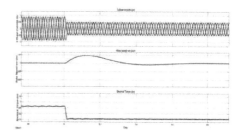

Figure 2: Variation in stator three- phase currents, rotor speed and electrical torque during 'switching- off' of 'Load2' at time t = 10 s

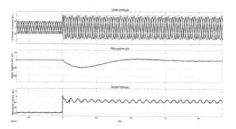

Figure 3: Variation in stator three- phase currents, rotor speed and electrical torque during switching on of 'Load3' at time t = 15 s

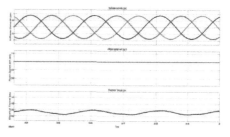

Figure 4: Variation in stator three- phase currents, rotor speed and electrical torque around time t = 25 seconds

electrical torque generated during switching off of 'Load2' at time t = 10 s. It is observed that the magnitude of the stator currents and the generated electrical torque reduces while switching off the load 'Load2'. The generated electrical torque reduces from 0.85 pu to 0.4 pu in response to this switching-off of 'Load2'. The rotor speed increases from 1 pu to 1.00875 pu and reduces to 0.9983 pu while quickly settling to 1 pu again because of the action of the PI controller. The ripple in rotor speed here comes out to be 0.01045 pu pp. Figure 3 shows the variation in stator three-phase currents, rotor speed and electrical torque generated during switching on of 'Load3' at time t = 15 s. It is observed that the magnitude of the stator currents and the generated electrical torque increases while switching on the load 'Load3'. The rotor speed reduces to 0.99083 pu from 1 pu and increases to 1.0025 pu and while quickly settling to 1 pu again because of the action of the PI controller. The ripple in rotor speed here comes out to be 0.01167 pu pp. The generated electrical torque increases from 1 pu to 1.05 pu and reduces to 0.75 pu while quickly settling around 0.85 again because of the action of the PI controller. The ripple in generated electrical torque here comes out to be 0.30 pu pp.

Figure 4 shows the variation in stator three-phase currents, rotor speed and electrical torque generated around time t = 25 seconds

It can be observed that the magnitude of the stator currents and the generated electrical torque remain approximately same at their initial values before the initiation of load variations. It is observed that the generated electrical torque remains within values of 0.865 pu and 0.84 pu giving a ripple of 0.025 pu pp.

6 Modelling and Simulation Results of the FOPI Controller in MATLAB Simulink

Figure 5 shows the modelling of the FOPI controller in MATLAB Simulink environment. The fractional PI controllers are used here to explore the possibilities of improving the performances of the PI controllers as discussed above. Figure 6 shows the modelling of the FOPI speed regulator/controller using MATLAB Simulink software. The proportional integral constant values are kept at the same previous values of 25 and 550 respectively. The fractional

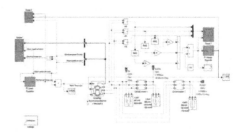

Figure 5: Modelling of the FOPI controller using MATLAB Simulink software

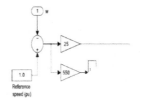

Figure 6: Modelling of the FOPI speed controller using MATLAB Simulink software

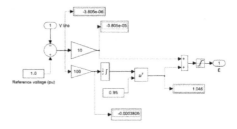

Figure 7: Modelling of the FOPI voltage controller using MATLAB Simulink software

integral power index was taken from literature (Subeekrishna and Aseem, 2019) and was kept at 0.95. Similarly, the FOPI voltage regulator/controller was modelled using the MATLAB Simulink software as shown in Figure 7.

The proportional integral constant values are kept at the same previous values of 10 and 100 respectively. The fractional integral power index was kept at same value of 0.95 (Subeekrishna and Aseem, 2019). The same simulation was repeated and the results are shown in Figures 8-10.

Figure 8 shows the variation in stator three-phase currents, rotor speed and electrical torque generated during switch-ing-off of 'Load2' at time t = 10 s with FOPI controller. It is observed that the magnitude of the stator currents and the generated electrical torque reduces while switching off the load 'Load2'. The gener-ated electrical torque reduces from 0.85 pu to 0.4 pu in response to this switching-off of 'Load2'. The rotor speed increases from 1 pu to 1.00875 pu and reduces to 0.9983 pu while quickly settling to 1 pu again because of the action of the FOPI control-ler. The ripple in rotor speed here comes out to be 0.01045 pu pp.

Figure 9 shows the variation in stator three-phase currents, rotor speed and elec-trical torque generated during switching on of 'Load3' at time t = 15 s with FOPI controller. It can be observed that the magnitude of the stator currents and the generated electrical torque increases while switching on the load 'Load3'. The rotor speed reduces to 0.99083 pu from 1 pu and increases to 1.00125 pu and while quick-ly settling to 1 pu again because of the action of the FOPI controller. The ripple in rotor speed here comes out to be 0.01042 pu pp. This is a reduction of ((0.01167 – 0.01042)/0.01167) × 100% = 10.711% as compared to the result of Figure 3. The gen-erated electrical torque increases from 1 pu to 1.05 pu and reduces to 0.775 pu while quickly settling around 0.85 again because of the action of the FOPI controller. The ripple in generated electrical torque here comes out to be 0.275 pu pp. This is a reduc-tion of ((0.30 – 0.275)/0.30) × 100% =

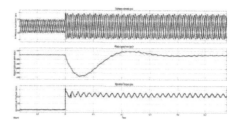

Figure 9: Variation in stator three- phase currents, rotor speed and electrical torque during switching on of 'Load3' at time t = 15 s with FOPI controller

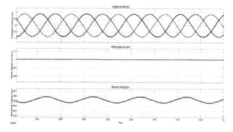

Figure 10: Variation in stator three-phase currents, rotor speed and electrical torque around time t = 25 s with FOPI controller

8.33% as compared to the result of Figure 3. Figure 10 shows the variation in stator three-phase currents, rotor speed and elec-trical torque generated around time t = 25 s with FOPI controller. It can be observed that the magnitude of the stator currents and the generated electrical torque remain approximately same at their initial values before the initiation of load variations. It is observed that the generated electrical torque remains within values of 0.865 pu and 0.84 pu giving a ripple of 0.025 pu pp.

7. Conclusion

It is concluded from the simulation results obtained in this work that the termi-nal-voltage magnitude and frequency of the synchronous generator remain constant at their rated values irrespective of variation in connected load (both real power and reactive power) with both the controllers,

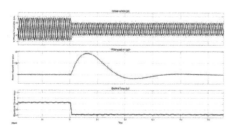

Figure 8: Variation in stator three- phase currents, rotor speed and electrical torque generated during 'switching-off' of 'Load2' at time t = 10 s with FOPI controller

PI and FOPI. However, the FOPI controllers reduce the electrical torque ripple by 8.33% in comparison to the PI controllers. Similarly, the ripple in the rotor speed reduces by 10.711% while using FOPI controllers. Thus, the FOPI speed and voltage controllers are found to be superior than their corresponding PI controllers.

References

Chapman, S. J. (2005). *Electric Machinery Fundamentals* (4th ed.). McGraw Hill Publication.

Elgerd, O. I. (1989). *Electric Energy Systems Theory* (TMH ed.). New Delhi: Tata McGraw Hill Publishing Company Ltd.

Fitzgerald, A. E., Jr. Kingsley, C., and Umans, Stephen D. (2003). *Electric Machinery* (6th ed.). New Delhi: Tata McGraw Hill Publishing Company Ltd.

Sanjenbam, C. D., and Singh, B. (2022). Universal active filter for standalone hydro-electric system with SPV and battery support. *2022 IEEE 1st Industrial Electronics Society Annual On-Line Conference (ONCON)*, pp. 1–6.

Sanjenbam, C. D., and Singh, B. (2023a). Modified adaptive filter based UPQC for battery supported hydro driven PMSG system." *IEEE Transactions on Industrial Informatics*, 19(7): 8018–8028. doi: 10.1109/TII.2022.3215950

Sanjenbam, C. D., and Singh, B. (2023b). PMSG based standalone distributed generation integrated universal active filtering system. *2023 IEEE IAS Global Conf. Emerging Technologies (GlobConET)*, pp. 1– 6.

Sarker, K. (2024). FC-PV-battery-Z source-BBO integrated unified power quality conditioner for sensitive load & EV charging station. *Journal of. Energy Storage*, 75: 109671.

Say, M. G. (1990). *Alternating Current Machines* (5th ed.). England: English Language Book Society/Longman.

Subeekrishna, M. P., and Aseem, K. (2019). Comparative study of PID and fractional order PID controllers for industrial applications. *International. Journal of. Engineering. Research & Technology. (IJERT)*, 7(01): 1–3, ISSN: 2278-0181.

Blockchain Smart Contract Fortification using Bytecode Analysis to Address Vulnerabilities

Mohammed Abdul Lateef, A. Kavitha

Jawaharlal Nehru Technological University, Hyderabad, India
E-mail: malateefz2101@gmail.com, athotakavitha@jntuh.ac.in

Abstract

Smart contracts and blockchain platforms have revolutionised various industries, offering decentralised and transparent execution of agreements. However, they are not immune to security lapses, and the presence of vulnerabilities has led to security issues. This research delves into the realm of smart contract security, employing bytecode analysis as a powerful tool to unveil and rectify vulnerabilities. By meticulously scrutinising the intricate low-level instructions and EVM opcodes, our objective is to unearth potential issues and provide actionable insights for enhancement. Through a compelling case study elucidating a smart contract designed for decentralised electoral integrity, we seamlessly integrate automated tool analysis with manual bytecode examination. Our overarching goal is to elucidate the nuanced process of leveraging bytecode analysis to effectively detect and mitigate vulnerabilities inherent in smart contracts, thus fortifying their robustness and reliability.

Keywords: Smart contracts, EVM opcodes, automated analysis, manual examination, vulnerability

1. Introduction

Blockchain tech, especially with Ethereum's smart deals (contracts), has changed lots of areas by making deals that run on their own with lots of auto-work. But these deals can be tricky because once they're set, they can't be changed, and everyone can see them. This means they need very careful checks for any problems, by people and with tools.

Smart deals work on a simple "if this, then that" way and have made things easier in places like banks and supply chains. But, their go-it-alone ability also brings risks that need sharp eyes. This study digs into the troubles with smart deals on Ethereum, aiming to understand what goes wrong, how bad it can be, and how to fix these troubles. This could help make blockchain more safe.

This research dives deep into the troubles with smart deals on Ethereum to grasp how they work, what goes wrong, and how to fix these troubles (Sui, 2023). Looking at how they're made and the big security issues, the aim is to make blockchain safer. With a close look, builders can make smart deals stronger, even when the rules around them are fuzzy, helping create a tougher blockchain world.

2. Literature Review

The birth of smart contracts was a major turning point in blockchain technology

Chaper 18 DOI: 10.1201/9781003581215

because it introduced self-executing agreements that are much faster than what is traditionally used. However, the more practical they became, THE more problems were discovered. Critics found everything from mistakes in coding to flaws throughout systems as flaws like these were being searched for (Xueshuo *et al.*, 2023). This increased scrutiny has shown how important it is to know about security when dealing with smart contracts on blockchains; therefore there should be research done continuously into this area so that risks can be identified such as reentrancy attacks which allow hackers access back into an already ongoing transaction after they've left once thereby stealing all its value since it keeps looping through forever until balance becomes zero even though overflow vulnerabilities were categorised still guiding towards making them stronger (Ma, 2023).

Smart contract vulnerability is a complicated and multifaceted problem that requires careful consideration and proactive measures (Vani, 2022). With new threats constantly emerging, collaborative efforts among different organisations should be encouraged in order to build strong security foundations around smart contract ecosystems (Fekih, 2023). A good way to avoid this would be using vulnerability assessments as well as setting up strict protocols based off them thus helping concerned parties traverse through issues arising from these weaknesses while ensuring trustworthiness within transactions conducted via block chains (Pise and Patil,2022).

3. SECURE-AMSVA —
Systematic Evaluation For Comprehensive And Unified Review Of Automated And Manual Smart Contract Vulnerability Assessment

This research is built around SECURE-AMSVA, which is a method used to identify weaknesses in Ethereum-based smart contracts. This formalised system is designed

to be comprehensive and meticulous so as to ensure that no potential security risk goes unnoticed. The methodology combines automated and manual techniques in order to carry out broad assessments for different kinds of vulnerabilities.

SECURE-AMSVA serves as an example of how thorough evaluations can be made towards ensuring safety measures for smart contracts built on Ethereum are effective. It is made up of tools that help with automation as well as human input which makes it holistic as shown in Figure 1. By doing this the approach attempts at benefiting from both methods thus increasing the range and intensity levels in which bugs may be found (Liu, 2022). Through using such automated programs alongside expert knowledge workers, efficiency gains can be achieved while still maintaining accuracy required during deep analysis aimed at establishing whether a given ethereum based application has any security holes or not.

- lKey Components of SECURE-AMSVA

Thorough Evaluation: Smart contracts are methodically evaluated for security flaws in SECURE-AMSVA. For this purpose, some automatic analysis instruments (like Mythril and Slither) are used which help to find known vulnerabilities as well as programming mistakes.

Unified approach: The program combines results of automated checks with manual ones so that users can see them all together as one picture about their contract's safety level (Luu, 2016). This allows them understand risks better and fix any weaknesses properly.

The Vulnerability Confirmation Engine (VCE) is what makes sure a smart contract vulnerability detection system is accurate and reliable. It acts as an interface between scanners' initial identification of potential vulnerabilities and developers' actionable remediation efforts (Samreen and Alalfi, 2021). In this article we do a deep dive into VCE technicalities, its methods, advantages, disadvantages are discussed here so that people will know more about it.

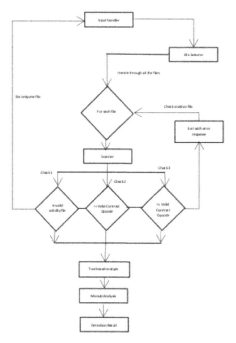

This is highly customisable and can be integrated into development pipelines for automatic vulnerability checks (Tikhomirov, 2018). Together, these tools offer a comprehensive approach to identifying and mitigating vulnerabilities in Solidity contracts, enhancing the overall security of blockchain applications.

4. Implementation and Results

The implementation requires the constant and immersive accent of the tools used for assessment. Both the tools i.e. Mythril and Slither play a crucial role in identification process.

All the 5 vulnerabilities are recognised by using tool based approach involving Mythril and Slither along with Manual analysis too as shown in Table 1.

5. Conclusion

Figure 1: Approach acquired to implement VCE

This research project embarked on a mission to fortify the security posture of Ethereum-based smart contracts by

Table 1: Dissemination of Smart contract vulnerability analysis

1. Bytecode Analysis					
Tool	Re-entrancy	Unchecked External call	Integer Overflow	Un-initialised variable	Access Control
Mythril	Yes	Incorrect assessment	Incorrect assessment	Incorrect assessment	Incorrect assessment
Slither	Not Applicable	Not Applicable	Not Applicable	Not Applicable	Not Applicable
Manual	Not Applicable	Not Applicable	Not Applicable	Not Applicable	Not Applicable
2. EVM Opcode Analysis					
Tool	Re-entrancy	Unchecked External call	Integer Overflow	Un-initialised variable	Access Control
Mythril	Not Applicable	Not Applicable	Not Applicable	Not Applicable	Not Applicable
Slither	Not Applicable	Not Applicable	Not Applicable	Not Applicable	Not Applicable
Manual	Yes	Yes	Yes	Yes	Yes

employing a rigorous and multi-faceted vulnerability assessment methodology, aptly named "SECURE-AMSVA" (Systematic Evaluation for Comprehensive and Unified Review of Automated and Manual Smart Contract Vulnerability Assessment). Through this comprehensive approach, the project sought to tackle the ever-evolving landscape of vulnerabilities that smart contracts face, including reentrancy attacks, integer overflows, unchecked external calls, uninitialised variables, and access control issues.

References

Fekih, R. B., Lahami, M., Jmaiel, M. and Bradai, S. (2023) "Formal Verification of Smart Contracts Based on Model Checking: An Overview," *2023 IEEE Int. Conf. on Enabling Technologies: Infrastructure for Collaborative Enterprises (WETICE)*, Paris, France, 1-6, doi: 10.1109/WETICE57085.2023.10477834.

Liu, Y., Xu, J., and Cui, B. (2022). "Smart contract vulnerability detection based on symbolic execution technology." In: Lu, W., Zhang, Y., Wen, W., Yan, H., Li, C. (eds) *Cyber Security. CNCERT 2021. Communications in Computer and Information Science*, 1506. Springer, Singapore. https://doi.org/10.1007/978-981-16-9229-1_12

Luu, L., Chu, D.-H., Olickel, H., Saxena, P., and Hobor, A. (2016). "Making smart contracts smarter." *Proceedings of the 2016 ACM SIGSAC Conference on Computer and Communications Security (CCS16)*, 254-269. doi.org/10.1145/2976749.2978309

Ma, C., Liu, S. and Xu, G. (2023). "HGAT: smart contract vulnerability detection method based on hierarchical graph attention network." *Journal of Cloud Computing*, 12: 93. doi: 10.1186/s13677-023-00459-x

Pise, R., and Patil, S. (2022). "A survey on smart contract vulnerabilities and safeguards in blockchain." *International Journal of Intelligent Systems and Applications in Engineering*, 10 (3s), 01–16.

Samreen, N. F., and Alalfi, M. H. (2021). "A survey of security vulnerabilities in Ethereum smart contracts." *CASCON20 Proc. 30th Annual Int. Conf. Computer Science and Software Engineering*, November 2020. arXiv. https://doi.org/10.48550/arXiv.2105.06974

Sui, J. Chu, L. Bao, H. (2023). "An opcode-based vulnerability detection of smart contracts." *Applied Sciences*, 13: 7721. doi: 10.3390/app13137721

Tikhomirov, E. Voskresenskaya, I. Ivanitskiy, Takhaviev, R., Marchenko, E. and Alexandrov, Y. (2018) "SmartCheck: static analysis of Ethereum smart contracts," *2018 IEEE/ACM 1st International Workshop on Emerging Trends in Software Engineering for Blockchain (WETSEB)*, Gothenburg, Sweden, pp. 9–16.

Vani, S., Doshi, M., Nanavati, A., and Kundu, A. (2022). "Vulnerability analysis of smart contracts." arXiv. doi.org/10.48550/arXiv.2212.07387

Xueshuo, X., Wang, H., Jian, Z., Fang, Y., Wang, Z., Li, T. (2023). "Block-gram: Mining knowledgeable features for efficiently smart contract vulnerability detection." *Digital Commun. Netw.*, ISSN 2352-8648, doi.org/10.1016/j.dcan.2023.07.009.

Fingers Identification Architecture for AIR CANVAS

A Case Study

Jitendra Kumar Gartia[1], Jemarani Jaypuria[2], Praveen Gupta[3], Rajesh Tanti[4],
Umashankar Ghugar[5]

[1]Department of CSE, PKACE, Bargarh, Odisha
[2]Department of Computer Science, GIET University Gunupur, Odisha
[3]GITAM (Deemed to be University), Visakhapatnam, India
[4]Department of CSE, OP Jindal University, Raigarh, Chhattisgarh, India
[5]Department of CSE, OP Jindal University, Raigarh Chhattisgarh, India
E-mail: yad2205@gmail.com, jemarani253@giet.edu, praveen2gupta@gmail.com,
rajeshtanti20017@gmail.com, ughugar@gmail.com

Abstract

The writing of content with the help of pen and paper has existed for a century. But with the development of new technologies, new methods can be invented. This paper is based on the gesture of finger movement. Suppose technology can sense the direction of the fingers. In this paper, OpenCV is applied to identify the movement of fingers. This paper aims to design and develop a system to draw anything on the screen. It is based on identifying the finger's movements and converts this motion to text. In recent years, few technologies have developed in this research area. This work uses computer vision to monitor finger movements. This detects our hand movements mainly and uses them for doing the task, i.e., without any touch or typing, the system will sense and interpret the same as a hand does automatically without any physical contact. This work tested finger movement from 5 to 10 fingers at different angles and tried to capture the action. It works fine to some extent, and with few modifications' efficiencies can be increased. In recent years, programming has moved towards Artificial Intelligence, and this concept is related to it. This is a concept that can be used for the development of the application in the coming years. It can be used to write text, operate systems, and perform tasks.

Keywords: Machine learning, OpenCV, air canvas, object detection, artificial intelligence

1. Introduction

Humans have been writing since 2500 BC; Neolithic people invented it. It has long-progressed tools with a time pace. It started with painting on the walls, then replaced by stone. Stones were replaced with cloth, and now it is using paper for a long time. Computers have been used for the last 60-70 years, and how writing has moved towards writing on a computer. The digital report can be done using keyboards, mouse, and other tools. but with the development of Artificial intelligence and cameras, efforts are being made to write using fingers. This paper integrates hand gesture detection with AI and Python programming to achieve natural human-machine interaction. Fingertip recognition is used to develop the system. Using the Python OpenCV technique, the fingertip is first detected, and then the fingertip trajectory is tracked and displayed on the screen. Python programming is used with the OpenCV libraries to develop the system. The contribution of this paper is as follows:

- To detect multiple fingers and identify their position on the air canvas.
- To identify the fingers at multiple angles/locations and the correctness of the count.

The paper has been organised in the following order, section-i describes the introduction part, section-ii describes the literature survey, section-iii describes the proposed work, in section-iii describes the experimental setup. The section describes the result analysis and testing part. Finally, section vi describes the future and scope of the research work.

2. Literature Review

Object tracking can be used to apply computer applications like computer vision and human-machine interaction. The literature suggests various applications for tracking algorithms. One group of researchers uses it to interpret signal language, and some see hand gestures; another group follows the text, recognises, and observes the body of an object's apparent movement, recognises characters based on finger tracking, and more. In this work, real-time hand gesture recognition is used with various methods. Few of the research work done on hand gestures (De Smedt et al., 2016) worked on fingerprint reorganisation using OpenCV and Mediapipe. A Framework for Hand Gesture Recognition and Spotting Using Sub-Gesture Modelling. They focused on the gesture recognition framework, gesture spotting framework and gesture spotting. Gestures are one of the ways of communication with a computer in an interactive environment. Recent advances in computer vision and machine learning have led to several techniques for modelling gestures in real-time environments (Yang et al., 2020).They have worked on gesture and

hand recognition using a Microsoft Kinect camera. Microsoft developed Kinect is a low-cost device with an RGB camera and a depth sensor. the company provided API for this framework. They worked on the three sections. The first section is hand detection, the second is finger identification and the third is gesture recognition. They implemented this using C# and Visual Studio 2015 on Windows 7. This system works on dynamic finger gesture recognition and gestures using fingers. This is an added advantage over other technologies. Raised Hand Touch Recognition (JRNN). Their software used sample sequences from still pictures to compare 5 to 9 different hand positions. Then he recorded when the contours of his hands began to crack during reinstallation. After the transition behavior of the set of places is determined.

3. Experimental Setup

In this paper code has been written by applying OPENCV and Mediapipe. This paper at the current stage is identifying the fingers and movement. this is a continued work and in the next step this model will get movement of fingers as input data and will get training under supervised environment and then will identify the alphabets of the English language. We are trying an effort to make words using two fingers. Making words are easy and fast using two fingers as compared to one figure.

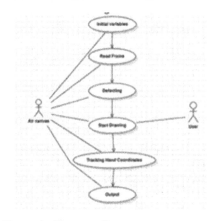

Figure 1: Use case diagram

OpenCV: is a free library using python programming. This can be used for images, object identification, real-time application development using images. This can be used by importing into python papers and creating objects and applying object-oriented concepts.

Mediapipe: is a framework developed for the use of video, and streaming of video. This is a solution for machine learning applications used with python and others. This is used for image detection, image processing, image analysis etc.

4. Testing and Result Analysis

We have considered three test cases each case having three phases for identifying the letters by using multiple fingers. Here have given some results in phase wise as below:

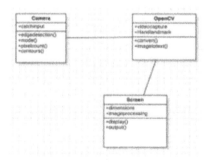

Figure 2: Class diagram

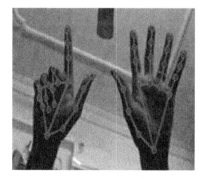

Figure 3: In this above figure the number of fingers is identified as 7 at angle-01.

Figure 4: the number of fingers is identified by 6 at angle-03

Figure 5: the number of fingers is identified as 7 at angle-02.

Figure 6: The number of fingers is identified as 7 at angle-01.

Conclusion

This research work is for the identification of the movement of fingers in the air. If accuracy is good and similarity between multiple iterations with different persons It can add some more features to the

Figure 7: the number of fingers is identified as 7 at angle-02.

Figure 8: the number of fingers is identified as 7 at angle 03

Figure 9: the number of fingers is identified as 8 at angle-01.

programming. the use of the fingers to write something on the air as canvas or it can be used to identify the words written using the fingers. It is found that at every time at different positions of multiple fingers system can track the fingers and count them successfully. In the future focus will be to identify text recognition by two fingers using machine learning approaches and the system will identify the word by movement of the fingers as a result less use of keyboards.

Figure 19.10: the number of fingers is identified as 8 at angle-02

Figure 19.11: the number of fingers is identified as 8 at angle 03

References

Chhabra, G. S., Sharma, A., and Krishnan, N. M. (2019, April). "Deep learning model for personality traits classification from text emphasis on data slicing." *IOP Conference Series: Materials Science and Engineering,* 495 (1): 012007. IOP Publishing.

De Smedt, Q., Wannous, H., and Vandeborre, J. P. (2016). "Skeleton-based dynamic hand gesture recognition." *Proc. IEEE Conf. Computer Vision and Pattern Recognition Workshops* (pp. 1-9).

Gupta, P., Rajpoot, C. S., Shanthi, T. S., Prasad, D., Kumar, A., and Kumar, S. S. (2022, October). "Image forgery detection using deep learning model." In *2022 3rd Int. Conf. Smart Electronics and Communication (ICOSEC)* (pp. 1256-1262). IEEE.

Hemalatha, S. (2020). "A systematic review on Fingerprint based biometric authentication system." *Int. Conf. Emerging Trends in Information Technology and Engineering* (ic-ETITE). IEEE, 2020. and Intelligent Systems (ARIS) (pp. 1-6). IEEE.

Performance Evaluation of Cloud with load balancing Algorithms

Malti Nagle, Prakash Kumar

JIIT Noida,
E-mail:nagle.malti083@gmail.com, Prakash.kumar@jiit.ac.in

Abstract

Cloud computing is the new emerging trend which contribute in process management and storage of huge amount of heterogeneous data. This needs efficient scheduling techniques for allocation of VMs and data centers. In this paper, cloud of extended cloudlets is simulated on Clousim simulator to perform scheduling algorithm on cloud with various PEs. Computational cost and response time are the parameter for comparison the performance of allocation of VMs to different cloudlet. It has been observed from the result that increasing the number of processor and apply efficient scheduling algorithm will successfully manage the distribution of cloudlet to VMs. Results of SJF scheduling is giving better result over the FCFS. The current trends of healthcare system requires the efficient scheduling algorithm to manage the modern system consist of historical data and data produced by IoT. In this paper, introduced the distribution of cloudlet to brokers with efficient load balancing.

Keywords: Cloud computing, load balancing, scheduling algorithm, FCFS, scalableSJF, computational cost, communication technology, CloudSim

1. Introduction

EMERGING healthcare with information technology is necessity of the next-generation of healthcare system. The next decade may well see a revolution in the treatment and diagnosis of diseases. Healthcare requires constant consideration to improvise their quality services with less operational cost. Any modernization of healthcare system that can ensure operational efficiency and service quality must include cloud computing as a core component. Another contest is enormous amount of medical data (in Tera Bytes and Peta Bytes) exist in medical organization that are heterogeneous in nature. According to jitendra Singh (2014) such data is the vital and basis for decision making for delivering accurate treatment. Cloud computing is the way out of "big data" issues. In order to overcome challenges of traditional model of healthcare study of Sagar Sharma et al. (2015) cloud computing is the best suitable platform to develop a system that can meet the requirement of modern healthcare. Cloud computing hype attraction to provide flexible, cost effective and collaborative infrastructure. It has been observed that most of the medical institutions still rely on traditional observation methods. Traditional models are more concerned with paper work and manual observations. That leads to error, in case of operations and sometimes inefficient for critical health issues. Upgrade healthcare with new technology enables novel approach to healthcare workflow. Medical activities and workflow of recent healthcare significantly underutilized. Computing resources can be accessible from remote location to the patients, doctors in any critical medical conditions. Medical data of patient accessed through mobile phones and tablets.

2. Literature Review

The rising thought of the Internet of Things (IoT) is quickly discovering its way all through our modern life, aiming to enhance the quality of life by associating numerous smart devices, technologies, and applications. Overall, the IoT would allow for the automation of everything around us. Study of kavita J., et al. (2018) have found that there is a potential application of IoT in to the healthcare related industry. To analyse performance of Cloud different scheduling algorithm is implemented at cloud platform

The significance of cloud performance monitoring tools by considering both the cloud provider as well as cloud user's perspective Prabal Verma, et al., (2018) has monitored the response time, CPU, storage usage, etc. in peak and off peak hours. Measured the performance of five (05) dominant cloud providers (Google, Rackspace, Salesforce.com, Amazon and terremark) for their performance by hosting a website. Study of Malik (2015) and Niharika (2017) measured the response time and latency of each of the above cloud providers suggested a new approach for infrastructure management to determine the performance in virtualized environment by monitoring infrastructure response time.

3. Methodology and Model specifications

Cloud computing facilitates opportunities to access physical and virtual resources at remote location. Continue development of cloud computing, causes numerous challenges for this technology. Tasks scheduling is one of challenges among all. Tasks scheduling is continuous process of mapping of

a tasks entered in cloud with the available resources in cloud computing environment. Scheduler always tries to look best match for VM mapping on the basis of response time and completion time. Different scheduling algorithm are available to assign the VM and cloudlets. Study of Naina Gupta (2017) among all other scheduling algorithm FCFS, SJF algorithms are implemented and simulation result is shown below.

3.1. Model Specifications

In the experimental setup different combination of VM and cloudlet in terms of CPU capacity, RAM size and bandwidth of disk and network are used. Number of cloudlet assign to datacenter are 100 quoted by Pantelopoulos, A., et al. (2010). Data center is a key component in the cloud computing and offers the various infrastructure to the users for instance.

4. Empirical Results

4.1 Configuration of Datacenter

Table 1 contains specifications used to create VM.

Simulation is carried out considering algorithm FCFS i.e. generally deals with allocation of process to first cloudlet then so on, while SJF execute with shortest job first then leads towards longest job. Algorithms are compared with regards of response time and actual CPU Time.

ScalableSJF – Algorithm

Step 1:	initialize the cloudlet
Step2:	create datacenter
Step 3:	assign broker to datacenter
Step 4:	Assign datacenter to PE
Step 5:	Repeat step 4 to create host list for different core processing element.
Step 6:	Submit host list to cloudlet
Step 7:	create list of cloudlet at broker
Step 8:	get length of cloudlet
Step 9:	compare the smallest cloudlet.
Step 10:	Add the smallest in the list

Table 20.1: Configuration of virtual machine

S. No.	Memory Size (Gb)	Storage Details(Gb)	Available Bandwidth	No. of Process	CPU mips
1	2	40	1000	2	1000
2	2	40	1000	4	1000
3	2	40	1000	8	1000

Step 11: sort the cloudlet list.
Step 12: Assign VM to host id by broker's priority list.
Step 13: Get the cloulet id and calculate the response time for scalable SJF.

The proposed paper consist of simulation of cloud. The CloudSim has been used to introduce the simulation setup for scheduling of cloud. Results has been carried out for various algorithm and shown in different tables and figures. The average Response Time, Max Response Time and Min Response Time for algorithm FCFS and SJF for 2, 4 and 8 core processor in shown in Figures 20.3, 20.4, and 20.5.

It revealed that increases the number of processor drastically reduces the Average Response Time. 8 core processor is found better to allocate the cloudlet and VM shown in (Figures 20.1 and 20.2). Quality services can be improvise by hosting applications on cloud and their performance should be checked i.e. response time, waiting time and execution time to analyse performance of cloud within the acceptance boundaries. Selection of type of scheduling algorithm that best fitted can be optimize the usage of cloud resources. Scheduling algorithms are extremely beneficial to distribute the load consistently. Response Time is calculated with the help of following expression:

Response Time = Cloudlet (finish time) – Cloudlet (Submission Time) 2, 4 and 8 core processor with different cloudlet for FCFS and SJF is described in (Figure 20.3, 20.4, and 20.5). It has been observed that increasing the number of cloudlet causes variation in response time of SJF but FCFS

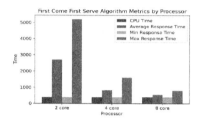

Figure 20.1: Average response time, max time, max response time for FCFS

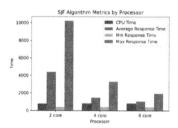

Figure 20.2: Average Response Time, Max Response Time, Min Response Time for SJF

is different in all cases. (Figures 20.3, 20.4, and 20.5). In 8 core processor at 62 cloudlet response time suddenly decreased and then till 78 cloudlet increased linearly. It has been observed that the maximum response time to process task is also minimum in 8 core processor.

5. Conclusion

Performance evaluation in terms of load balancing in cloud computing (CC) is a significant matter and shall be controlled at various level. Study of scheduling revealed that average response time (RT) is a noteworthy aspect that monitored the overall performance of cloud. Response time

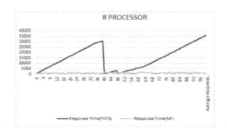

Figure 20.3: 8 core processor for FCFS and SJF

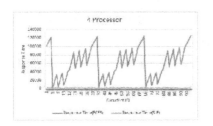

Figure 20.4: 4 core processor for FCFS and SJF

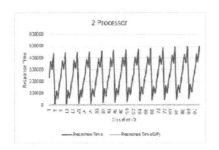

Figure 20.5: 2 core processor for FCFS and SJF

(RT) can be evaluated using load balancing algorithms. In current pape, experiment has been conducted to examine the dependency of RT in consideration of broker service policy along with datacenters. Results obtained revealed that among all processing elements allotment i.e. 2, 4 and 8 core processor; 8 core processor has best performance over the other 2 core and 4 core processors. Whereas, to examine the datacenter performance, when the number of cloudlet are increased, response time (RT) reduced in case of SJF proportionally upto a certain level before reaching to

almost constant value. The result has been compared with FCFS algorithm and found significantly better. Average response time shall be decrease by assigning the suitable Scheduling algorithm with appropriate number of processor. Consequently, performance of cloud can be improved significantly.

References

Jitendra, S., (2014). Study of Response Time in Cloud Computing, IEEB, 5(1): 36-43.

Sagar Sharma. et al., (2018), Toward Practical Privacy-Preserving Analytics for IoT and Cloud-Based Healthcare Systems, in healthcare informatics and privacy, IEEE Internet Computing, 1:1089-7801.

Kavita Jaiswal et al., (2018). An IoT-Cloud based smart healthcare monitoring system using container based virtual environment in Edge device, Proceeding of ICETIETR.

Prabal Verma et al., (2018). Cloud-centric IoT based student healthcare monitoring framework, J Ambient Intell Human Compute, (9): 1293-1309.

Maulik Parekh et al., (2015). Designing a Cloud based Framework for Health-Care System and applying Clustering techniques for Region Wise Diagnosis, Elsevier, ScienceDirect, 2nd International Symposium, Procedia Computer Science, 50(1): 537 – 542.

Niharika Kumar, (2017). IoT Archtecture and System Design for Healthcare System, ICOSTSN, 1118-1123.

Naina Gupta et al., (2017). Study and implementation of IOT vased smart healthcare System, ICEI, pp 541-546, 2017

Pantelopoulos, A., et al., (2010). A survey on wearable sensor-based systems for health monitoring and prognosis, IEEE Trans. Syst. Man Cybern. C, Appl. Rev., 40(1):1–12.

Automation of Genetic Diagnosis for Detecting Chromosome Structure

P. Devika[1,2*], B. Bhaskara Rao[2], M. Nagapavani[3], Ashok Sarabu[4],
V. Surya Narayana Reddy[4]

[1]Department of CSE(CyS,DS) and AI&DS, VNR Vignana Jyothi Institute of Engineering and Technology, Hyderabad, India, devikapotta23@gmail.com
[2]Department of Computer Science and Engineering, GITAM School of Technology, GITAM Vishakhapatnam, India, bbodu@gitam.edu
[3]Department of Humanities and sciences,Hyderabad Institute of Technology and Management, Hyderabad, India, naga.pavani84@gmail.com
[4]Department of Computer Science and Engineering, BVRIT HYDERABAD College of Engineering for Women, Hyderabad, India, sarabu.ashok@gmail.com, veeramreddysurya@gmail.com

Abstract

Chromosome is protein structure found inside nucleus where each cell has 23 pair of chromosomes. Every pair if well-structured is said as regular chromosomes. Many ways are approached to detect the structure of chromosomes. In this paper we are coming with new idea of adopting Yolo Technique in order to detect the normal chromosome structure. An attention-based model that will automatically learn to describe the content of the photos was just introduced. We also go over the deterministic and stochastic training methods we'll use for this model, including how we'll maximize the variation lower bound and take into account hyperparameters like Precision, Recall, and F1 to compare the results to the methods used to find the pair of chromosomes.

Keywords: Yolo, hyper parameters, chromosome

1. Introduction

The automation of genetic diagnostics has received significant scientific attention recently. Years of training are needed to identify genes and chromosomes so that clinicians can diagnose with a low risk of error because the majority of chromosomes are normal, the imbalance of normal and abnormal data in chromosomal abnormality detection that understanding the appearance of abnormal chromosomes becomes more challenging for students. Doctors will therefore need additional information and years of training to become experts in this area. A segment of chromosomal DNA that is missing, extra, or irregular and is brought on by an aberration, disease, or mutation is referred to as a chromosomal abnormality. The prevalence of Down syndrome (DS) occurs in approximately 14.2 out of every 10,000 live births, with an expected total of 5400 cases in the coming years, according to studies conducted by Gert de Graaf. According to estimates, by 2007, the implementation of the medical initiative connected to DS in several nations resulted in a 30% reduction in the incidence of DS births. This underscores the

Chapter 21 DOI: 10.1201/9781003581215

vital importance of prenatal screening and the identification of chromosomal illnesses as they intersect significantlyreduce the occurrence of birth defects (Devi *et al.,* 2022).The first study to appear on the computer vision scene was the YOLO study by Joseph Redmon *et al.*, which was published in 2015 which subtly caught curiosity of other computer experts. Before YOLO was created, region proposal networks (RPNs) were employed by CNNs, like the R-CNN.RPNs first created proposal bounding boxes for the input image, and then they used those bounding boxes to run a classifier.

In order to eliminate duplicate detections and subsequently improve the bounding boxes, post-processing is then used. (Balagalla *et al.*, 2022)

Basically, it wasn't suitable to train the R-CNN network's stages individually. It was even tough and very slow to optimise the R-CNN network. The main goal of the author is generally to create comprehensive neural network model for all phases after following a pre-unique neural network featuring multiple convolution layers with an input image that contains (or does not contain). The YOLO system simultaneously processes every aspect of an image and generates predictions for all items at once rather than focusing on particular portions of the image one at a time.Inserting grid cell of default size of SxS (7x7) is the main objective of YOLO v1 (Song *et al.*, 2022). A grid cell is responsible for object detection when an item's center reaches that grid cell (Figure 1).

2. Methodology

Our motive is to automate the process of finding abnormalities in the chromosome. While trying out the different ways and testing different libraries for building such system, we came to know about the YOLO library. As per our planned procedure for building such system, YOLO proved out to be a very helpful tool. The planned methodology for building a system which could detect the abnormality is heavily

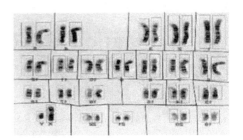

Figure 1: Inputsplit

dependent on a tool which could help to distinguish a chromosome from an image. As per our methodology the first step is to provide input to the system, the input can be provided as an image which 23 pairs of chromosomes. After the input is provided the first step which our system will do is to calculate the distance between each chromosome and then internally split the image into 23 pairs each image containing a pair of chromosomes. In case of abnormality where there are more than 2 chromosomes in any pair, the input image will split in such a manner that the abnormal pair will divide that many number of chromosomes. (Leung *et al.*, 2022) The calculation for splitting the image in 23 pairs will be on the basis of distance between each chromosome. The distance between each chromosome will be compared. Numbering will be assigned to each chromosome, then the distance between each chromosome will be calculated. Largest distance between each adjacent chromosome will be noted down.

(Nimitha *et al.*, 2018) The split point of the images will be the top 22 distanced chromosome. By this way the input will be divided into 23 pairs. Now, we will have each separate pair of chromosomes. Now in the next function each pair image will pe passed one by one.

2.1. Procedure of Implementation

2.1.1. *Environment*

A free (IDE) Integrated Development Environment) supported from Google called Google Colab (Google Collaboratory) supports AI research and education (AI). Collaboratory offers the same coding environment as Jupyter Notebook, and Tensor Process Unit (TPU) and Graphic Process Unit (GPU) usage are both free (TPU). The most widely used libraries for Deep Learning research were PyTorch, Keras, and TensorFlow, which were already installed on Google Colab. Deep learning and machine learning cannot be used with regular PCs since they need a system with high processing power and speed (usually driven by a GPU). To assist AI researchers, Colab offers the Tesla V100, one of the most powerful GPUs currently available, as well as TPU (TPUv2) on cloud. Even Colab offers sufficient storage and RAM that is at least 16 GB.Preparing the dataset: The redundant parameters can now be removed.

2.1.2. Creating Label Text Files

In PyTorch, YOLOv5 reads the data contained within the box from a text file(.txt) rather than a CSV file (.csv). To combine all of the data from bounding boxes in an image into a single text file, the bounding boxes in that image must be clustered. The GWHD/label directory will house all label text files after writing. Positive_images are pictures with items, while negative_images are pictures without anything. For the YOLOv5 model to access and use those label text files during training, they have to match those negative images. Due to the absence of any objects, these images lack boxes that surround. Consequently, their label text files will be empty of content.

2.1.3. Creating the data.yaml File

A yaml file is used by the PyTorch YOLOv5 model to access and use the photos as an input that contains summary of the dataset. Because it lacks a data. yaml file, the supplied dataset needs to be initialised. The data.yaml file is often created using Notepad or Notepad++, saved in yaml file, and then published to Drive. However, it will immediately be recorded here in Colab. To replace the blank yaml file, a method called Core magic from iPython must be imported.

2.1.4. Establishing the Infrastructure

The procedure of selection of YOLOv5 model samples that are built upon past hypotheses were unpredictable. These architectures will be read from the yaml file by the PyTorch YOLOv5 model, which will then generate them within the train.py file. Furthermore, this allows the design to accommodate various issues with object detection.This study's objective seeks to assess the efficiency of the YOLOv5 algorithm, the model won't be momentarily altered or enhanced with additional algorithms or optimisation techniques. The original architecture will be applied in its place. 80 classes are defined as a result of utilising this prototype YOLOv5 architecture to train the COCO dataset. For the wheat dataset, adjusting the number of classes is necessary. Auto-learning for the anchor box has been incorporated since then.

2.1.5. Training Model

The model will be trained using the command line by compiling the file train.py and its programmable arguments. To specify the input photo size, use img. To make training easier, the image's original 1024 by 1024 dimension was shrunk. (Redmon et al., 2015) After conducting numerous experiments on various computers,52 vision researchers determined that the maximum input size that can be used without significant information loss is 416x416 pixels. Batch: A single epoch, or amount of time during which the model learns all of the weights, increases significantly when thousands of photos are fed into the neural network simultaneously allowing the batch size to be determined. In order to allow for individual training of each batch, the dataset is frequently split into numerous groups of n pictures. . The outcomes from each training batch are stored in RAM and combined after all batches have finished. Because the weights that are learnt from the batches are saved here, using more batches will result in a greater RAM use.A total of 86 batches, each with a batch size of 32 and a training set of 2738 pictures, will be created.Epochs: To specify how many training epochs there are. All input images must be learned, or trained, by an epoch. One epoch will be in charge of training every batch since the dataset is separated into many batches.. The frequency of epochs in model training, typically exceeding 3000, reflects how often the model learns from input data and updates its weights towards correct labels. This decision often relies on intuition and past experiences. Additionally, the data. yaml file, located in the dataset summary, provides the path to the dataset. The model utilises this path to access the validation directory for evaluation after each epoch. Cfg: Provide the route for configuring our model. In accordance with the help of this command line, the train.py programme may compile and construct the architecture for training input pictures, which was previously defined in the model yaml file. Utilising a weight that has already been trained helps speed up exercise. In the event that it remains empty, the model will automatically initialise training weights at random.Name: the folder name for the outcome. A directory holding all of the training results will be generated by the model. cache: For faster training of cache Pictures.The training process took an average of 10 seconds for 86 batches, while the evaluation procedure took an average of 9 seconds for 22 batches over one epoch. The dataset had 3422 photos, and the execution time for 100 epochs was 35 minutes and 41 seconds. 93.5% of the map is from the most recent period. The weight for the highest accuracy does not necessarily correspond to the weighting result from the preceding era. TensorBoard was created as an add-in as a result, making the training process extremely visible.The model runs through one epoch in around 20 seconds, and it takes 100 epochs for the model to achieve an accuracy of 93%This shows that, with no further optimisation methods, the model remains fast and precise while utilising the YOLO v5 original architecture. Pt files are the most recent and

best weight available.The weight yielding the maximum accuracy for the most recent epoch is stored in the pt file. Because both files are under 14MB in size, they can be easily included as a pretrained weight into AI systems (in online or mobile applications) while still maintaining 93.5% accuracy.

2.1.6. Inference with Trained Weight

(Zhang *et al.*, 2018 Using trained weights, the wheat head of any image may be located. If a wheat head is detected, a bounding box will be created to enclose the identified object,indicating the potential presence of a wheat head.Within the GWHD dataset, there are roughly ten images of outdoor wheat that are non-duplicated from the given 3422 training images (Ding *et al.*, 2019). The bounding boxes in the real world are not labelled on these images either. You could imagine them as photographs of farmers taken in their fields that were then put to the test using weights that had already been taught to the model for finding wheat heads. Using trained weights, object detection can be accomplished using the same technique that was used to train the aforementioned model. The programme displayed in Figure 5 is used to create the detect.py file, which reconstructs the training architecture. Then With an accuracy of 93.5%,the taught weights will be utilised to anticipate objects and limit boxes for them.The anticipated bounding boxes that enclose the objects (wheat heads) will be added to the photographyonce detection is finished. Usually, even though their presence is obvious, In the lower left and centre of one of the ten projected photographs number 58, there is some wheat that isn't supposed to be there.

3. Discussion

Primary objective of this study is to automate the generation of Karyograms by identifying each chromosome's class in a microscopic image of the human chromosomes.We require a Karyogram image, which is normally created manually, in order to examine human chromosomes for anomalies. Additionally, building a machine learning classifier for abnormality diagnosis using the original chromosomal images is exceedingly challenging and will yield very poor recall, precision, and accuracy scores.This is how we want to go about solving problems. First, we use a range of data augmentation techniques to improve the 75 original and karyogram images that we obtained from a facility in China. To augment the image count and mitigate overfitting in our model, as well as classifier, and improve recall, precision, and accuracy of our classifier, we developed the scripts to carry out various types of data augmentations, such as 90 and 180degree rotations, reflections/mirroring, random change in brightness, noise injections, and random cropping.Second, we develop a set of convolution neural network classifiers for chromosomal detection and categorisation using the improved Karyogram images.

4. Result

To assess whether a classifier is effective, the evaluation metric used is essential. Selecting the incorrect metric for a model's

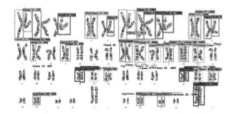

Figure 2: Predictions of the given sample image with the help of Faster RCNN Inception. V2

Figure 3: It specifies the Predictions of the given sample image using SSD Mobilenet V2

evaluation could lead to the selection of the incorrect model or a flawed overall understanding of the model's anticipated performance in real-world circumstances.Our challenge is to identify chromosomes and even the class of each one by

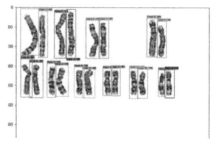

Figure 4: It shows Predictions from image with just twenty chromosomes using inception V2

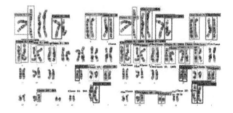

Figure 5: It shows Predictions from image using the RFCN Resnet101.

Figure 6: It demonstrates the outputs produced by the YOLO v5 model trained over a span of 250 epochs.

performing chromosome identification on upgraded Karyogram photos, which serve as actual chromosomes.We have a problem where some classes, like classes X and Y, are underrepresented since they are less important than other classes. The selection of performance metrics for evaluation will be significantly impacted by this. Let's use our dataset as an example, which has 420 classes of Y and 1200 classes of X. We also give the levels of confidence for each projection. With the exception of YOLO v5, which performed better with a lower epoch number of 200; more than 200 resulted in a very slight improvement, and it takes a long time for the model to train, the predictions made using the models Faster RCNN Inception v2, RFCN Resnet 101, SSD Mobilenet v2, and YOLO v5 are shown in the figures.

The model also records the outcomes of two weightings in a pt file. lastly, as depicted in Figures 5 and 6.

Forecasts generated by SSD Mobilenet v2, Faster RCNN Inception v2, and RFCN Resnet101 exhibit inferior performance compared to YOLO v5. YOLO v5 surpassed all other models in terms of the quantity of recognised chromosomes (bounding boxes) per image. We explored potential software vulnerabilities that might have contributed to this poor performance, however testing

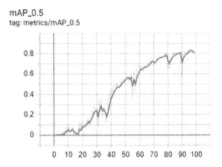

Figure 8: Metrics/precision map of YoloV5

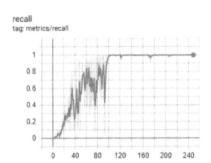

Figure 7: Metrics/recall map of YoloV5

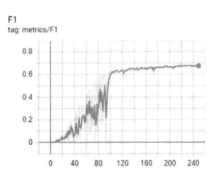

Figure 9: F1/metrics map of YoloV5

Table 1: The model identified every chromosome and correctly identified each chromosome's class for all test results, resulting in faster performance and more precise results. The models Faster RCNN, RFCN, and SSN, on the other hand, produced results that were comparable. We found that the model needed more epochs to obtain good accuracy and recall outcomes, and that its training time was significantly slower than YOLO v5's

Model	Batch Size	Epoch No	Recall	Precision	F1
Yolo v5	16.0	250.0	00.98	00.83	0.89
Faster RCNN Inception V2	14	5000	0.67	0.59	0.62
RCNN RESENET 101	12	5000	0.68	0.72	0.69
SSN MOBILENET v2	12	10000	0.60	0.69	0.64

on photos with fewer chromosomes led to improved bounding box results, as can be seen in the figure above.

5. Performance Evaluation

Precision, F1 score and Recall are the performance indicators used to assess our models . The classes of each chromosome are the main focus of our experiment; as a result, the F1 score is particularly intriguing since it provides an appropriate summary on categorised techniques on the basis how well they predict those classes.

Below is a list of the cross-validation findings utilising our data. We used a range of hyperparameters and selected the most effective ones.

The model does not become better when we go beyond a certain number of epochs, and at that time, we can conclude that we shouldn't increase the epoch number above the given level because the model's training time will increase and the outcomes will not be enhanced. The size of the batch has a big impact on how quickly the model trains. The recall, accuracy, and F1 results of each model are therefore compared in Table 1.

P. Devika et al., 2021. In contrast, the output from the Faster RCNN, RFCN, and SSN models was comparable to that of YOLO v5.(D. Sreeram et al., 2021) However, following extensive testing, we discovered that the model took significantly longer to train compared to YOLOv5, and that more epochs were necessary to get satisfactory accuracy and recall outcomes.

6. Conclusion and Future Scope

Within this system we have integrated automated detection of abnormal chromosomes. The only required input is the image of whole 46 chromosomes. The whole process revolves around detection of chromosome, which is being done with the help of yolo library. With the help of yolo library we are detecting chromosome and counting the number. This automated process will save a lot of time which would have been spent on the manual process of detection of chromosome.

(Qin et al., 2019) Even if YOLOv5 hasn't made much progress, there's still a lot of debate on its nomenclature and advancements within the field of computer vision. Nevertheless, YOLOv5 performs at least as well as YOLOv4 regarding speed and precision when the moniker is taken aside. With the integrated PyTorch framework, it is likely that YOLOv5 will acquire more contributions and have higher potential for growth in the future because it is more community-based and user-friendly than the Dark net framework.

Second, we trained each model with dozens of various hyper parameters to obtain the perfect outcomes. We compared the outcomes of each model in the end and discovered that YOLO v5 performed the best. In contrast to previous models, Yolo v5 had remarkable recollection., accuracy, as well as F1 scores which were 0.98, 0.83, and 0.89, respectively. Additionally, it was able to recognise and categorize every one of the dataset's 46 chromosomes.

References

Balagalla, U. B., Samarabandu, J., Subasinghe, A. (2022). "Automated human chromosome segmentation and feature extraction: current trends and prospects [version 1; peer review: 1 approved with reservations]." *F1000Research*, 11(ISCB Comm J): 301. https://doi.org/10.12688/f1000research.84360.1.

Devi, N. V. S. N., Murale, M., Prasanth, R. K. Nilesh, D. (2022). "Multi-label classification for diagnosis of tuberculosis from chest X-ray images." *International Conference on Advances in Computing, Communication and Applied Informatics (ACCAI)*, 1-7. https://doi.org/10.1109/ACCAI53970.2022.9752472.

Devika, P., Rao, B. B. and Saralaya, A. "A Survey on Automation of Chromosome Based Genetic Diagnosis Using

Machine Learning," 2021 *IEEE 6th International Conference on Computing, Communication and Automation (ICCCA)*, Arad, Romania, 2021, pp. 166-170, doi: 10.1109/ICCCA52192.2021.9666336.

Ding, W., Chang, L., Gu, C., Wu, K. (2019). "Classification of chromosome karyotype based on faster-rcnn with the segmatation and enhancement preprocessing model." *12th International Congress on Imageand Signal Processing, BioMedical Engineering and Informatics (CISPBMEI)*. IEEE, 1–5.

Leung, H. Y., Yeung, M. H. Y., Leung, W. T., Wong, K. H., Tang, W. Y., Cho, W. C. S., ... Wong, S. C. C. (2022). The current and future applications of in situ hybridization technologies in anatomical pathology. *Expert Review of Molecular Diagnostics*, 22 (1), 5–18. https://doi.org/10.1080/14737159.2022.2007076.

Nimitha, N., Arun, C., Puvaneswari, A. S., Paninila, B., Pavithra, V. P., Pavithra, B. (2018). Literature survey of chromosomes classification and anomaly detection using machine learning algorithms. *Published under licence by IOP Publishing Ltd, IOP Conference Series: Materials Science and Engineering, Volume 402, 2nd International conference on Advances in Mechanical Engineering (ICAME 2018)*, 22–24 March 2018, Kattankulathur, India.

Redmon, J., Divvala, S., Girshick, R., Farhadi, A. (2015). "You only look once: unified, real-time object detection." ArXiv./abs/1506.02640.

Sreeram, D., Peneti, S., Tejaswi, P., Chandra, N. S. and Yadav, R. M. "Retracted: Helmet Detection using Machine Learning Techniques," *2021 6th International Conference on Communication and Electronics Systems (ICCES)*, Coimbatre, India, 2021, pp. 1250-1254, doi: 10.1109/ICCES51350.2021.9489096.

Song, S., Bai, T., Zhao, Y., Zhang, W., Yang, C., Meng, J., Ma, F., Su, J. (2022). "A new convolutional neural network architecture for automatic segmentation of overlapping human chromosomes." *Neural Process. Letters*, 54, 1(Feb 2022), 285-301. https://doi.org/10.1007/s11063-021-10629-0.

Qin, Y., Wen, J., Zheng, H., Huang, X., Yang, J., Song, N., Zhu, Y.-M., Wu, L., Yang, G.-Z. (2019). "Varifocal-net: a chromosome classification approach using deep convolutional networks." *IEEE Transactions on Medical Imaging*, 38 (11), 2569-2581.

Zhang, W., Song, S., Bai, T., Zhao, Y., Ma, F., Su, J., Yu, L. (2018). "Chromosome classification with convolutional neural network based deep learning." *2018 11th International Congress on Image and Signal Processing, BioMedical Engineering and Informatics (CISP-BMEI)*. IEEE, 1–5.

Exploring Cardiac Dynamics

An Investigation of Machine Learning Approaches in Heart Attack Prediction

Sasmita Rout[1], Amaresh Sahu[2], Hemanta Kumar Bhuyan[3], Rojalin Mohapatra[4], Partha Sarathi Sahoo[5], Swayumjit Ray[6]

[1]Lecturer, Department of Computer Science, Raghunath Jew Degree College Cuttack, India, Email: routsasmita48@gmail.com
[2]Associate Professor, Master in Computer Application. Ajay Binay Institute of Technology, Cuttack, India, Email: amaresh_sahu@yahoo.com
[3]Associate Professor, Department of Information Technology and Computer Applications, Vignan's Foundation for Science, Technology & Research, Guntur, India, Email: hmb.bhuyan@gmail.com
[4]Ph.D Student, Computer Science and Engineering, Gandhi institute of Technical Advancement, Bhubaneswar, India, Email:rosalinmohapatra000@gmail.com
[5]Student, Department of Computer Science, Ajay Binay Institute of Technology Cuttack, India, Email: partha.orcl@gmail.com
[6]Student, Information Technology and Management, Ajay Binay Degree College Cuttack, India, Email: swayumjitray@gmail.com

Abstract

The heart is a complex organ prone to issues, affecting its ability to pump blood. Timely observation, care, and dietary adjustments can minimize complications post-heart attack. Machine learning algorithms like logistic regression, GNN, light GBM, stacking CV, Naive Bayes, XGB, decision tree, KNN, support vector machine, and random forest assess heart attack likelihood. Procedures include correlation matrix examination, feature visualisation, and AUC analysis. Results show light GBM model with 99% accuracy as a promising candidate for predicting heart attacks, suggesting its potential as the optimal model.

Keywords: Decision Tree, GNN, heart attack prediction, KNN, logistic regression, light GBM, machine learning, Naive Bayes, random forest, stacking CV, XGB

1. Introduction

The rise in heart disease underscores the need for early detection and monitoring. Utilising machine learning, a predictive system has been developed to analyse patients' medical histories, aiding in identifying those at risk. Algorithms like logistic regression and K-nearest neighbours enhance precision in classifying and predicting heart disease patients. Logistic regression offers simplicity and interpretability, while Graphic Neural Network (GNN) captures intricate data relationships. These techniques enable personalised risk assessments and real-time insights, transforming proactive cardiovascular healthcare. By integrating machine learning, medical professionals can effectively manage cardiovascular risks, thereby improving patient outcomes and healthcare strategies.

Chaper 22 DOI: 10.1201/9781003581215

2. Literature Review

Extensive research has been conducted on predicting heart disease, employing various data mining and machine learning algorithms on datasets from heart patients. Despite numerous advancements, challenges persist in this domain. Recent studies techniques for heart disease analysis of various type of classification algorithms like Navie Bayes and Decision Tree were used for heart disease prediction. Chitra, R, and Seenivasagam, V. in 2013 provided an overview of common prediction techniques. Kaur, B., & Singh, W. in 2014 developed a heart diagnosis system based on Support Vector Machines (SVM) with minimal sequential optimization, reporting SVM's superiority over Radial Basis Functions. Prasad, R., et al. in 2019 focused on logistic regression for heart disease prediction, showcasing its outperformance compared to other algorithms. Khan, S. N., et al. in 2019 conducted a comparative assessment of machine learning algorithms, highlighting SVM's efficacy in heart disease prediction. Various algorithms like Light GBM, Naive Bayes classifiers, stacking CV classifier, K-nearest Neighbors, Decision Trees, SVM, and Random Forest Classifier offer distinct advantages based on data characteristics. Despite their differences, these algorithms collectively contribute to improving heart attack prediction accuracy, thereby aiding diagnosis and preventive measures.

3. Methodology and Model Specifications

Our study follows a structured methodology aimed at comprehensively evaluating the execution of different machine learning classifiers in heart attack prediction. The methodology can be summarised into the following key steps:

3.1. Data Preprocessing

The Pandas library was used to load heart attack datasets. Depicted in Table 1. The

Table 1: Description of dataset

```
<class 'pandas.core.frame.DataFrame'>
RangeIndex: 1025 entries, 0 to 1024
Data columns (total 14 columns):
 #   Column    Non-Null Count  Dtype
---  ------    --------------  -----
 0   age       1025 non-null   int64
 1   sex       1025 non-null   int64
 2   cp        1025 non-null   int64
 3   trestbps  1025 non-null   int64
 4   chol      1025 non-null   int64
 5   fbs       1025 non-null   int64
 6   restecg   1025 non-null   int64
 7   thalach   1025 non-null   int64
 8   exang     1025 non-null   int64
 9   oldpeak   1025 non-null   float64
 10  slope     1025 non-null   int64
 11  ca        1025 non-null   int64
 12  thal      1025 non-null   int64
 13  target    1025 non-null   int64
dtypes: float64(1), int64(13)
```

CONCISE SUMMARY OF THE DATAFRAME

dataset was split into features (X) and target labels (y). The function train_test_split of Scikit-learning was used to divide the data further into training and testing sets. The ratio of train set and test set is 80:20.

3.2. Classifier Selection and Training

We had utilised 10 various machine learning classifiers in this study:

Logistic regression, Graph Neural Networks(GNN), LightGBM Classifier, K-Nearest Neighbors, Stacking CV Classifier, Naive Bayes Classifier, XG Boost Classifier, Decision Tree Classifier, Support Vector Machine (SVM), Random Forest.

3.3. Performance Evaluation:

Reports of classifications are generated for each classifier, providing precision, recall, F1-score, and support metrics.

Confusion matrices are plotted for visualising the classification results using the plot_confusion_matrix function. Metrics are calculated to assess the accuracy, precision, recall and F1 score of each classifier.

3.4. Statistical Analysis

We calculate macro and weighted average accuracy metrics to understand the overall performance of the classifiers.

By analyzing the class-wise accuracies, we provide insights into the impact of class distribution on classifier performance.

3.5. Visualisation

A comparative bar plot displays the accuracy of each classifier, with macro and weighted average accuracy indicated.

3.6. Discussion and Insights

The results and visualisations from performance evaluation are scrutinised to reveal the strengths and weaknesses of each classifier, while discussing trade-offs between interoperability, accuracy, and computational complexity. This rigorous methodology aims to offer a thorough comprehension of machine learning classifiers' capabilities and limitations in heart attack prediction, facilitating informed decision-making in practical scenarios.

3.7. Empirical Results

Our study encompasses a detailed analysis of ten machine learning classifiers for heart attack prediction: Logistic regression, Graph Neural Networks (GNNs), LightGBM, StackingCV Classifier, Naive Bayes, Boost Classifier, The decision tree, The KNN algorithm, The Random Forest Classifications and The Support Vector Machine Algorithm (SVM).

3.8. Performance Evaluation

The classification reports for each classifier are given in Table 2.

3.9. Confusion Matrices

The following figure shows the best confusion matrices among all the classifiers. Each cell represents the number of instances that were classified into a specific class. Depicted in Figure 1.

3.9. ROC Curves

ROC curve for Logistic regression, Graph Neural Networks (GNNs), LightGBM,

Table 2: Classification report

Classifier	Precision	Recall	F1-Score
Logistic regression	1.00	0.98	0.99
Graph Neural Networks (GNNs)	0.98	0.91	0.94
Light GBM	1.00	0.96	0.98
Stacking CV Classifier	0.97	0.94	0.95
Naive Bayes	0.98	0.91	0.94
XG Boost Classifier	0.92	0.79	0.85
Decision tree	0.88	0.86	0.87
The KNN classifier	0.97	0.90	0.93
The Support Vector Machine (SVM)	0.92	0.97	0.95
The Random Forest Classifier	0.88	0.81	0.84

StackingCV Classifier, Naive Bayes Etc. are depicted in Figure 2.

3.10. Accuracy Comparison

The accuracy comparison is shown in the figure by using a bar graph which demonstrates the accuracy of each classifier. Depicted in Figure 3.

4. Conclusion

Various methods have been explored, with machine learning emerging as a valuable tool for predicting heart disease, a prevalent socioeconomic issue. As development in machine learning continues to progress, it is conceivable that new methods will enhance its effectiveness in healthcare. The algorithms used in this study have demonstrated noteworthy performance with current variables. In conclusion, machine learning has the potential to reduce physical and emotional harm by accurately identifying cases of cardiovascular illness.

Acknowledgement

We'd like to thank the researchers and institutions that supplied the heart attack datasets utilised in this study. We thank the following people for creating the heart attack datasets: A kaggle Data-set by (Bhat, 2020) of Heart Attack Prediction.

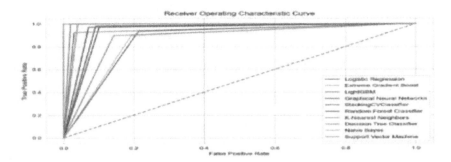

Figure 1: The confusion matrices of GNN, DT, RF, SVM, NB, STACKING CV, KNN, LR and XGBoost.

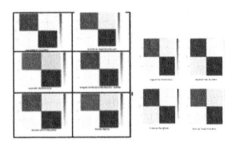

Figure 2: ROC Curve shows performance of all classification models threshold

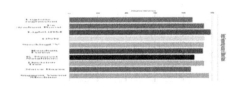

Figure 3: Accuracy comparison

References

Bhat, N. (2020), "A kaggle datasets for Heart Attack prediction." https://www.kaggle.com/nareshbhat

Chitra, R., Seenivasagam, V. (2013). "Review of heart disease prediction system using data mining and hybrid intelligent techniques." *ICTACT Journal on Soft Computing*, 3 (04): (July):605-609. DOI: 10.21917/ijsc.2013.0087

Kaur, B., Singh, W. (2014). "Review on heart disease prediction system using data mining techniques." *International Journal on Recent and Innovation Trends in Computing and Communication*, 2(10) (October): 3003-3008. https://www.academia.edu/9327781/Review_on_Heart_Disease_Prediction_System_using_Data_Mining_Techniques.

Khan, S. N., Nawi, N. M., Shahzad, A., Ullah, A., Mushtaq, M. F. (2019). "Comparative analysis for heart disease prediction." *International Journal on Informatics Visualization*, 1 (4-2): 227-231, DOI: http://dx.doi.org/10.30630/joiv.1.4-2.66.

Prasad, R., Anjali, P., Adil, S., Deepa, N. (2019). "Heart disease prediction using logistic regression algorithm using machine learning." *International Journal of Engineering and Advanced Technology*, 8(3): (February):659-662. https://www.ijeat.org/wp-content/uploads/papers/v8i3S/C11410283S19.pdf

Ramalingam,V.V., Dandapath, Ayantan, & Karthik Raja M. (2018), "Heart disease prediction using machine learning techniques: a survey." *International Journal of Engineering & Technology*, 7(2.8):684-687. DOI:10.14419/ijet.v7i2.8.10557.

Sahu, A., Pattnaik, S. (2017). "Feature selection using evolutionary functional link neural network for classification." *International Journal Of Advanced And Applied Sciences*, 6(4) , (December): 359-367. DOI:10.11591/IJAAS.V6.I4.PP359-367

An Investigation of the Applicability of Machine Learning Methods to the Forecasting of Cardiovascular Disease

I. Sundara Siva Rao[1], Salim Amirali Jiwani[2], Barkat Amirali Jiwani[2], Rakesh Nayak[3], Umashankar Ghugar[3]

[1]Department of Computer Science and Engineering, GITAM University, Visakhapatnam, India, isro75@gmail.com
[2]Department of CSE, Vaagdevi College of Engineering, Bollikunta, India, salimj06@gmail.com, barkatjiwani86@gmail.com
[3]Department of CSE, OP Jindal University, Raigarh, India, nayakrakesh8@gmail.com, ughugar@gmail.com

Abstract

Heart disease is recognised as one of the leading causes of mortality around the globe in the modern era. A greater degree of precision, perfection, and accuracy is required to predict heart illnesses. This is because even a little error may result in the death of an individual, and heart diseases are also related to several risk factors. To effectively address the issue, it is necessary to have a prediction system that can provide accurate and trustworthy information on illnesses. Machine learning offers a method for predicting any event by using training from natural occurrences as its source of information. Logistic Regression, K-nearest neighbors, Support Vector Machines, Decision Trees, and Random Forests are some of the supervised machine-learning classification methods we developed in this study. We have also calculated the accuracy of these algorithms by using pre-existing datasets from the Cleveland database of the University of California, Irvine repository of heart disease patients.

Keywords: KNN, support vector machines, logistic regression, random forests, decision trees

1. Introduction

The impact of cardiovascular disease has been rapidly expanding over the last several years on a global scale. A large number of researchers have been studying the causes of heart disease to pinpoint the exact risk factors and determine their relative importance. Some people call heart disease a "silent killer" because it may kill a person without any outward signs of harm. To reduce the number of complications, it is crucial to identify high-risk patients with cardiovascular disease early on so that they may decide whether to adopt lifestyle changes. Heart disease and other cardiovascular illnesses have been the top killers worldwide throughout the last decade. Coronary artery disease and stroke cause 80% of these deaths (Seckeler and Hoke, 2011). Typically, the countries with the greatest death tolls are those with moderate or low incomes. Personal and occupational habits, as well as a genetic predisposition, are among the many risk factors that might lead to the onset of cardiovascular disease. Heart disease risk factors include a variety of lifestyle choices and physiological variables, including smoking, heavy alcohol and caffeine consumption, stress, and insufficient physical activity. The many classification techniques that are used are compared and studied. For this study, I consulted the dataset housed in the UCI repository. The model for classifying heart disease uses random forests, decision trees, and classification support vector machines (SVM) (Gaziano *et al.*, 2010).

1.1. Datasets

The dataset that is now available to the public on the Kaggle website is derived from cardiovascular research that is currently being carried out on residents of Framingham, Massachusetts. The Python Panda library technique is used to transform it into a data frame.

1.2. Organisation of the Paper

There are a total of six parts to this article. The first section included the introduction. Section 2 discussed complementary literature. The suggested model is laid out in Section 3. The strategies for machine learning are presented in Section 4. Following the presentation of the data and the discussion paper in Section 5, the research is brought to a close in Section 6.

2. Literature Survey

Heart disease is a major concern for public health as it is the leading cause of mortality globally. A favorable treatment outcome and the prevention of further complications are substantially improved upon with early detection of heart illness. Recently, machine learning and hybrid machine learning models have been used to predict cardiovascular disease based on many risk variables. This review of the literature will examine a few recent studies that have attempted to forecast cardiac illness using machine learning or a hybrid approach.

Data from studies titled "Predicting the risk of heart disease using machine learning techniques" are summarised here (Weng *et al.*, 2017). Thirty studies exploring the use of machine learning for risk prediction of cardiovascular disease were thoroughly examined and meta-analyzed in this research. Researchers found that machine learning algorithms performed better than traditional risk prediction models in predicting the development of cardiovascular disease. To enhance the prediction of cardiac events, a mixed-learning technique was used. Predicting the occurrence of cardiovascular disease was the goal of this research, which aimed to assess several hybrid machine-learning methods.

The researchers found that the most accurate model for predicting heart disease was a hybrid one that combined the random forest technique with an artificial neural network. "Using machine learning to predict cardiovascular risk factors" (Ramalingam *et al.*, 2018). This paper presents logistic regression, decision trees, and artificial neural networks for cardiac disease prediction. These results demonstrate the potential application of machine learning and hybrid machine learning models for heart illness prognostic prediction. It is possible that early-stage cardiovascular disease prevention efforts could benefit from models that have shown strong prediction ability across several risk factors.

3. Proposed Model

This research predicts the chance of acquiring heart disease as the probable cause of computerised heart disease prediction to help medical professionals and patients. This study analyses a dataset and applies multiple machine-learning techniques to accomplish its goal. The research goes on to show certain characteristics are more important than others in predicting improved accuracy. As not all qualities are likely to have a significant impact on the result, this might save money by reducing the number of trials a patient undergoes. Numerous pieces of missing and noisy data are present in the

Figure 1: Data processing

real-world data. To circumvent these problems and create strong predictions, these data are pre-processed (Fatima and Pasha, 2017). Our suggested model's sequential chart is shown in Figure 1.

Data cleansing often reveals missing values and noise. The data has to be noise- and missing-values-filled to achieve a useful and accurate result. To make the data more understandable, the transformation modifies its format. It requires operations on aggregate, normalisation, and smoothing. Integration It is necessary to integrate the data before processing it, as it may not be obtained from a single source but from several sources. It is necessary to format the data to reduce it effectively, and this process is difficult. After that, the data is divided into a test set and training set, and different algorithms are used in each set to get the accuracy score (Pahwa and Kumar, 2017).

4. Machine Learning Approaches

Many supervised classification algorithms have been used which are as follows (Lavecchia, 2015): Logistic Regression, Support Vector Machine (SVM). Decision Tree, Random Forest Algorithm.

5. Result Analysis and Discussion

Researchers want to find cardiovascular disease-prone persons. We tried these supervised machine learning classification algorithms: The UCI repository employed K-nearest neighbour, decision tree, random forest, and Logistic Regression. Several classifiers were tried. The research used Intel Core i7 8750H processors with up to 4 GHz and 18 GB RAM. Training and test sets were made from the categorised dataset. We preprocess data and measure accuracy using supervised classification methods including decision trees, Random Forests, K-nearest neighbour, and Logistic Regression. Various algorithms' percentage accuracy scores are shown. We also compared the effectiveness of the suggested model in predicting the occurrence of heart disease. The models are evaluated using accuracy, precision, recall, and F1 scores since our research involves a classification issue. The terms TP, FP, TN, and FN will be defined here. A positive event that the model accurately predicts is called a true positive (TP), while an inaccurate prediction is called a false positive (FP). If the model accurately predicts a negative.

5.1. Dataset and Element Characteristics

The information is sourced from the UC Irvine data warehouse. The Heart Disease Dataset is the name given to this specific dataset housed in the machine learning repository at UCI.David Aha and other UCI Irvine students established this resource in 1987. Four organisations contributed to the heart disease dataset (Otoom *et al.*).

1. Cleveland Clinic Foundation.
2. Hungarian Institute of Cardiology,
3. Budapest.
4. V.A. Medical Centre, Long Beach, CA.
5. University Hospital, Zurich, Switzerland.

In fig 2 to fig 4 , we have represented the simulations model using the heart disease dataset.

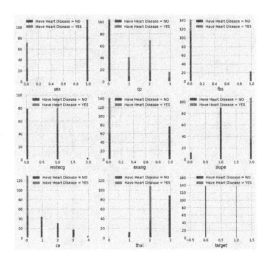

Figure 2: Simulations on heart disease dataset

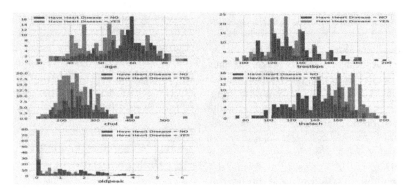

Figure 3: Simulations on heart disease dataset

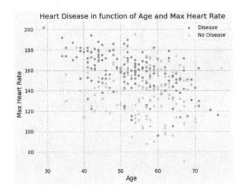

Figure 4: Simulations on heart disease dataset

5.2. Models Building

In table 1, we have compared the results of several methods. Training is error-free, and tests are 91.8% accurate. The first configuration utilises max depth and default estimators. The forest contains 100 trees since n estimators are 100. max depth is None, thus internal nodes need less samples. Despite theoretical 100% accuracy, test accuracy is 85.25 percent. Overfitting may be to blame. Reshuffled the dataset and tried numerous random state values from 1 to 2000 to fix any remaining concerns with the training data not being generalised. Random state 1826 yields 91.80%

Table 1: Results of several methods

Methods	Train accuracy	Test accuracy	precision	recall	F1 score
Logistic Regression	85.88%	87.25%	0.89	0.80	0.84
SVM	90.26%	88.89%	0.92	0.82	0.86
Naïve Bayes	84.47%	88.25%	0.90	0.80	0.86
Random Forest	99%	95.80%	0.95	0.91	0.92

test accuracy. For instance, 100 trees make sense in the forest. Since the model only optimises for training data, underfitting happens when there aren't enough trees to represent the data. When there are too many trees, a model becomes too intricate and sensitive to data, causing overfitting. Random Forest excels in handling feature-rich datasets, balancing variation, and ignoring data noise. The best of these five models is Random Forest.

6. Conclusion

Develop data mining tools for heart disease prediction. We strive for accurate prediction with fewer features and tests. Studying 14 vital traits only. K-nearest neighbour, Logistic Regression, SVM, decision tree, and random forest classified data. The model used preprocessed data. This model's best algorithms are K-nearest neighbor. This study is limited, thus more complex and coupled models are required to predict early cardiac disease.

References

Fatima, M., Pasha, M. (2017). "Survey of machine learning algorithms for disease diagnostic." *Journal of Intelligent Learning Systems and Application*, 9: 1-16.

Gaziano, T. A., Bitton, A., Anand, S., Abrahams-Gessel, S., Murphy, A. (2010). "Growing epidemic of coronary heart disease in low-and middle-income countries." *Current Problems in Cardiology*, 35 (2): 72-115.

Lavecchia, A. (2015). "Machine-learning approaches in drug discovery: methods and applications." *Drug Discovery Today* 20 (3): 318-333.

Pahwa, K., Kumar, R. (2017). "Prediction of heart disease using hybrid technique for selecting features." *4th IEEE Uttar Pradesh Section International Conference on Electrical, Computer and Electronics (UPCON)*. IEEE, 500-504.

Patel, J., Tejal Upadhyay, D., Patel, S. (2015). "Heart disease prediction using machine learning and data mining technique." *Heart Disease*, 7 (1): 129-37.

Ramalingam, V. V., Dandapath, A., Raja, M. K. (2018). "Heart disease prediction using machine learning techniques: a survey." *International Journal of Engineering & Technology*, 7 (2.8): 684-687.

Seckeler, M. D., Hoke, T. R. (2011). "The worldwide epidemiology of acute rheumatic fever and rheumatic heart disease." *Clinical Epidemiology*, 3: 67.

Weng, S. F., Reps, J., Kai, J., Garibaldi, J. M., Qureshi, N. (2017). "Can machine learning improve cardiovascular risk prediction using routine clinical data?" *PLoS One*, 12 (4): e0174944.

Otoom, Ahmed F., et al. "Real-time monitoring of patients with coronary artery disease." International Journal of Future Computer and Communication 4.3 (2015): 207.

Simulation and Analysis of Power Quality Improvement FLC Based DVR for Mitigating Voltage Sags in Grid Connected Power Systems

Sritam Parida[1], Maheswar Prasad Behera[1], Manoj Kumar Sahu[2]

[1]Department of Electical Engineering, IGIT, Sarang, BPUT, Rourkela, India, sritamparida@igitsarang.ac.in, mpbehera@igitsarang.ac.in
[2]Department of Electical Engineering, CAPGS, BPUT, Rourkela, India, capgs.mksahu@bput.ac.in

Abstract

A dynamic voltage restorer (DVR) is a power conditioner belongs to the Flexible AC transmission system (FACT) device family. The concept of DVR developed to maintain an uninterrupted three-phase load voltage level under any sudden grid voltage distortion condition. Grid voltage distortion is a power quality (PQ) issue cause due to voltage sag, voltage swell, interruption and flickering. The requirement for electrical power increasing day by day due to rapid development of industry and their automation technology. The use of power semiconductor devices in the network and integration of green energy (GE) sources such as solar, wind energy enhance system reliability, but the system face some challenge in maintain the power quality and system stability. In order to compensate such trouble in recent era FACTS controller such as DVR, DSTATCOM etc. developed. The DVR play an active role in mitigating PQ issues. This research work proposed a test model, which deals with the protection of sensitive loads from power quality disturbances using a DVR. A control algorithm adapts soft computing logics i.e. a fuzzy logic controller proposed to generate a reference filter voltage to compensate the load voltage during any power quality issue. The proposed model and results simulated in MATLAB/SIMULINK environment and the performance of model verified.

Keywords: Power conditioner, PQ, sag, swell, FACTs devices, DVR, 3phase faults, fuzzy controller, voltage source inverter, Injection transformer, harmonic reduction.

1. Introduction

The power system broadly divided in to four parts i.e. power production, power transmission, power distribution, power utilisation. Power system fed domestic as well as industrial load on distribution side (Dugan *et al.*, XXXX). Normally sensitive load (hospital's operation theaters, processing plants, data processing units etc.) affects more due to low power quality. Power quality affects both utility and consumer as well (Tan *et al.*, 2023). DVR is a FACTS device adopted widely for mitigating nonstandard voltage in distribution side as shown in Figure 1. All the simulation of proposed test system conducted in

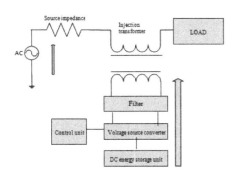

Figure 1: Basic topology of a DVR

Table 1: Characteristics table of voltage profile

Sr.no	Parameter	Objective
1	Voltage variation	Within tolerance level10%
2	Wave shape	Perfect sinusoidal
3	Voltage	Balanced in three phase

MATLAB/SIMLINK environment to know the effectiveness of DVR based system.

Based controller discussed to control the DVR (Moghassemi *et al.*, 2021). This paper proposed a DVR structure, which mainly consists of a VSC connected to the system through a line injection transformer (Sabin and Sundaram 1996, Subrahmanyam *et al.*, 2018). The test system, consist of a grid connected system supply electrical power to a sensitive load.

Previously conventional controller based on PI logics is used to control the DVR topology (Hingorani, 1995; Kadandani and Maiwada, 2015). But PI controller has a narrow range of controllability and un reactive to abrupt alternation of error. Hence various soft computing techniques used to overcome these issues. In this proposed research work, fuzzy

Table 2: Cause of power quality issues

Sr.no	Source of Disturbance	Parameter
1	Natural phenomena	Storm, Lightening, Animal activity
2	Utilities	Capacitor, Load switching
3	Load end	High load demand, Sensitive nonlinear load, Improper earthing

penetration, unbalanced load distribution in power system network strongly affect the quality of power. Some basic cause of PQ stated in Table 2.

There are various issues arises due to poor power quality, which are stated below-

1. The data loss or damage of data complete loss of power supply of various sensitive loads.

2. The operational problem related to microprocessor based and micro controller based control equipment, which are responsive to voltage disturbance.

3. Shut down of equipment if the available voltage is not within the range of voltage limit for duration more than tolerance time.

4. Resonance occurs due to harmonics, which in turn results in mal operation of protective device.

2. Power Quality Overview

Power quality is termed as the concept of maintain the various parameters of system such as voltage, current, frequency, power etc. within in its acceptable range irrespective of any fault or disturbances (S. Akhtar & et.al). We can say supply power to have good quality considering only voltage prospective if it has following characteristics noted in Table 1.

Due to addition of various nonlinear load to the existing power system, increase demand of power electronics devices in various drives renewable energy source

Thus, the power quality needs to measure for the above-mentioned problems due to voltage level variation. There are two main power quality problems related to voltage disturbance i.e. voltage sag and voltage swell. Both issues cause large system unbalance and shutdown of sensitive equipment. According to voltage reduction standard of IEEE std. 1159-1995, the voltage sag normally comes in a range of 10% to 90% of the actual operating voltage value (magnitude 0.1pu to 0.9pu). The duration of voltage sag lies in between 0.5 cycle to 1minute. Similarly, the voltage swell normally comes in a range of 110% to 180% of the actual operating voltage value (magnitude1.1puto1.8pu).The duration of voltage sag lies in between 0.5 cycles to 1 minute. Comparative to voltage swell voltage sag is the common type of PQ problem, hence it cause more attention.

3. Fuzzy Logic Controller

The brain of human easily understood the information provided by various part of human body similarly fuzzy theory deal with the processing of information supplied from the system (Chankhamrian *et al.*, 2014, Shakil *et al.*, 2013). The computation done by linguistic value specified by membership function (AI-Mathnani *et al.*, 2007, Chankhamiran and Bhumkittipich, 2011). The rule base consist of if then rules is the key component of fuzzy inference system (FIS). The block diagram explaining controller based on fuzzy logic place in Figure 2. The fuzzy controller (FLC) performs input parameters of the fuzzification, such as voltage variation, wave shape, voltage. define the rule, inference mechanism, defuzzification.

FLC required fuzzy value to work hence crisp value converted in to fuzzy value, this process is called fuzzification . The input parameter to the fuzzy controller is crisp value .Using Fuzzification process the crisp value converted to fuzzy value which represented by the membership functions .Fuzzy logic designed by fuzzy rule based system. The rule base uses the given linguistic value

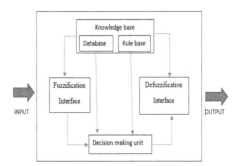

Figure 2: Block diagram of FLC

as its antecedent and consequents. The antecedent express inequality. The consequent is the output if antecedent satisfied. The rule base system define any logic by using IF –THEN rule as IF antecedent and THEN consequent.

The output parameter of rule base system is a fuzzy value, which given to a fuzzy inference system(FIS) as input parameter (PanduSathishBabu, Nagappan Kamaraju). The FIS system convert fuzzy input to fuzzy output using a knowledge base. In this research work, we use Mamdani type inference system. The output parameter of FIS is fuzzy value; hence, a Defuzzification is required to convert this fuzzy value to crisp value. In this test model, centroid type Defuzzification method used.

4. Control Startegyof DVR

The FLC has two input parameter. They are the error which is mention as E (k) and change of error defined as CE (k) at predefined specific sampled times k. The FLC has one output. FLC is used to compute the output as change in phase angle $\emptyset c$. Rule base table on which controller is designed as given in table3.

The five fuzzy sets are (i) NH-Negative High, (ii)NL-Negative Low, (iii) ZR-Zero, (iv) PL- Positive Low, (v) PH-Positive High.

The operation, which performed by the control circuit, are as follows:-

a. Power quality fault detection
b. Calculation of requirement of compensating voltage

Table 3: FLC rule base

Controller O/P NH		Change Inerror			
		NL	ZR	PL	PH
ERROR	NH	NH	NH	NL	NL	ZR
	NL	NH	NL	NL	ZR	PL
	ZR	NL	NL	ZR	PL	PL
	PL	NL	ZR	PL	PL	PH
	PH	ZE	PL	PL	PH	PH

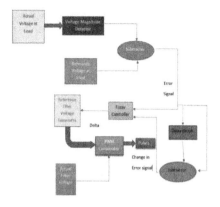

Figure 3: Proposed control strategy of the DVR

c. Generation of gate pulse for converter switch using PWM technic during fault condition.

d. Turnoff the converter switch after clearing of fault

As shown in flow chart, the 'voltage magnitude detector block' of controller, calculates the magnitude of load voltage in PU. The controller set the reference load voltage at 1PU. Then a mathematical Subtract or block is there in the controller, which calculates the difference in voltage magnitude by comparing actual load voltage and reference load voltage. During no fault condition the actual value and reference value both are same .so the error become zero and there is no action for controller (S. Behera &et.al). However, during any fault condition there is differences in voltage, which name das error signal, occur. That error signal and the change in error signal (find out by using a delay circuit) become the input to fuzzy based controller.

The control strategy of DVR which is shown in Figure 3.

The FLC depending on inputs, generate the essential reflection in phase angle. Using this change in phase angle a reference filter voltage generate using the Equations (1)–(3) which are summarised as;-

$$Varf = Vmr\sin\{wt + \emptyset c\} \qquad (1)$$

$$Vbrf = Vmr\sin\{(wt + \emptyset c) - 120^0\} \qquad (2)$$

$$Vcrf = Vmr\sin\{(wt + \emptyset c) + 120^0\} \qquad (3)$$

The PWM controller block compares the actual filter voltage with the reference filter voltage to generate the required gate pulses. Then the obtained gate pulses fed to the converter switch for smooth operation of DVR.

5. Test Model of the Proposed System

The test model of the proposed system has been designed in MATLAB Simulink simulation environment to create a grid-connected system, which supply electrical power to sensitive load through step up and step down transformer connected in transmission line. The PQ problem i.e. voltage sag also designed in the proposed model for a particular time interval .An efficient DVR topology along with its fast response controller designed which connected with the existing power network through an injection line transformer. The block diagram of proposed test model display in Figure 4.

As shown in figure the sensitive load are responsive to voltage sag. In this proposed test system from observation point of view primarily a voltage sag is introduced for a time interval of 0.1s to 0.2s .During this period the load voltage under sag with and without DVR circuit are drawn with the help of MATLAB software .and observed.

From result it shown that, the DVR injected voltage during the disturbance and the FFT analysis of load voltage with or without DVR during fault analyzed to observe the THD percentage.

The specification used in the proposed test model given above in Table 4.

6. Results of proposed system

Various wave form of load voltage under test extracted from MATLAB SIMULINK of proposed model.

The load voltage of sensitive load connected with grid system without any fault

shown in Figure 5 .The figure shows a three-phase load voltage in PU varies from 1 PU to –1 PU .The Figure 6 shows load voltage magnitude of proposed test system in PU.

A sag introduced in the system for a period of 0.1s to0.2s shown in Figure 7. It is shown that a 20% voltage dip occurred in test system for0.1s.

The Figure 8 shows load voltage magnitude of proposed test system with a voltage sag during a period of 0.1s to 0.2s.

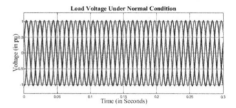

Figure 5: Load voltage of proposed system without fault

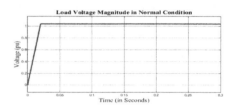

Figure 6: Load voltage magnitude of proposed system without fault

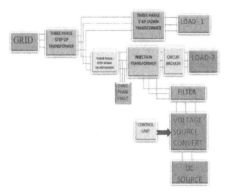

Figure 4: The proposed system of test model topology

Table 4: Specification used in proposed test model

Sr.no	Parameter		Ratings
1	Frequency		50HZ
2	Utility voltage		13KV
3	Transmission Parameters	Resistance	0.001ohms
		Inductance	0.005hennery
4	Transformer (250MVA)	Step up	13/115KV
		Step down	115/11KV
5	Load-1		10MW
6	Load-2		5MW
7	DC voltage source		9.5KV
8	DC link capacitor		750µF
9	Injection line transformer		250MVA 11KV/11KV

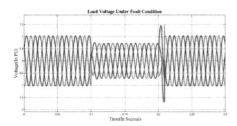

Figure 7: Load voltage of proposed system with sag

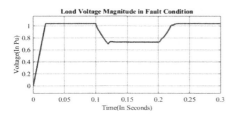

Figure 8: Load voltage magnitude of proposed system with sag

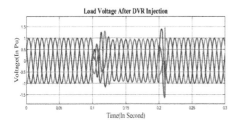

Figure 9: Load voltage of proposed system with sag with DVR

The DVR inject the required amount of voltage to safely the sag voltage .The load voltage wave form with DVR during the sag is shown in Figure 9.

The load voltage magnitude of proposed test system with DVR during a voltage sags during a period of 0.1s to 0.2s shown in Figure 10.

7. Conclusion

A detail analysis of DVR system as a power conditioning devices has been analysis with the help of MTLAB/SIMULINK/ SIMULATION environment .Very low cost, easy to implement, require less

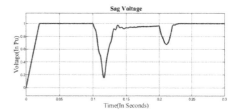

Figure 10: Load voltage magnitude of proposed system with sag with DVR

Table 5: Parameters for THD calculation

Srno	Parameter (In%)	
1	THD during voltage sag	5.78
2	THD during voltage sag with DVR	0.98

computation to find out error, simple control logic in compared to other devices are some basic advantage of dynamic voltage restorer (DVR).The control strategy system has been developed with the help of fuzzy based system with pulse width modulation (PWM) with pure technique. The FLC controller calculates error between load side voltages of the test model system, with its reference for compensating voltage sags. From result of the simulation results it is observed that show the DVR performance is showing efficient to maintain the voltage level irrespective of any power quality issue. It gives the load voltage balance and constant at the nominal value. In addition to it, the proposed topology also effectively balances the system disturbances and keeps harmonics under the specified limit of the IEEE as shown in Table 5.

Reference

Al-Mathnani, A. O., Mohamed, A. W., Ali, M. A. M. (2007). "Photovoltaic based DVR for voltage sag mitigation." *SCORED*, 5. pp. 1-6, DOI: 10.1109/SCORED. 2007.4451393,January 2008.

Akhtar, S., Saha, A., Das, P. (2012). "Modelling, simulationand comparison of various FACTS devices in powersystem." *International Journal of Engineering Research & Technology (IJERT)*, 1 (8), pp 1-13 ISSN: 2278-0181.

Ayadashaker, M. T. H., Sohrabi, F., Gunsel, I. S. (2017). "Fuzzy based controller for DVR in the presence of DG." *ICSCCW*, 120, pp. 684-690, ISSIN 1877-0509. DOI: 10.1016/j.procs.2017.11.296, January 2017.

Babu, P. S., Kamaraju, N. (2007). *Power Quality Enhancement using Dynamic Voltage Restorer*. Lambert Academic Publishing, Chapter 1.

Behera, S., Dash, S. k., Sahu, M. K., Sahu, I., Parida, S. (2023). "Design and development of a new soft-switching buck converter." *2023 International Conference on Power Electronics and Energy (ICPEE)*. IEEE, 1-6.

Chankhamiran, W., Bhumkittipich, K. (2011). "The effect of series connected transformer in DVR application." *Eco-Energy and Material Science and Engineering Symposium*, Vol. 9. https://doi.org/10.1016/j.egypro.2011.09.033.

Chankhamrian, W., Winittham, C., Bhumkittipich, K., Manmai, S. (2014). *Load Side Voltage Compensation of Small Hydropower Grid Connected System using DVR Based on PV Source*. Elsevier.

Donsion, M. P., Aguemes, J., Rodriguez, J. M. (2007). "Power quality benefits of utilizing facts controller." *IEEE Transactions*. pp. 26-29. https://doi.org/10.1109/EMCECO.2007.4371637.

Dugan, R. D., Mcgranaghan, M. F. F., Beatty, H. W., Santos, S. (XXXX). *Electrical Power System Quality*, 2nd edition. Tata McGraw Hills Publications.

Hingorani, N. G. (1995). "Presenting custom power." *IEEE Spectrum*, 41-48.

Kadandani, N. B., Maiwada, Y. A. (2015). "Anoverview of FACTS controllers for power quality improvement." *The International Journal of Engineering and Science (IJES)*, 4 (9): 09-17. ISSN (e): 2319-1813. ISSN (p): 2319 – 1805.

Moghassemi, A., Padmanaban, S., Ramachandramurthy, V., Mitolo, M., Benbouzid, M. (2021). "A novel solar photovoltaic fed trans ZSI-DVR for power quality improvement of grid connected PV systems." *IEEE Spectrum*. https://doi.org/10.1109/NAPS52732.2021.9654507.

Omar, R., Rahim, N., Sulaiman, M. (2009). "Modellingand simulation for voltage sag/swell mitigation usingDVR." *Journal of Theoretical and Applied Information Technology*, 464-470.

Sabin, D. D., Sundaram, A. (1996). "Quality upgradesdependability." *IEEE Spectrum*, 34-41.

Shakil, S., Srivastava, K. K., Pandey, A. V. (2013). "Power quality enhancement and sag mitigation by DVR." *International Journal of Science and Research*, 4, (6).

Subrahmanyam, K. B. V. S. R., Vedik, B., Kumar, M. P., Dhanraj, K. (2018). "A study on the issues of power quality in power systems." *International Journal of Engineering & Technology*, 7 (3-24): 525-528. https://doi.org/10.14419/ijet.v7i3.24.22806.

Tan, K.-H., Chen, J. H., Lee, Y.-D. (2023). "Intelligent controlled dynamic voltage restorer for improving transient voltage equality." *IEEE Spectrum*. https://doi.org/10.1109/access.2023.3293823.

A Comparative Analysis of Working Memory Dynamics through EEG Methodology

Bichitra Mandal, Subasish Mohapatra

OUTR Bhubaneswar, Bhubaneswar, India, bichitra012@gmail.com, subasish.
mohapatra@gmail.com

Abstract

The Working memory, a foundational cognitive system holds a significant importance in various cognitive tasks. Electroencephalography (EEG) provides precise temporal resolution, allowing for the study of neural changes in working memory tasks. This study uses EEG data to observe brain activities linked to working memory and employs diverse experimental paradigms such as task manipulations, stimuli variations and analytic approaches. This comparative analysis focuses on key methodological aspects and emphasizes the value of EEG in studying cognitive processes.

Keywords: Working Memory (WM), electroencephalography (EEG), cognitive processes

1. Introduction

In cognitive neuroscience, understanding Working Memory (WM) is crucial in studying human cognition. WM is essential for cognitive tasks like reasoning, choice selection and analytical thinking (Kemps *et al.*, 2000). It allows temporary storage and manipulation of information. Its dynamic nature presents a challenge in studying its complex operations and neural correlates. Electroencephalography (EEG) stands out as a versatile tool, offering real–time insights into the neural dynamics underlying WM processes. It records the brain's electrical activity with time precision which helps in investigating the temporal dynamics of WM revealing how it is maintained, updated and manipulated (Mohamed *et al.*, 2018, Grissmann *et al.*, 2017).

This comparative analysis examines WM dynamics using EEG methodology across various experimental paradigms. This study aims to analyze neural signatures of different WM tasks to reveal its similarities and differences in neural mechanisms. It combines EEG study to enhance our understanding of WM dynamics for future research and cognitive applications.

The subsequent sections of the paper are structured as follows. In section II, the literature survey is discussed. The EEG methodology is presented in section III followed by challenges and limitations of the paper in IV. In section V the comparison analysis of the study is described and finally in section VI conclusion of the paper is outlined.

2. Literature Survey

The literature includes numerous studies that have been conducted to detect various cognitive states using EEG signals. The majority of existing methods employ machine learning algorithms to develop models capable of predicting cognitive states from EEG signals as shown in Table 1.

Bashiri *et al.* (2015) utilizes coherence as a feature between EEG signals and as a result five coherence feature predicts WM-attention performance with 90% accuracy during the distraction in the encoding stage and 35 features predict with 80% accuracy during the distraction in the maintenance stage, and another 35 features predict with 85% accuracy during the interruption in the maintenance stage.

Mohamed *et al.* (2018) proposed an approach to measure WM using EEG signals and machine learning algorithms. They found the best accuracy for attention is **84%** using SVC, KNN and GP, while the highest accuracy achieved for WM is **81%** utilizing Linear SVC. Kaushik *et al.* (2022) proposes a comparative study based on machine learning techniques, and found that LSTM classifier achieved the highest accuracy of 95.86% and 95.4% in predicting attention and distraction states.

Zhang *et al.* (2020) developed a fast and accurate algorithm for processing EEG signals and predicting WM ability and demonstrate a strong linear relationship between actual observations with an R-square value of 0.72. Islame *et al.* (2023) proposes a method to analyze the mental workload of humans using EEG signals recorded during a simultaneous capacity experiment (SIMKAP). The Neural Network (Narrow) classification algorithm demonstrated outstanding performance with an accuracy of 86.7 %, precision of 84.4%, F1 score of 86.33%, and recall of 88.37%.

Zygierewicz *et al.* (2022) has designed four neural network models and found that Contrastive model employing gated multilayer perceptron (gMLP-MoCo), attained an accuracy of 65.29% along with Matthew's correlation coefficient of 0.288.

Charbonnier *et al.* (2016) directly compares several connectivity measures for

Table 1: Related work on working memory performance

Authors'	Novel Approach	Best Accuracy	Dataset
Mohammad Bashiri *et al.*	Logistic regression	Encoding stage (Acc) = 90% Maintenance stage (Acc) = 80% during distraction Maintenance stage (Acc) = 85 % during interruption	19 healthy participants, permitted by the local ethics committee of Universiti Teknologi PETRONAS
Zainab Mohamed *et al.*	SVM, KNN, Decision trees, random forest, neural networks	SVM gave highest Accuracy WM= 84% Accuracy FA = 81 %	86 subjects accepted by the Research Office in Zewail City.
Pallavi Kaushik *et al.*	LSTM	Accuracy 95.86% and 95.4% in delta and theta bands	24 male Tibetan monks hailing from Sera Jey monastic university, India
Yuanyuan Zhang *et al.*	Functional Linear model	R square value =72%	145 university students aged 18 to 22

(*Continued*)

Table 1: (*Continued*)

Md. Ariful Islame *et al.*	SVM, Neural network (bilayered), Ensemble subspace, Ensemble bagged trees, Neural network (narrow)	Neural network (narrow) provides highest accuracy = 86.7 %, precision = 84.4%, F1 score = 86.33%, and recall = 88.37%.	45 Subjects has been taken from an open accessed pre-processed EEG dataset
Jaroslaw Zygierewicz *et al.*	Neural network models	gMLP-MoCo is the best model with accuracy = 65.29% and a Matthews correlation coefficient of 0. 288	87 healthy participants from local universities and work agencies.
S Charbonnier *et al.*	Cross – correlation, spatial covariance, spectral coherence and phase locking value	Covariance having best accuracy = 60.64%	20 healthy right handed volunteers approved by the French ethics committee.
Sriniketan Sridhar *et al.*	EEDM and DNN	Accuracy = 97.62%	18 subjects, counting 14 normal subjects and 4 diagnosed with mild cognitive impairment

workload estimation and suggests that covariance in the beta band is a promising feature for this purpose. The best results were achieved using covariance in the beta band, with an average accuracy of 60.64%. Sridhar *et al.* (2023) proposed that an ensemble empirical mode decomposition (EEMD) and deep neural network (DNN) methods are effective techniques for cognitive load prediction in both normal and mild cognitive impairment (MCI) subjects. The proposed method achieved high prediction accuracies, averaging 97.62%.

3. EEG Methodology

The electrical signals produced by neurons in the brain can be thoroughly recorded through a non – invasive neuroimaging technique called EEG (Islame *et al.*, 2023). Because of its exceptional temporal resolution, it is commonly used to study WM dynamics and allows researchers to examine the rapid

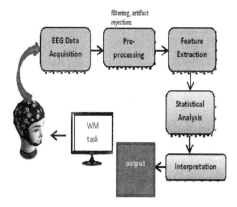

Figure 1: Block diagram of WM

neural processes involved in WM tasks (Bell and Cuevas, 2012; Lenartowicz *et al.*, 2021, Baddeley, 2012). The block diagram presented in Figure 1 describes how EEG methodology is used for accessing WM dynamics.

Participants are first exposed to specific stimuli or tasks like visual or auditory cues to stimulate their WM assessment using EEG Methodology

They react to various stimuli and carry out cognitive processes as per the guidelines of the task. EEG electrodes on the scalp record brain activity during tasks (Mohamed *et al.*, 2018). Then the raw EEG data are refined like removing disturbances, discarding anomalies like eye blinks and muscle movements, and establishing a consistent baseline reference point.

Once pre-processed, EEG data unveils valuable features like ERPs, frequency elements, and brain area interconnections (Mohamed *et al.*, 2018). These features are key for understanding WM dynamics which then undergo thorough statistical analysis to identify differences and connections in experimental conditions, groups, or cognitive variables. Then the data results are analyzed and interpreted within the research question and theoretical frameworks, revealing insights on neural mechanisms in WM processes. This helps in excavate our knowledge of cognitive processes, potentially impacting fields like neuroscience, psychology, and cognitive science.

4. Comparative Analysis

As described in Table 2, the studies reveal patterns and discrepancies in WM dynamics, varying in experimental approaches and participant samples.

Bashiri *et al.*, Mohammad *et al.*, Zhang *et al.*, Charbonnier *et al.*, used an n-back task while Kaushik *et al.*, utilized a debate task. Islame *et al.*, employs the SIMKAP task while Charbonnier *et al.*, used digit-span task. All the studies focus on young adults aged 18-30 years of age and used high – density EEG systems with a minimum of 32 electrodes. It was found that SVM an LSTM algorithm excel in decoding EEG data for WM, closely followed by NN. EEG methods can deepen our grasp of WM dynamics investigating how changing task factors like cognitive load and complexity, impact WM in experiments and how it can improve our grasp of cognitive processes and resource allocation. The application of machine learning and network analysis methods to EEG data can unveil brain activity patterns impacting WM dynamics.

Table 2: Comparisons of methodologies employed across several studies

Developed Model	Objective	Accuracy
Coherence model (Mohammad Bashiri *et al.*)	Investigate brain connectivity through EEG data on WM	90%
cognitive skill detection model(Zainab Mohamed *et al.*)	Predicts levels of WM using EEG signals	84%
Long short term memory model (Pallavi Kaushik *et al.*)	Evaluate the performance on EEG data using ML and DL methods	95.86%
functional linear model (Yuanyuan Zhang *et al.*)	To extract features from EEG signals	72%
Neural network (Md. Ariful Islame *et al.*)	Predict WM performance using EEG signals	86.7%
gLMP_MoCo (Jaroslaw Zygierewicz *et al.*)	To identify and classify thyroid nodules using DL	65.29%
EEG connectivity workload index (S Charbonnier *et al.*)	Compare several connectivity measures from EEG signals	60.64%
DNN (Sriniketan Sridhar *et al.*)	Predict WM load using EEG signals	97.62%

5. Challenges and Limitations

Based on the Table 2 analysis, all the models studied WM dynamics with EEG has limitations of small data sizes and limited cognitive tasks, affecting data availability.

Small data sizes can lead to over fitting in complex models, resulting in inflated performance estimation and difficulty in generalizing to new data. EEG signals can vary due to noise and individual differences. Studying WM with limited cognitive task may not fully capture its complexity. The results from narrowly or limited defined tasks may not be applicable to broader contexts. To address these constraints, it is crucial to employ precise experimental design and increase the sample sizes. Furthermore, integrating EEG with other methodologies can significantly enhance the research outcomes.

6. Conclusion

The methodology of EEG provides significant understanding into the dynamics of WM. The study compares WM in different conditions and populations, revealing insights into the involved mechanisms and neural correlates. EEG data gives insights into brain activity during memory tasks aiding the understanding of cognitive processes and contributing to cognitive neuroscience.

References

Baddeley, A. (2012). Working memory: Theories, models, and controversies. *Annual review of psychology*, 63, 1-29

Bashiri, M., Mumtaz, W., Malik, A. S., Waqar, K. (2015). "EEG-based brain connectivity analysis of working memory and attention." *2015 IEEE Student Symposium in Biomedical Engineering & Sciences (ISSBES)*. IEEE, 41-45.

Charbonnier, S., Roy, R. N., Dolezalova, R., Campagne, A., & Bonnet, S. (2016). *Estimation of working memory load using EEG connectivity measures*, 4, 122-128.

Grissmann, S., Faller, J., Scharinger, C., Spüler, M., Gerjets, P. (2017). "Electroencephalography based analysis of working memory load and affective valence in an n-back task with emotional stimuli." *Frontiers in Human Neuroscience*, 11: 616.

Islame, M. A., Sarkar, A. K., Hossain, M. I., Ahmed, M. T., Ferdous, A. I. (2023). "Prediction of attention and short-term memory loss by EEG workload estimation." *Journal of Biosciences and Medicines*, 11 (4): 304-318.

Kaushik, P., Moye, A., Vugt, M. V., Roy, P. P. (2022). "Decoding the cognitive states of attention and distraction in a real-life setting using EEG." *Scientific Reports*, 12 (1): 20649.

Kemps, E., De Rammelaere, S., Desmet, T. (2000). "The development of working memory: Exploring the complementarity of two models." *Journal of Experimental Child Psychology*, 77 (2): 89-109.

Lenartowicz, A., Truong, H., Enriquez, K. D., Webster, J., Pochon, J. B., Rissman, J., Bilder, R. M. (2021). "Neurocognitive subprocesses of working memory performance." *Cognitive, Affective, & Behavioral Neuroscience*, 21 (6): 1130-1152.

Mohamed, Z., El Halaby, M., Said, T., Shawky, D., Badawi, A. (2018). "Characterizing focused attention and working memory using EEG." *Sensors*, 18 (11): 3743.

Sridhar, S., Romney, A., Manian, V. (2023). "A deep neural network for working memory load prediction from EEG ensemble empirical mode decomposition." *Information*, 14 (9): 473.

Zhang, Y., Wang, C., Wu, F., Huang, K., Yang, L., Ji, L. (2020). "Prediction of working memory ability based on EEG by functional data analysis." *J. Neuroscience Methods*, 333: 108552.

Zygierewicz, J., Janik, R. A., Podolak, I. T., Drozd, A., Malinowska, U., Poziomska, M., Rogala, J. (2022). "Decoding working memory-related information from repeated psychophysiological EEG experiments using convolutional and contrastive neural networks." *Journal of Neural Engineering*, 19 (4): 046053.

Zero Forcing Conditioning of Energy Efficiency in a Contemporary Quasi-Massive MIMO System

Prajna P. Nanda[1], Bikramaditya Das[1], Suvendu N. Mishra[1], Ashima Rout[2]

[1]Department of ETC Engineering, VSSUT, Burla, Sambalpur, India
[2]Department of ETC Engineering, IGIT, Sarang, Dhenkanal, India
E-mail: prajnaparamitananda@gmail.com, adibik09@gmail.com, susoveny@ieee.org,
ashimarout@igitsarang.ac.in

Abstract

This paper aims to optimise the energy efficiency in a quasi-massive MIMO system. Here, we have modified the number of active terminal users in the downlink and uplink as well as the number of antennas in the base station in comparison with massive MIMO system. We propose a new parameter for computation of energy-efficiency along with the already existing parameters. We have utilised Zero Forcing conditions for simulation in both multi-cell & single-cell scenario. Our simulation suggests maximisation of the energy efficiency in both imperfect and perfect Channel State Information conditions.

Keywords: CSI, energy efficiency, quasi-massive MIMO

1. Introduction

Human beings are relentlessly using mobile phones in a 24 x 7 environment. Accordingly, energy consumption is increased as far as a typical modern Base Station (BS) is concerned. In the past, people used the 2G, 3G, and 4G technologies, which allowed up to eight numbers of antenna ports in BS (Rao et al., , 2020). In the present paper, we have explored the consequences on energy consumption with a prospective escalation in the number of antennas in BS. There is an inherent problem associated with the escalation in number of antennas related to energy consumption. For this concern, we have tried to optimise the energy efficiency with an additional parameter in zero forcing (ZF) process using a time division duplexing (TDD) technique. This optimisation studies a quasi-massive MIMO configuration, in which the BS has M (M=100) antennas and the terminal has K (K=60) active consumers. The ratio between M and K can be regarded as a little constant number, individual signals from each BS are pre-coded using a linear transmission procedure (Arshad et al., 2020, Hoydis et al., 2013). We have simulated a standard quasi-massive MIMO system by analyzing the entire power consumption during transmission that includes a new parameter along with previously reported other parameters (Bjornson et al., 2015). Various simulation outcomes include both imperfect and perfect Channel State Information (CSI), concerning ZF processing. Our results reveal the following inferences

1. The parameters M and K are correlated with each other for higher rate of data transmission.

Chapter 26 DOI: 10.1201/9781003581215

2. Number of antennas increases with the circuit and transmitter power.

Because ZF processing uses active interference-suppression at a reasonable complexity, it achieves the maximum energy efficiency (EE). This result also follows the results of Bjornson *et al.* (2015) and Bjornson *et al.* (2013).

2. Literature Review

Energy efficiency is also one of the benefits of 5G. When equated with 4G set-ups, 5G systems use 90% less energy (Chataut *et al.*, 2020). The core technology of 5G mobile communications depends upon massive MIMO technology (Yang *et al.*, 2021). This massive MIMO greatly increases the system accessibility rate by utilising the same time-frequency responses to serve many terminal users by placing hundreds of antennas at the BS (Yang *et al.*, 2021). Compared to MIMO, massive MIMO offer more benefits, such as reduced power consumption and higher network capacity due to operation of bigger antenna apertures (Arshad *et al.*, 2020, Rao *et al.*, 2020). Nevertheless, the equal amount of radiofrequency (RF) chains are needed if all antennas are employed for data transmission, which raises the system's cost and lowers the EE (Yang *et al.*, 2021). Consequently, to fulfill the increasing demand for green energy, the EE performance of massive MIMO must be improved which in turn leads to energy-efficient communications (Yang *et al.*, 2021). To obtain maximum EE and optimal area throughput, it is essential to analyze a massive MIMO configuration for genuine modelling of overall power consumption during both downlink and uplink communications (Rao *et al.*, 2020).

3. Methodology

We have considered a single cell and quasi-massive MIMO structure in downlink and uplink for simulation purposes. Co-located arrays with M antennas help to facilitate communication between the

Uplink τ^{ul}K	Uplink transmission	Downlink τ^{dl}K	Downlink transmission

Figure 1: TDD coherence block (U) for bidirectional transmission (uplink and downlink transmission) (Yang *et al.*, 2021)

BS and K numbers of active terminal users. The active terminal users are chosen in a round-robin manner. Figure 1 shows the active terminal users and the TDD protocol. The parameters τ^{dl} and τ^{ul} represent the pilot multiplexing factors in downlink and uplink respectively.

We have considered a TDD operated quasi-massive MIMO system. In this situation, it is assumed that the propagation path loss is same for each user and each base station antenna. For single user and multi user setting (Yang *et al.*, 2021),

$$EE = \frac{TOTAL\,SYSTEM\,CAPACITY\,(C_{TOTAL})}{TOTAL\,POWER\,CONSUMPTION\,(P_{TOTAL})} \quad (1)$$

$$C_{TOTAL} = Rate\,of\,uplink\left(C_{\tau^{ul}}\right) + Rate\,of\,downlink\left(C_{\tau^{dl}}\right) \quad (2)$$

For ZF processing

$$C_{TOTAL} = K\left(1 - K\frac{\left(\tau^{ul} + \tau^{dl}\right)}{U}\right)\underline{C} \quad (3)$$

$$\underline{C} = B log\left(1 + \rho\left(M - K\right)\right) \quad (4)$$

B = the system's total bandwidth (in Hz)
ρ = the signal to noise plus interference ratio (SNIR)

Total power consumption comprises of two numbers of components, viz. average power of the power amplifier (P_{TX}) and circuit power (P_{CP}).

$$P_{TOTAL} = P_{TX} + P_{CP} \quad (5)$$

$$P_{TOTAL} = \left(P_{TX}^{ul} + P_{TX}^{dl}\right) + P_{CP} \quad (6)$$

Circuit power consumption includes entire amount of power consumed by all the signal processing circuits. We have added an extra parameter P_{EBU} in the already proposed expression of circuit power (Bjornson *et al.*, 2015) for the efficient use of circuit power in our quasi-massive MIMO systems.

$$P_{CP} = P_{FIX} + P_{CE} + P_{C/D} + P_{LP} + P_{TCP} + P_{BH} + P_{EBU} \quad (7)$$

Here, P_{EBU} is the emergency backup power required for short term power fluctuations. Fixed power P_{FIX} is required for site cooling. The power necessary for channel estimation process is P_{CE}. The power necessary for encoding and decoding units is $P_{C/D}$. The power necessary for linear processing is P_{LP}. The power consumption in transceiver chain is P_{TCP}. P_{BH} is the power necessary for backhaul.

ZF processing is used for simulation in total transmission (Bjornson *et al.*, 2015, Rao *et al.*, 2020, Yang *et al.*, 2021). It involves both uplink and downlink transmission. ZF processing solves the optimisation problem of EE. ZF processing modifies equation (1) in to equation (8).

$$EE^{(ZF)} = \frac{K\left(1 - K\frac{\left(\tau^{ul} + \tau^{dl}\right)}{U}\right)\underline{C}}{P_{TX} + P_{CP}^{(ZF)}} \quad (8)$$

$$P_{CP}^{(ZF)} = P_{FIX} + P_{CE} + P_{C/D} + P_{LP}^{(ZF)} + P_{TCP} + P_{BH} + P_{EBU} \quad (9)$$

We have simulated the above equation (9) by considering our new proposed parameter P_{EBU}. Its value is the sum of other six numbers of parameters (Bjornson *et al.*, 2015, Rao *et al.*, 2020,) in right hand side of equation (9). After a conservative calculation, we found its value as 22 W/ (Gbit/s). Figures 2–5 show the simulation results.

4. Result and Discussion

We have considered the outputs of CSI (perfect and imperfect) with the help of 60 numbers of active terminal users and 100 numbers of antennas. Figure 2 shows the global EE-optimum to be 26.3379 Mbit/joule. Similarly Figure 3 shows the global EE-optimum is 21.8412 Mbit/joule. Definitely, a quasi-massive MIMO configuration is ideal. Figures 2 and 3 have concave, smooth surfaces, and their combined effect is nearly optimum EE. Figure 4 displays the maximum EE for various antenna, and Figure 5 displays the equivalent RF

power (RF power per BS antenna). Here, we demonstrate how the system is affected by inter-cell interference by lowering the throughput, which in turn lowers transmitter power usage and, ultimately the EE. It's interesting to observe that the maximum pilot reuse factor (τ^{ul} = 4) provides the maximum area throughput and EE. There is evidence to support the need for active

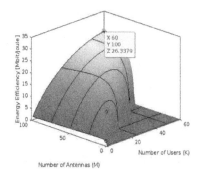

Figure 2: EE (in Mbit/Joule) of single cell in perfect CSI by the use of ZF processing

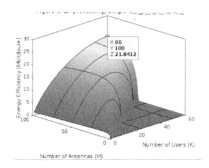

Figure 3: EE (in Mbit/Joule) of single cell in imperfect CSI by the use of ZF processing

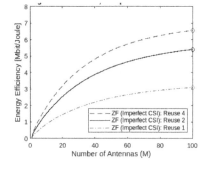

Figure 4: EE value comparison in a multi-cell using CSI(perfect and imperfect)

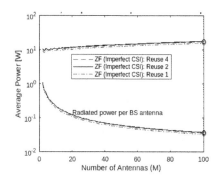

Figure 5: Comparing the RF power in a multi-cell environment using CSI (perfect and imperfect)

pilot-contamination reduction in multi-cell quasi-massive MIMO system.

5. Conclusion

The present study examined how to choose K and M for single cell and multi cell quasi-massive MIMO systems in order to optimise the EE. Our power consumption model is nearly accurate and clearly explains how M and K affect the overall power consumption. We consider ZF processing under perfect CSI conditions, which is validated by simulations conducted for alternative processing techniques in symmetric multiple cell set-ups through imperfect CSI. There is a finite global optimum for the EE (in Mbit/Joule), since it is a quasi-concave function of M and K. Based on the results, we propose that the EE-optimal approach with current technology is to deploy 80–100 antennas to support a reasonably larger number of K. The outcomes show that the emitted power per antenna is reducing with M. This suggests that instead of utilising traditional industry-quality high-power equipment at the base stations, quasi-massive MIMO can be implemented by employing low-power consumer-friendly transceiver devices.

References

Arshad, J., Rehman, A., Rehman, A. U., Ullah, R., and Hwang, S. O. (2020). "Spectral efficiency augmentation in uplink massive MIMO systems by increasing transmit power and uniform linear array gain." *Sensor*, 20 (4982): 1-15.

Bjornson, E. and Jorswieck, E. (2013). "Optimal resource allocation in coordinated multi-cell systems, foundations and trends in communications and information theory." 9 (2-3): 113-381.

Bjornson, E., Sangunetti L., Hoydis J. and Debbah, Merouane (2015). "Optimal design of energy –efficient multi-user MIMO systems: is massive MIMO the Answer? arXiv:1403.6150v2[cs.IT]: 1-16.

Chataut, R. and Akl, R. (2020). "Massive MIMO systems for 5G and beyond networks-overview, recent trends, challenges, and future research direction." *Sensors*, 20 (2753): 1-35.

Hoydis, J., Brink, S. T., and Debbah, M. (2013). "Massive MIMO in the UL/DL of cellular networks: how many antennas do we need? *IEEE Journal on Selected Areas in Communications*, 31 (2): 160-171.

Kassam, J., Castanheira, D., Silva, A. Dinis, R. and Gameiro, A. (2023). "A review on cell –free massive MIMO systems." *Electronics*, 2023, 12 (1001): 1-18.

Kolomvakis, N., Bavand, M., Bahceci, I.. and Gustavsson, U. (2022). "A distortion nullforming precoder in massive MIMO systems with nonlinear hardware." *IEEE Wireless Communication Letter*, 11 (9): 1775-1779.

Rao, M. A., Jehangir, A., Mustafa, S., Sohail, M. N., and Ateeq, U. R. (2020). "Energy efficiency augmentation in massive MIMO systems through linear precoding schemes and power consumption modelling." *Wireless Communications and Mobile Computing, Hindawi*, 2020: 1-13.

Yang, J., Zhang, L., Zhu, C., Guo, X., and Zhang, J. (2021). "Energy efficiency optimization of massive MIMO systems based on the particle swarm optimization algorithm." *Wireless Communications and Mobile Computing, Hindawi*, 2021: 1-11.

Energy Efficient Wireless Sensor Networks using Support Vector Machine

Sidhartha Sankar Dora, Prasanta Kumar Swain

Department of Computer Application, Maharaja Sriram Chandra Bhanja Deo University, Baripada, India
E-mail: lpurna@gmail.com, prasantanou@gmail.com

Abstract

Energy-efficient WSNs aim to balance the trade-off between energy consumption and network performance, enabling long-term, autonomous operation in resource-constrained environments. The role of kernel functions in designing energy-efficient wireless sensor networks (WSNs) using Support Vector Machines (SVMs) lies primarily in optimising the classification performance of the SVM model while considering the energy constraints of sensor nodes. In this paper different kernel functions are used to manage power utilisation, temperature of different rooms in different time periods of smart home. In order to simulate a wireless sensor network, we have used the datasets of the smart home from the public data repository. Our simulation findings show that by employing the right number of features and SVM kernel functions, the suggested model may be used to extend the life of the wireless sensor networks and accomplish our goals. The RBF kernel function produces best result than other functions.

Keywords: WSN, SVM, lifetime, kernel functions, dataset

1. Introduction

Wireless Sensor Networks are group of sensor nodes that communicate wirelessly to collect data from target area and pass the data to the sink where it can be verified. Wireless Sensor Networks (WSNs) are employed by Heinzelman et al., (2000) in many applications such as forest fire detection and climate prediction). In 1990, Support Vector Machine (SVM), a powerful machine learning intelligent technique, was introduced. SVMs are implemented differently than other machine learning algorithms.

In this paper, feature selection technique is used to manage energy in such a way that the life time of WSNs can be enhanced. Different kernel functions of SVM are also used. Smart home can be classified on the basis of energy consumption in different weather condition and categories.

This is how the remaining paper is arranged. The pertinent works are discussed in Section 2. In Section 3, the suggested system is described. In Section 4, the process is elucidated. Section 5 focuses on the findings of the experiment, and Part 6 concludes.

2. Literature Review

The necessity of employing an energy-efficient management plan was emphasized as a means of extending network lifetime. (Wang et al., 2018) have created a localisation algorithm to address the localisation problem that is based on the polar coordinate system and SVM. (Panda et al., 2022) have developed a dependable method based on machine learning approaches that forecasts improved signal strength for optimal communication in subterranean sensor networks. SVM is considered one of the most powerful techniques in the field of data mining. Many different scientific applications have used it successfully (Purnami *et al.*, 2011). The authors claim that LSVM efficiently strikes a compromise between memory requirements and the proportion of processed input that is correctly classified. Furthermore, the sensor node's microcontroller can implement Linear-SVM (Barnawi and Keshta, 2014; Magno *et al.*, 2010). Within the field of machine learning, It was discovered that SVM with a linear kernel was a suitable technique that could learn and acquire the necessary knowledge with less training. (Narayan, & Daniel, 2022) have designed model by using regression technique which improves network life time.

3. Proposed System

This research suggests an intelligent model that makes use of SVM to manage energy in WSNs efficiently. We can evaluate accuracy and performance using a smart home dataset. There are two sections to the data set. 30% of the data are used to evaluate the suggested model, while the remaining 70% are used for training. The formula below provides the Lifetime Extension Factor (LTEF), according Alwadi and Chetty (2012).

The four stages of our suggested model's operation are depicted in Figure 27.1.

4. Methodology

The machine learning classification algorithm support vector machine with different kernel functions are applied for reduction the volume of data to get better accuracy, consume less energy and extend life of WSNs. The support vector machine and its kernel functions are explained below.

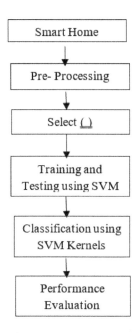

Figure 1: Proposed model

4.1 Support Vector Machine

To handle classification and regression problems, support vector, a kernel-based learning approach, is typically utilised. It is a method of supervised learning. Cortes and Vapnik first put up the idea in 1995 (Faisal *et al.*, 2013).

4.2 Kernel functions

SVM techniques use a set of mathematical operations called kernels. In order to transform data input into the necessary format, a kernel is needed. Different types of kernel functions are used by different SVM algorithms. These functions come in sigmoid, polynomial, radial basis function (RBF), linear, and nonlinear varieties, among others.

4.3 Linear Kernel

Primarily one-dimensional, this is the most basic type of kernel. When there is an abundance of features, it ultimately becomes the finest one. Since most text classification problems can be partitioned linearly, the linear kernel is usually preferred for these problems.

Usually one-dimensional in nature, this is the most basic type of kernel. Linear Kernel Formula

$$F(x, xj) = Sum(x, xj)$$

The data you are attempting to categorise is shown above as x, xj.

4.4 Polynomial Kernel

The data you are attempting to categorised is shown above as x, xj. It provides a broader representation of the linear kernel. It is not as well-liked as other kernel functions because of its decreased accuracy and efficiency. The formula of polynomial kernel is written below.

$$F(x, xj) = (x, xj + 1)d$$

"d" denotes the degree in this instance, while ". " shows the product of the two values.

F (x, xj) represents the boundary used to decide how to partition the given classes.

4.5 RBF Kernel

In the svm kernel, it is one of the most favoured and utilised functions. For non-linear data, it is typically used. When no prior information of the data is available, it aids in proper separation.

$$F(x, xj) = \exp\left(-gamma * \|x - xj\|^2\right) fa$$

The range of values for gamma is 0–1. You have to specifically input the code's gamma value. A setting of 0.1 for gamma is ideal.

5. Result Analysis

5.1 Dataset

A brief summary of each of the smart home datasets utilised in the study is given in this section. The csv file contains measurements from a smart metre for 350 days over the course of one minute, together with weather information for that particular area.

5.2 Simulation Environment

Anaconda3, an Intel Core i5 processor operating at 3.50 GHz with 8GB of RAM, and Microsoft Windows 10 are used for the simulation tasks. We have used following parameter values which are shown in Table 27.1.

We have observed from Table 27.2 and Figure 27.2, the accuracy and lifetime depend on number of feature selection and kernel function. When the features increase, the lifetime decreases and accuracy also changes. It is cleared from Table 27.3, the accuracy changes when kernel function changes. The RBF kernel function produces highest accuracy.

Table 1: Simulation parameters

Parameters	Value
% Training	70%
% Testing	30%
Kernel for SVM	RBF

Table 2: Accuracy and life time

Features	Accuracy	Lifetime
10	99.9	3.2
15	99.2	2.1
20	99.6	1.6
25	99.8	1.28
30	99.8	1.06
32	99.9	1

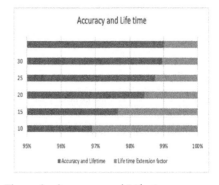

Figure 2: Accuracy and Life time

Table 3: Kernel and accuracy

Kernel	Accuracy
RBF	99.9
Sigmoid	98.5
Polynomial	98.7
Linear	99.8

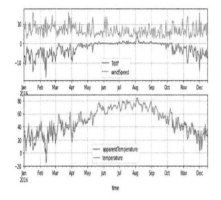

Figure 3: Weather condition

The temperature will vary from time to time in different months. The Figure 27.3 shows different temperatures and weather condition on the basis of time. So, power consumption also changes in different months.

6. Conclusion

An intelligent model for effective energy management in WSNs is presented in this paper. SVM will be used in real-time applications in the future.

References

Alwadi, M. and Chetty, G. (2012). "Feature selection and energy management for wireless sensor networks." *IJCSNS International Journal of Computer Science and Network Security*, 12(6).

Barnawi, A. Y. and Keshta, I. M. (2014). "Energy management of wireless sensor networks based on multi-layer perceptrons." In *Proceedings of the 20th European Wireless Conference, European Wireless*. VDE (1-6), Barcelona, Spain.

Faisal, S., Javaid, N., Javaid, A., Khan, M. A., Bouk, S. H. and Khan, Z. A. (2013). "Z-SEP: zonal-stable election protocol for wireless sensor networks." *Journal of Basic and Applied Scientific Research*, pp. 1 - 9.

Heinzelman, W., Chandrakasan, A., and Balakrishnan, H. (2000). "Energy-efficient communication protocol for wireless micro sensor networks." *System Science. In Proceedings of the 33rd Annual Hawaii International Conference* (pp. 4-7). https://www.kaggle.com/. Datasets

Magno, M., Brunelli, D., Zappi, P. and Benini, L. (2010). "Energy efficient cooperative multimodal ambient monitoring." *In Smart Sensing and Context* (pp. 56–70), Springer.

Narayan, V., and Daniel, A. K. (2022). "Energy efficient protocol for lifetime prediction of wireless sensor network using multivariate polynomial regression model." *Journal of Scientific and Industrial Research*, 81: 1297-1309.

Panda, H., Das, M., and Sahu, B. (2022). "Received signal strength prediction model for wireless underground sensor networks using machine learning algorithms." *Journal of Information and Optimization Sciences*, 43 (5): 949-962.

Purnami, S. W., Zain, J. M. and Heriawan, T. (2011). "An alternative algorithm for classification large categorical dataset: k-mode clustering reduced support vector machine." *International Journal of DatabaseTheory and Application*. 4 (1): 19-30.

Wang, Z., Zhang, H., Lu, T., Sun, Y., and Liu, X. (2018). "A new range-free localisation in wireless sensor networks using support vector machine." *International Journal of Electronics*, 105 (2): 244-261.

Power control of a DFIG based wind turbine using bio-inspired algorithm optimised fuzzy logic controller

Satyabrata Sahoo

Nalla Malla Reddy Engineering College, Hyderabad, India
E-mail:jitu_sahoo@yahoo.com, sahoo.eee@nmrec.edu.in

Abstract

The growing importance of renewable energy, specifically wind power, in meeting clean energy demands needs effective power control for wind turbines. This paper represents a new methodology to power control for DFIG-based wind turbines. It exploits a bio-inspired algorithm optimised fuzzy logic controller, such as particle swarm optimisation (PSO), to enhance FLC performance. The optimisation process focuses on enhancing reference tracking accuracy and achieving optimal power control.

Keywords: wind turbine, DFIG, fuzzy logic controller, bio-inspired particle swarm optimisation

1. Introduction

The global requirement for clean and sustainable energy has forced wind power to the forefront of the renewable energy solutions. Universally, wind power capacity touched 992 GW by 2023, increasing at an annual rate of 12 % over the last 5 years, emphasizing its crucial role in addressing climate change. India has notably added to this growth, boasting an installed wind power capacity exceeding 44 GW by 2023 (Romera and Richards, 2023).

The intermittent quality of wind power poses challenges to grid integration. Effective control approaches are vital for optimizing power generation, augmenting grid stability, and ensuring the reliability of wind turbine system. Several control methods such as PI controllers, model predictive control, and Fuzzy logic controllers (FLCs) are widely exercised. Though traditional methods are favored for their affordability and simplicity, they struggle to handle the nonlinearities and uncertainties. Innovative technique like Model Predictive Control (MPC) (Soliman et al., 2010) poses resilience to wind speed vacillations but face challenges in selecting appropriate control and prediction horizons. Fuzzy logic controllers show promise in tackling the uncertainties, and incorporating optimisation algorithms like Particle Swarm Optimisation (PSO) can further upgrade the control performance (Alsakati et al., 2022).

This article is structured as follows: The next segment presents the dynamical numerical modeling of DFIG-based wind turbine systems. Following that, the controller design, incorporating Particle Swarm Optimisation and Fuzzy Logic Controller principles are discussed. Subsequently, the simulation results and their analysis are presented, followed by the conclusions.

Chapter 28 DOI: 10.1201/9781003581215

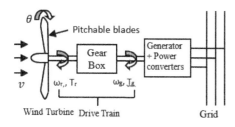

Figure 28.1: Outline of wind power turbine

2. System Dynamic Model

Figure 28.1 explains a systematic diagram of a grid-integrated wind turbine along with its allied components. The system is separated into the segments including the actuator dynamics, aerodynamic system, drive train, wind and generator model. (Sahoo *et al.*, 2016).

The real power obtained by the wind power turbine from the wind is:

$$P_a = \frac{1}{2}\rho\pi R^2 v^3 \qquad (1)$$

where ρ denoted air density, R stands for radius of the swept region and v is the operational wind speed. The rotor power, , is calculated by

$$P_t = P_a C_p(\lambda,\theta) \qquad (2)$$

$$C_p(\lambda,\theta) = 0.5176\left(\frac{116}{\lambda_i} - 0.4\theta - 5\right)e^{-21/\lambda_i} + 0.0068\lambda \qquad (3)$$

$$\frac{1}{\lambda_i} = \frac{1}{\lambda + 0.08\theta} - \frac{0.035}{\theta^3 + 1} \qquad (4)$$

The name $Cp(\lambda,\theta)$, recognised as the power efficiency, has conceptua highest value of 0.593, denoted as the Betz limit and varies based on tip speed ratio λ and the pitch angle θ. The tip speed ratio λ is described as the ratio of the operative wind velocity to the blad tip rate.

$$\lambda = \frac{R\omega_r}{v} \qquad (5)$$

The rotor speed is written as ω_r. By regulating the blade pitch angle θ, the power

coefficient C_p (λ,θ) is changed, there by influencing the Power output P_e. This fundamental rule highlights the control of wind power by regulating the blade angle.

$$P_e = T_g\omega_g \qquad (6)$$

The generator speed is ω_g and Torque of the generator is T_g.

The scintific developing of a DFIG is baseed on its dq alike prototype, with the real power P_e and reactive power Q_e created through DFIG (Sahoo *et al.*, 2022) is given by

$$\begin{cases} P_e = 1.5\left(v_{ds}i_{ds} + v_{qs}i_{qs}\right) \\ Q_e = 1.5\left(v_{qs}i_{ds} + v_{ds}i_{qs}\right) \end{cases} \qquad (7)$$

where v the voltage, i the current, the indices q and d specify the elements of quadrature and the direct axis and s signifies the stator quantities. The equation of induction generator specified by

$$\omega_g = \frac{p}{J_r S}\left(T_g - T_t\right) \qquad (8)$$

$$T_g = pL_m\left(i_{dr}i_{qs} - i_{qr}i_{ds}\right) \qquad (9)$$

where L_m is the mutual inductance.

Time constants are essential factors in the acurate modeling of the system, as they vary between electrical and mechanical systems. Therefore, the generator system is presented as

$$\dot{T_g} = -\frac{1}{\tau_t}T_g + \frac{1}{\tau_t}T_{gr} \qquad (10)$$

here T_{gr} is output of the actuator and represents the generator torque reference value. Pitch actuators are utilised to regulate the blade's positioning. The simulation model for actuator is

$$\begin{cases} \frac{d\theta}{dt} = -\frac{1}{\tau_\theta}\theta + \frac{1}{\tau_\theta}\theta_d \\ \theta_{min} \leq \theta \leq \theta_{max} \end{cases} \qquad (11)$$

The parameter θ_d signifies the pitch angle reference value, while θ_{min} and θ_{max} denote the blade pitch angle minimum and maximum value. The drive train equation

$$T_{tw} = k_d \beta_{tw} + B_d \left(N\omega_t - \omega_g \right) \quad (12)$$

$$T_{tw} = k_d N\omega_t - k_d \omega_g$$
$$-\left(\frac{N^2 B_d}{J_t} + \frac{B_d}{J_g} \right) T_{tw}$$
$$+\frac{NB_d}{J_t} T_t + \frac{B_d}{J_g} T_g \quad (13)$$

J_t and J_g represent the inertia cnstant of turbine & generator respectively. k_d and B_d signifies the shaft stiffness & damping coefficient. N & β_{tw} is the gear ratio and twist angle of the shaft.

The speed of wind $v(t)$ is comprised of 2 different parts: a low-frequency component $v_l(t)$ and turbulence component $v_t(t)$, each with different frequencies represented as:

$$v(t) = v_l(t) + v_t(t) \quad (14)$$

3. Controller Design

3.1 Particle Swarm Optimisation

Particle Swarm Optimisation (PSO) is a robust optimisation algorithm motivated by collective behaviour perceived in nature, like the bird gatherings and schooling of fish. Presented by James Kennedy and Russell Eberhart in 1995, PSO has been commonly adopted across various domains including engineering, finance, and machine learning. It simulates the movement of a group of particles within a multidimensional search space. In PSO, particles represent potential solutions within a swarm, and their velocities determine how they move through the search space. Through iterative refinement, particles adjust their positions based on their individual best-known positions and the collectively optimal positions identified by neighboring particles. This process leads to convergence upon the optimal solution (Shami *et al.*, 2023). The relationship between particle position x_i and velocity v_i is mathematically expressed as follows:

$$\begin{cases} x_i^{k+1} = x_i^k + v_i^k \\ v_i^{k+1} = w v_i^k + c_1 r_1 \left(P_{besti} - x_i^k \right) \\ \quad + c_2 r_2 \left(G_{besti} - x_i^k \right) \end{cases} \quad (15)$$

where w denotes weight of inertia, C_1 and C_2 are the coefficients of acceleration, P_{besti} represents the personal best of particle i, while G_{besti} represents its Global best. Besides, r_1 and r_2 are random numbers ranging from 0 to 1.

3.2 Particle Swarm Optimised Fuzzy Logic Controller

A hybrid control system recognised as a particle swarm optimised fuzzy logic controller unites the merits of both fuzzy logic and particle swarm optimisation, offering a potential solution for numerous applications. Fuzzy Logic presents an intuitive approach to representing and reasoning about intricate systems in a manner like to human thinking. Meanwhile, particle swarm optimisation provides robust and efficient search algorithms, facilitating the finding of optimal solutions. In this setup, the fuzzy logic controller determines control action built on the system's current states, while the particle swarm optimisation always explores for the optimum parameters of the fuzzy logic controllers. This addition yields an intelligent and adaptive control system capable of effectively managing complex and non-linear systems with significant precision and efficiency.

The fuzzy rules building block (Aguas-Marmolejo and Castillo, 2013) are classically crafted using expert information in a heuristic manner, a process that can be time-intensive and may not result optimal outcomes. PSO donates a compelling alternative for automatically subtle tuning MF parameters, resulting in improved output power performance, which is displayed in Figure 28.3 (Silva, 2021). In this context, utilizing gbell MFs, the depiction

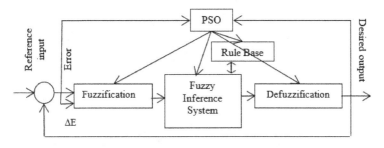

Figure 28.2: PSO based FLC optimised control block diagram

of a gbell MF with its parameters is detailed in (16)

$$\mu\left(pitch\ angle\right) = \frac{1}{1 + \left(\dfrac{\left|pitch\ angle - c\right|}{a}\right)^{2b}} \quad (16)$$

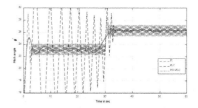

Figure 28.4: Pitch angle comparison

4. Results and Discussion

In this part, it is explore the use of PI, fuzzy logic and particle swarm optimised fuzzy controllers in a MATLAB / SIMULINK simulation to optimie the performance of a DFIG wind turbine during high wind

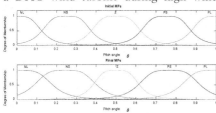

Figure 28.3: PSO optimised MFs of FLC

Table 1: Optimised rules for Fuzzy controller

Error	NL	NS	Z	PS	PL
Change in error					
NL	SC	BC	BC	VBC	VBC
NS	SC	SC	SC	NC	NC
Z	BC	BC	NC	BC	VBC
PS	VBC	NC	NC	BC	VBC
PL	VBC	VBC	BC	VBC	VBC

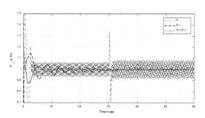

Figure 28.5: Power output comparison

speeds, exceeding the evaluated threshold. The specifications of the machinery are sourced from (Sahoo and Panda, 2022). The wind turbine's functionality requires a stepwise elevation of wind speed to 14 m/s within the first 30 seconds, succeeded by a further increase to 16 m/s, maintained for a total operation time of 60 seconds.

Figures 28.5 and 28.6 illustrate a evaluation of the wind turbine blade angle and output power, respectively, utilising PI, fuzzy logic and PSO based fuzzy logic controllers. Across all scenarios, the improved fuzzy logic controller with PSO demonstrates superior performance with regard to oscillation, tracking, stabilisation and over shoot when contrasted with both the classical and fuzzy logic controller.

5. Conclusion

This paper introduces an innovative method for power control in a wind turbine runs through DFIG. This involves the design, comprehensive discussion, and implementation of a novel fuzzy logic controller enhanced through particle swarm optimisation (PSO-FLC). Our objective is to enhance power control in a wind turbine runs through DFIG while addressing the challenges posed by varying wind speeds on the DFIG rotor speed. Additionally, employing the PSO-FLC controller has a positive outcome on the quality of power from the wind turbine by reducing oscillations in output power. In conclusion, the research outcome indicates that the particle swarm-optimised fuzzy logic controller, especially when combined with PSO, establish to be more effective and adaptable in controlling the power and speed of a DFIG centred wind turbine compared to traditional controllers such as conventional fuzzy logic controllers and classical Propertional- Integral (PI) controllers.

References

Aguas-Marmolejo, S. J. and Castillo, O. (2013). "Optimization of membership functions for type-1 and type 2 fuzzy controllers of an autonomous mobile robot using PSO. In *Recent Advances on Hybrid Intelligent Systems. Studies in Computational Intelligence*, 451: 97–104 Springer, Berlin, Heidelberg.

Alsakati, A. A., Vaithilingam, C. A., Alnasseir, J., Naidu, K., and Rajendran, G. (2022). "Transient stability enhancement of grid integrated wind energy using particle swarm optimization based multi-band PSS4C." *IEEE Access*, 10: 20860-20874.

Romera, M. P. and Richards, A. (2023). "Global wind energy report in the world wide web." https://gwec.net/wp-content/uploads/2023/03/GWR-2023_interactive.pdf (accessed October 12, 2023).

Sahoo, S., Subudhi, B., and Panda, G. (2016). "Pitch angle control for variable speed wind turbine using fuzzy logic," *2016 International Conference on Information Technology (ICIT 2016)* (pp. 28-32). IEEE

Sahoo, S., Rajsekhar, E., and Puhan, P. S. (2022). "Design and simulation of a GA optimized variable speed DFIG based wind turbine using MATLAB. In Sustainable Energy and Technological." *International Conference on Intelligent Controller and Computing for Smart Power (ICICCSP)* (pp. 01–04). IEEE

Shami, T. M., Mirjalili, S., Al-Eryani, Y. *et al.* (2023). "Velocity pausing particle swarm optimization: a novel variant for global optimization." *Neural Comput and Applications*, 35: 9193-9223.

Silva, J. L. D. (2021). "Fuzzy Logic Control with PSO Tunin. Fuzzy Systems." *Theory and Applications of IntechOpen.* 96297: 1–14.

Soliman, M., Malik, O. P. and Westwick, D. T. (2010). "Multiple model MIMO predictive control for variable speed variable pitch wind turbines." *2010 American Control Conference* (pp. 2778-2784). IEEE.

Implementation and Evaluation of Software Defined Wide Area Networking (SD-WAN) Infrastructure Using Cisco System Technologies

Rasmita Kumari Mohanty, Kolipaka Rushik, Devarakonda Sridevi, Modhyala Sai Sumanth Reddy, P. Devika

Department of CSE-(CYS & DS) and AI &DS, VNR Vignana Jyothi Institute of Engineering and Technology, Bachupally, Hyderabad, India
Email: rasmita.atri@gmail.com, rushikkolipaka@gmail.com, devarakondasridevi0502@gmail.com, sumanthreddy5112@gmail.com, devika_p@vnrvjiet.in

Abstract

This paper describes how Software-Defined Wide Area Networking (SD-WAN) was implemented and tested using technology from Cisco. The research paper focuses on using SD-WAN solutions to simplify and streamline the operation of enterprise networks as the requirements change. The main goal is to evaluate how well the new infrastructure works in terms of its performance, its ability to handle growth, and how well it performs compared to the old WAN architecture like Multiprotocol label switching (MPLS). The evaluation process that was undertaken consisted of analyzing the network throughput and latency in addition to evaluating the user experience. Through this assessment, it was clear that the adoption of SD-WAN resulted in improved agility as well as reduced costs. The results show a significant increase in network performance, simple administration, and efficient utilisation of resources. This research emphasises the importance of Cisco SD-WAN technologies in enhancing network performance, which allows companies to remain competitive in the ever-changing networking scenario.

Keywords: Cisco SD_WAN, MPLS, SD-WAN, WAN

1. Introduction

Our project is dedicated towards exploring and implementing the exciting, advanced technology of Software-Defined Wide Area Networking (SD-WAN) in order to cater to the evolving technology and business requirements of the fast-paced modern world where the companies need more flexibility and protection while communicating. We are using the best technology from Cisco Systems to make a network that can securely and quickly adjust to the needs of a business worldwide. imagine it happening in a virtual setting using EVENG lab, which is a great platform on Google cloud thank to u can practice almost anything So, our company has its main building located in India, and it also has some other offices in Dubai and Londn (Gedel et al. 2024).

At the heart of our virtual network are three superhero-like controllers: VMange,

VBond, and VSmart. VManage is like the main authority figure who keeps an eye over all the operations. Vbond is our security guard, he complies as to the rules allowed, he formerly checks if someone is allowed within or bad guy out Then, VSmart acts like a smart brain, and makes the quickest decisions about how the data can travel so that it gets to the exact end point faster.

But, Our project's security is a supreme concern. Our institute has implemented some advanced security measures to safeguard our network from unauthorised access. The communication between our controllers is secure thanks to encryption and VBond makes sure only authorised devices can access it. Every outpost has a digital firewall that protects the network from unauthorised access. Such technology exists to protect against any security breaches. We use Virtual Private Networks (VPNs) over our MPLS and internet routers to ensure everything is securely connected. Our network design has a several different components. These components include provider routers, which allow us to easily connect with other networks, MPLS, which produces a private path between our offices, and regular old internet routers for accessing the internet. This blend allows us to evaluate our SD-WAN's ability to manage different situations.

This project is aiming to convey the tale of how Cisco SD-WAN technology simplifies working in complex situations. Yeah, cyber security requires not only fast and adaptable thinking but acting like a superhero to defend against cyber threats. In this presentation, we are demonstrating how this technology can help businesses in several ways. Firstly, it can help them save costs, and secondly, make their operations smarter and secure. We will also discuss how businesses can manage the challenges of the digital world using this technology.

2. SD-WAN architecture

Software-Defined Wide Area Networking (SD-WAN) is a specialised network architecture that's designed to be agile, scalable, and secure so it can effectively connect and manage distributed networks. And those key components make sure that the network performance works well across the entire wide area. Let's delve into each part of the SD-WAN architecture in detail:

2.1 Edge Devices:

SD-WAN Routers or Edge Devices: Branch offices or remote locations could be referred as devices deployed at the network's edge. This is regardless whether they are physical or virtual. Secure communication between sites is maintained by them through the process of encapsulation and decapsulation of traffic. This means that SD-WAN routers can support a variety of connections through different modes of transportation.

2.2 SD-WAN Controllers

VManage (Management Plane): vManage is responsible for configuring, managing, monitoring, and troubleshooting. It allows administrators to visually define policies,devices manage, and monitor the health of the network. The Restful API of vManage allows for programmatic access, which is useful for automation and integration purposes.

VBond (Orchestration Plane): vBond is responsible for getting SD-WAN devices up and running, verifying their identity, and making sure they are properly connected to the network. It communicates with the controller board after first authenticating the device. The procedure guarantees that the gadgets can connect to one another securely.

VSmart (Control Plane): The control plane is composed of VSmart controllers, which aid in intelligent routing and decision-making. After displaying the network architecture and evaluating the link's state, it routes traffic appropriately. Additionally, it modifies itself in response to the latest policies and circumstances. Dynamic decision-making enables apps to function well and dependably.

2.3 SD-WAN Data Plane

Overlay Tunnels: When sending data packets from the sender to the recipient, the data plane is an essential part of the system. Data packet encapsulation is the method used by SD-WAN to create virtual connections between distant devices. In this manner, secure data can be transferred via unprotected networks. It offers an extra degree of protection to guarantee efficient data transfer and communication. We can create tunnels using the GRE and IPsec protocols, which enable secure and essential data transport.

2.4 Transport Networks:

MPLS (Multiprotocol Label Switching): MPLS is a widely used technique in SD-WAN configurations to enable communication between several networks. The output from this connection is reliable, secure, and efficient in terms of latency. Using SD-WAN in addition to MPLS connections can increase the flexibility of transportation choices and enable us to adjust and modify them as necessary.

Internet Links: Using internet connections, SD-WAN can offer affordable connectivity. The solution's goal is to control and optimise traffic flow by deciding on the optimal path at any given time in compliance with a predetermined set of guidelines or policies. Links to the internet are often utilised for insignificant activities like social media surfing, video watching, and online gaming.

2.5 Security Features:

1. Encryption: SD-WAN uses encryption to protect data transfer over the network. This can help keep unsolicited parties from accessing our personal data.

2. Firewall and Security Policies: Because edge devices have firewalls built to control and filter transmitted traffic, they can defend the network from hacker attacks and unauthorised access. In other words, vManage is used to specify and carry out the security protocols for every device connected to the network. The network as a whole is kept at a constant degree of security in this way. One technique that can be used to safeguard and isolate specific types of traffic is micro-segmentation.

3. Micro-Segmentation: One of the main benefits of SD-WAN is small segments, in which the network is split up into several sections. Restricting an attacker's ability to move laterally within a network increases the likelihood of successful attacks by making it harder for defences to be overrun.

2.6 Quality of Service (QoS) and Traffic Optimisation:

Application-Aware Routing: By utilising SD-WAN controllers, we can gain extensive insight into applications operating on a network. This can be helpful if they need to manage traffic and specific applications during the high demand periods.

WAN Optimisation: SD-WAN makes use of WAN optimisation technologies to improve the efficiency of the network. It enables quicker processing of data and efficient use of resources It also aids in reducing data transfer which is essential for applications that demand quick response time. SD-WAN Architecture is show in below Figure 1.

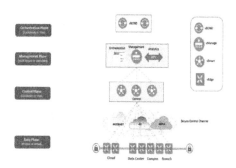

Figure 1: SD-WAN architecture

3. Literature review

Swamy, S. N. L. K. (2023) has proposed Performance Analysis of an SD-WAN Infrastructure Implemented Using Cisco System Technologies . It contains results from tests conducted on-site of the new SDWAN architecture in which features such as traffic shaping, load balancing and high availability are considered. The results demonstrate just how good the network architecture is and point clearly toward the advantages of SDWAN being superior to MPLS technology. If there are malfunctions or if the system is attacked by hackers, handedness will ensure that centralization In such cases These security vulnerabilities may result from network intelligence within SD-WAN. Tiana, D et al. 2023 has proposed this work "SD-WAN: an Open-Source Implementation for Enterprise Networking Services".In this study, which was an early adopter of SD-WAN that used Open Daylight as the basis for its controller and implemented it on top of a layer-2 network built with OpenvSwitch. A set of services for automatically selecting paths using policy rules and monitoring network is proposed (OvS). Indra, D. et al.2024 has proposed this work "SD-WAN: how the control of the network can be shifted from core to edge" .The purpose of this paper is to explain how SD-WAN works by using a centralized control function to effectively manage and route network traffic across many Wide Area Networks (WANs). The SD-WAN edge solution only offers control over CPEs, or network edge devices, which are positioned at the perimeter of the enterprise's network rather than over WAN network devices like switches and routers. Baba Yara, et al. 2024 has proposed this work "Resiliency in SD-WAN with eBPF Monitoring: Municipal Network and Video Streaming Use Cases"Two SD-WAN testbeds are shown in the research report; one is replicated in a lab and the other is deployed in a city's municipal network in Italy. Using a novel BPF-based monitoring technique, the research article aims to show how SD-WAN can ensure service availability and resilience in the event of network disruptions. According to the report, Blidborg, E. (2022) proposed SD-WAN is a technology that can completely change how WAN services are used and has a significant impact on WAN environments and CC services. Salazar-Chacón, G. (2022) proposed this work " SD-WAN Threat Landscape" This work's fundamental thesis is that all conventional network and SDN dangers, along with newly introduced vendor-specific product threats, are included in the SD-WAN threat model. The purpose of the study is to apply a practical strategy to comprehend SDWAN dangers. It covers the fundamental features and components of SD-WAN, investigates the attack surface, investigates the security of different vendor features, and analyses the risks and weaknesses present in SD-WAN systems. Leivadeas, A et al. 2023 Provide any suggestions or fixes to the problems to enhance SD-WAN monitoring . Depending on the vendor, setup and network circumstances, there may be differences in the techniques and resources used to monitor SD-WAN performance. Not all the variables influencing SD-WAN monitoring may be taken into consideration in the comparative study of the techniques and resources. Mohanty, R. K., et al. 2023 has proposed this work "Hybrid Networking SDN and SD-WAN: Traditional Network Architectures and Software Defined Networks Interoperability in digitization era" Determine the essential elements and standards for assessing hybrid networking systems' compatibility in the age of digitization. Use both quantitative and qualitative methods of data analysis to find recurring themes, trends, and problems with hybrid networking interoperability. Mohanty, R. K.,et al. 2023 has proposed this work " Analyzing the Performance of SD-WAN Enabled Service Function Chains Across the Globe with AWS ". Mohanty, R. K., et al. 2022 has proposed this work " Comparative analysis of Cyber security mechanisms in SD-WAN architectures: A preliminary results "The article lists the cyber-attack vectors that can target SD-WAN, including man-in-the middle assaults, data interception, denial of service, and illegal access. Using commercial and open-source solutions, they create an SD-WAN architecture with two branches and a data centre in a simulation scenario they present using the GNS3 software. The simulation scenario is oversimplified and ignores the variety and complexity of SD-WAN deployments in the real world. Tahenni, A., & Merazka, F. (2023) proposed the security protocols are not uniform and can differ based on the SD-WAN system provider or vendor. The study does not cover every potential avenue for cyber-attacks or security risks that SDWAN could encounter, including ransom ware, phishing, and malware.

4. Network model

The objective of this project is to design and evaluate a network infrastructure that will enable the distribution and transfer of data across numerous sites via wired and wireless links. The goal is to combine Cisco ISR routers, switches, and SD-WAN controllers to build a network topology that can handle heavy traffic loads and effectively distribute traffic among numerous devices. The purpose of the configuration tasks is to close the gap between the local area network (LAN) and the wide area network (WAN) and enhance the capabilities of the network infrastructure. In addition, the configuration duties include traffic steering, providing quality of service (QoS), and performance optimisation for the network—all of which are aspects of software-defined networking (SD-WAN). The configuration duties also include handling network resource management and communication security requirements. To confirm the SD-WAN solution's dependability, robustness, and redundancy under varied conditions, extensive testing will be conducted. This project comprises multiple elements that would enable a thorough examination of its execution. These elements include documentation, cost-benefit analysis, user training, and performance monitoring. By combining these elements, we can assess the implementation's

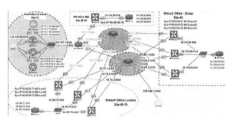

Figure 29.2: SD-WAN topology

effectiveness and identify opportunities for development. Essentially, a thorough report detailing the entire process will be provided, along with test results and suggestions for enhancing SD-WAN infrastructure.

5. Proposed model

Cisco Systems' implementation and evaluation of an SD-WAN (Software-Defined Wide Area Network) infrastructure will bring users a rapid, flexible, secure network with easy management. For this project, we have configured a high-performance network topology within the EVE-NG environment which contains all of the essential elements from Cisco's SD-WAN solution. The topology is modeled on a multibranch enterprise network, with cloud-based controllers having various different branch offices connected through both MPLS and internet services shown in below Figure 2.

5.1 Topology Overview:

5.1.1 Cloud Controllers:

- **Manage:** Thisis the central management controller for all parts of the SD-WAN infrastucture. It offers a single point of centralisation for configuration, monitoring and troubleshooting.

- **Bond:** As orchestrator for the overlay network, the vBond oversees making connections secure and providing contact points to help you set up control connections.

- **Smart:** These are the intelligent routers that provide dynamic routing and traffic engineering within the SD-WAN network.

5.1.2 Management Cloud:

- The Management Cloud is the center for managing and monitoring all of an enterprise's SD-WAN controllers. This helps communication and cooperation between controllers.

5.1.3 Head Office - India Branch:

- **EdgeRouter:** At the India branch, a ViptelavEdge router is placed behind an existing Cisco IOS. This branch works as an interconnect node from the provider router, MPLS router and cloud controllers.

5.1.4 Provider Router:

- The router connects various elements within the cloud and thence to its India city. The connection between the controllers, switch and vEdge router at its head office is attached there.

5.1.5 MPLS Router:

- The MPLS router sets up a connection between the provider and your own branch of this network. Any data sent will flow through it directly to another portion of the network, securely and quickly moving from one location to site.

5.1.6 Dubai Branch:

- This branch employs three vEdge routers, a switch and another Cisco IOS router. The vEdge routers are plugged into the switch, which is then linked to the Cisco IOS router. All of the three vEdge routers are connected to both MPLS and an internet router.

5.1.7 London Branch:

- Like the Dubai branch, all these are three Edge routers and a switch using Cisco IOS software. Switch are connected to the vEdge routers, and these in turn connect with the Cisco IOS router. All three vEdge routers have links to both an MPLS and an internet router.

5.2 Key Terms:

- **SD-WAN (Software-Defined Wide Area Network):** A technology that uses software-defined networking concepts to intelligently route and optimise network traffic across the wide area network, improving connectivity.

- **Manage, vBond, vSmart:vManage** serves as the management controller, vBond is an orchestrator and vSmart plays the role of intelligent routing device. These three are all parts of Cisco's SD-WAN solution.

- **EdgeRouter:** A router to support SD-WAN solution implementation, offering secure and reliable links between branch offices and data centers.

- **MPLS (Multiprotocol Label Switching):** An efficient packet-forwarding protocol often used in wide area networks to improve network performance and reliability.

- **Cisco IOS Router:** A Cisco router running the Internetwork Operating System, which provides all necessary features for routing and security as well as network management.

Configuring and onboarding provider routers, MPLS, and Internet routers in EVE-NG for SD-WAN involves multiple steps.

5.3 Provider router configuration

This configuration is provides in below Figure 3.

Internet Router configuration : This configuration is provided in Figure 4.

Figure 29.3: Configuring and onboarding PE-Router

Figure 29.4: Configuring and on boarding MPLS

-enable
-configure terminal
-hostname ProviderRouter
-interface GigabitEthernet0/0
 -ip address <IP_Address><Subnet_Mask>
 -no shutdown
-exit

MPLS configuration
-enable
-configure terminal
-hostname MPLSRouter
-interface GigabitEthernet0/0
- ip address <IP_Address><Subnet_Mask>
-no shutdown
-mplsip
-exit

Internet Router configuration
-enable
-configure terminal
-hostname InternetRouter
-interface GigabitEthernet0/0
 -ip address <IP_Address><Subnet_Mask>
 -noshutdown-exit

6. Conclusion

We observed great advancements in how fast data is transmitted, how safe our information is, and how well our overall network operates. By utilising vManage's monitoring functions, we were able to observe significant advancements in our network performance. The information we obtained allowed us to take control and be ready for potential issues before they became major problems, thereby streamlining our network management process. The clear diagram displayed the architecture in a simplified manner which aided in comprehending the network more efficiently. Also, the ability to visualise the design helped in implementing it successfully. Analytics and Quality of Service (QoS) policies have optim'd application performance and network management. The security dashboard and the proactive alerting system ensured that we were always prepared to protect our system effectively from potential threats. They also proved their efficacy by providing real-time information on all security-related incidents. By using SD-WAN solution, the network operations across the diverse branches and cloud controllers were observed to have brought about a significant change which can be viewed by comparing the before and after scenarios. Looking forward, the future scope of this work involves continuous monitoring and optimisation to adapt to evolving network demands and emerging technologies. To make the SD-WAN solution more resilient and adaptable, we need to investigate advanced security features, automation capabilities, and how it can work with emerging networking paradigms. Moreover, using machine learning algorithms for forecasting and self-repairing mechanisms can increase the effectiveness of the network. This project is a basic one but the basic one is fundamental for add any further technical advancement in the fields of networking technology.

References

Baba Yara, F., Davis, C., Grigoris, F., and Kantak, P. (2024). "Risk from the inside out: understanding firm risk through employee news consumption." *Kelley School of Business Research Paper*, (2023-4354256).

Blidborg, E. (2022) "An overview of monitoring challenges that arise with SD-WAN".

Gedel, I. A., and Nwulu, N. I. (2024). "Low latency 5G IP transmission backhaul network architecture: A techno-economic analysis. *Wireless Communications and Mobile Computing*, 2024(1), 6388723 , pp. 2-7.

Indra, D., Umar, F., Fattah, F., Azis, H., and Manga, A. (2024). "7 The Microcontroller-Based Technology for Developing Countries in the COVID-19 Pandemic." *The Spirit of Recovery: IT Perspectives, Experiences, and Applications during the COVID-19 Pandemic.*

Salazar-Chacón, G. (2022). "Hybrid networking SDN and SD-WAN: traditional network architectures and software-defined networks interoperability in digitization era." *Journal of Computer Science and Technology* 22 (1): e07-e07.

Leivadeas, A., Pitaev, N., and Falkner, M. (2023) "Analyzing the performance of SD-WAN enabled service function chains across the globe with AWS." In *Proc. 2023 ACM/SPEC Int. Conf. Performance Engineering*, 125-135.

Mohanty, R. K., Sahoo, S. P., and Kabat, M. R. (2023). "Sustainable remote patient monitoring in wireless body area network with Multi-hop routing and scheduling: a four-fold objective based optimization approach." *Wireless Networks*, 1-15.

Mohanty, R. K., Sahoo, S. P., and Kabat, M. R. (2023, June). Sustainable remote patient monitoring in wireless body area network with Multi-hop routing and scheduling: a four-fold objective based optimization approach. *Wireless Networks*, 29(5), 2337-2351.

Mohanty, R. K., Sahoo, S. P., and Kabat, M. R. (2022, October). "A Network Reliability based Secure Routing Protocol (NRSRP) for Secure Transmission in Wireless Body Area Network," 2023 *8th International Conference on Communication and Electronics Systems (ICCES)*, Coimbatore, India, 2023, pp. 663-668,

Swamy, S. N. L. K. (2023). "A Study on Security Attributes of Software-Defined Wide Area Network" (2023). *Culminating Projects in Information Assurance.* 138.

Tahenni, A. and Merazka, F. (2023). SD-WAN over MPLS: A Comprehensive Performance Analysis and Security with Insights into the Future of SD-WAN. *arXiv preprint arXiv*:2401.01344.

Tiana, D. G., Permana, W. A., Gutandjala, I. I., and Ramadhan, A. (2023, December). "Evaluation of software defined wide area Network Architecture Adoption Based on The Open Group Architecture Framework (TOGAF)," 2023 *3rd International Conference on Intelligent Cybernetics Technology & Applications (ICICyTA)*, Denpasar, Bali, Indonesia, 2023, pp. 278-283

e-Ensemble for Diagnosis Gene Expression Data using Hybrid FeatureSelection

Chapala Maharana[1,2], Ch. Sanjeev Dash[1,2], Bijan Bihari Mishra[1,2]

[1]Faculty, CSE, Parala Maharaja Engineering. College, Berhampur, India
[2]Associate Professor Silicon Institute of Tech, Bhubaneswar, India
E-mail: chapala.maharana@pmec.ac.in, sanjeevc@silicon.ac.in,
misrabijan@gmail.com

Abstract

Genes are closely associated with diseases. They contain valuable information about the physical status. Death rate can be reduced to a large extend with proper analysis of such datasets. Advanced machine learning (ML) methods named ensemble methods, expert systems, artificial intelligence are used for such health care issues. In this paper we have proposed such an advanced robust extended ensemble ML model for medical benefits.

KeyWords: AltWOA, Blending, Deep Learning, Ensemble Learning, Feature Selection, Microarray, Stacking

1. Introduction

Maharana *et al.* (2020) studied the machine learning applications built of the independent identical base classifiers. The feature selection technique is adopted in model building for faster and accurate results shown by Maharana *et al.* (2020). The different traditional ensemble methods are max voting, average voting and weighted average voting ensemble classifiers. If the ensemble learning is used for the deep learning then it is called deep ensemble model. In our work we have proposed a method of two phases for efficient diagnosis of health care microarray data. We have used extended ensemble method for early diagnosis of the dangerous diseases with advanced feature selection method. The first phase includes statistical methods i.e. Luka, Passiluukka, and the filter methods are fuzzy entropy measure and optimisation algorithm which is inspiredby animal behavior altruism whale optimisation algorithm (AltWOA) developed by Kundu *et al.* (2022). The second phase is the prediction model designed as an extended ensemble approach with heterogeneous base classifiers and combination of voting classifier, stacking classifier and blending classifier.

2. Related Work

Adulaziz et al. (2020) proposed an effective ensemble boosting learning method which is used for efficient classification on with selective features. The EBL-RBFNN has 98.4% accuracy for classifying the cancer gene expression datasets. A robust ensemble learning approach is used by Abasi et al. (2021) used stabilized lasso (SCOPE) to determine core genes from large sets of genes. Nair et al. (2019) used ElStream utilizing the majority voting technique. Kosinna et al. (2022) studied review of ensemble methods for gene

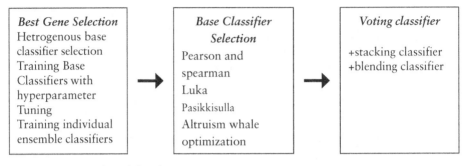

Figure 1: Proposed model architecture

expression datasets. A clustering ensemble algorithm using optimized multi objective particle (CEMP) was proposed by Liu and Wang (2023). An ensemble framework was proposed by Liu and Zou (2021) which used adaptive slice kNNs (scASK) for high dimensionality data analysis. Bi clustering method was used by Akinbo and Daramola (2021). The proposed model by Padilha and Carvalho (2018) using five ensemble classification methods that proved with higher performance results. Ensemble classification framework (ENSCF) was used as the ensemble decision method (2022). Low rank subspace ensemble clustering framework (LRSEC) was used by Khoshgoftaar and Fazelpour (2016) in ensemble method to analyze scRNAseq dataset. A randomized ensemble method was used by Koul and Manvi (2020) for feature selection from Leukemia cancer gene expression data. An ensemble-based classification algorithm was proposed by Hoangp et al. (2022). Dorabiala et al. (2023) used a new ensemble combination method. The performance metrics of it are accuracy, precision, recall, and f1 score. They were found as 95.5%, 0.96, 0.95 and 0.95 respectively. Saha et al. (2016) proposed ensemble method. Anh et al. (2018) proposed a few algorithms Ganaie et al. (2021) have shown deep learning ensemble learning approach for medical applications. Xie et al. (2020) used voting system for better classification using ensemble model. Zamri et al. (2023) have shown many feature selection methods. Ren et al. (2020) have shown improved ensemble method.

3. Proposed Methodology

Our contribution is in selecting appropriate gene selection method and finding the combing rule of classifiers for a robust model which takes least computational time. Fig. 1 shows proposed model. Figure 1. Proposed Model Architecture

4. Result Evaluation

We have found the e-ensemble classifier which takes 0.001 unit of time both for mutation RNA and gene expression datasets whereas voting classifier takes 12.63, 18.87 unit of time, stacking classifier takes 69.16, 186.83 unit of time and blending classifier takes 0.87, 1.81 unit of time for building the model with evaluation. Our model takes 0.001unit of time which is the least. Figure 2 shows ROC curve.

$$\text{Voting } Y\text{voting} = \text{argmax } (Y_i)$$
$$[i=1 ,... k \text{ classifier}]$$

$$\text{Stacking } Y\text{stack} = \sum_{1}^{k} Yk$$

$$\text{Blending } Fl\text{blend} = \sum_{i=1}^{n} w(i) Fl(i)$$

$$Y\text{e\_ensemble} = Y\text{voting}$$
$$(Y\text{voting}+Y\text{stack}+Fl\text{blend})$$

5. Conclusion

The overall performance of stacking classifier is found better than other ensemble classifiers but it takes maximum time than our proposed method.

Table 1. Performance Metrics

dataset altwoa (exec time 0.001 ns)	(training/ preci cancer gene gxpression	recall	f1	support	sensitivity	specificity	cohen's	features	Acc	kappa
class 0	213/ 20530	1.00 / 0.99	0.990	0.97	0.98	75	0.99	0.97	0.98	0.001
class 1		1.00	1.00	1.00	76	1.00	1.00	1.0		
class 2		1.00	0.99	0.99	87	1.00	0.98	0.99		
class 3		0.97	1.00	0.99	67	0.99	1.00	0.99		
class 4 mutation RNA data (exec time 0.001 ns)		1.00	1.00	1.00	70	1.00	1.00	1.0		
class 0	223/	1.00/ 0.61	0.52	0.62	0.57	236	0.72	0.63	0.71	0.001 662
class 1		0.61	0.58	0.60	192	0.81	0.57	0.69		
class 2		0.63	0.74	0.68	173	0.88	0.63	0.71		

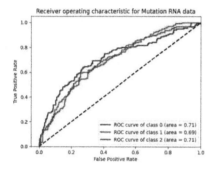 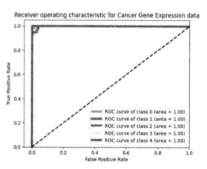

Figure 2: ROC curve

6. Future Scope

Zhongbin et al. and our's proposed algorithm shows a path to challenges of ensemble method. Its implementation with different health care dataset can be analyzed. The challenges in ensemble classifier design are lack of sample size, high dimensionality, class imbalance, dataset invariance, different ML algorithms implementation, deep learning algorithms implementation, hybrid ensemble learning algorithms, hyper tuning of the model.

References

Abdulaziz, K. Osman, A.H. Moetque, H., (2020). "An effective of ensemble boosting learning method for breast cancer virtual screening using neural network," under Grant no. G:1521-830-1440, 10.1109/ ACCESS 2976149, Deanship of Scientific Research.

Abbasi, A. Javed, A. R. Chakraborty, C. Nebhen, J. Zehra, W. and Jalil, W., (Apr, 2021). "ElStream: an ensemble learning approach for concept drift detection in dynamic social big data stream learning." *Open Access Journal*, doi: 10.1109 /ACCESS.2021.3076264, IEEE.

Akinbo, R. S. and Daramola, O.A., (2021). *A Book Chapter, Ensemble Machine Learning Algorithms for Prediction and Classification of Medical Images, Intech open on book Machine Learning - Algorithms, Models and Applications*, doi: http: //dx.doi.org/ 10.5772 / intech open .100602.

Anh, V. Thu, H. and Geurts, P., (2018) "dyn-GENIE3: dynamical GENIE3for the inference of gene networks from time series expression data" *Scientific Reports* doi: 10.1038/s41598-018-21715-0.

Ganaie, M. A., Hu, Tanveer, M., Suganthan, P. N., (2021). *Ensemble Deep Learning a Review Dept of mathematics, IIT, Indore, School of Electronic Engg & Tech University*, Singapore, 350674063 Elsevier.

Hoangp, Q. H., Nguyenp, P. L., Do, T.D. Tran, A. V., and Nguyen, T. A., (2022). "Leukemia classification using informatics," 26 (8), doi: 10.1109 /JBHI .2022.3169542.

Khoshgoftaar, T. M., Fazelpour, A., Rittman,D. J., and Napolitano, A., (2016). "Ensemble vs. data sampling: which option is best suited to improveclassification performance of imbalanced bioinformatics data," 9, doi: 10.1109 /ICTAI. 2015.106 , IEEE.

Kossinna, P.,Cai, Lu, W. X. Shemanko,,Zhang, C.S., (2022). "Stabilized Core gene and pathway election uncoverspan-cancer shared pathways and a cancer specific driver Pathum," doi: 10.1126/ sciadv .abo28 46, Research article on Computer Science.

Koul, N. and Manvi, S. S. (2020). "Ensemble feature selection from cancer gene expression data using mutual information and recursive feature elimination." doi: 10.1109/ ICAECC 50550 .2020.9339518, IEEE.

Kundu, R. Chattopadhyay, S. Cuevas, E., and Sarkar, R. (2022). "AltWOA: altruistic whale optimizationalgorithm for feature selection on microarray datasets." *Computer in Biology and Medicine* 144, doi:10.1016, Elsevier.

Liu, Q., Zhao, X., and Wang, G., (2023). "a clustering ensemble method for cell type detection by multi objective particle optimization," 20, doi: 10.1109 /TCBB .2021 .3132400, IEEE ACM Transaction.

Liu, B. F., Wu, X., and Zou, X., (2021). "scASK: a novel ensemble framework for classifying cell types based on single-cell rna-seq data," 25 (8), doi:10.1109/ JBHI. 2021. 30509 63, IEEE.

Maharana, C., Mishra, B. B., and Dash, C. S. (2020). "A topical survey applications of machine learning in medical issues," 17: 5010-5019, doi. org/10.1166/ jctn. 20.20.9334, JCTN.

Maharana, C., Mishra, B. B, and Dash, C. S. (2020). "Featureselection for classification using extreme learning machine," 7 (6): 769-775 2006103, www. Jetir .org (ISSN-2349-5162), JETIR. Multidisciplinary Rapid Review.

Nair, A. J., Rasheed, R., Maheeshma, K.M., Aiswarya, L. S.. and Kavitha, K. R., (2019)."An ensemble-based feature selection and classification of gene expression using support vector machine, K- nearest neighbor, decision tree," doi:10.1109/ ICCES 45898. 2019. 90020 41, ICCES

Padilha, V. A. and de Carvalho, A. C. P. L. F. (2018). "A comparison of hierarchical Bi clustering ensemble methods," doi: 10.1109/BRACIS.2017.38, IEEE

Ren, Y., Yang, Z., Zhang, H., Liang, H. Y., Huang, H.,andChai, H. (2020). "A genotype based ensemble classifier system for non small cell lung cancer," (8):128509-128517. IEEE.

Xie, W., Zheng, Z., Zhang, W. Huang, L., Lin, Q., and Wong, K.C. "SRG-Vote: predicting mirna-gene relationships via embedding and LSTM Ensemble," *Journal of Biomedical and Health*, doi10.1109/JBHI.2022.3169542, IEEE.

Zamri, N. A., Aziz, N. A., Bhubaneswari, T., Aziz, N. H. A., Ghazali, A. K., (2023). "Feature selection of microarray data using simulated kalman filter with mutation in artificial intelligence and big data for helth care and big data," 11 (8): 2409, doi.org/ 10.3390 /pr11082409. MDPI.

Zhongbin, S., Qinbao, S., , Xiaoyan Z., Heli S., Baowen X., and Yuming Z., (2014). *A Novel Ensemble Method for Classifying Imbalanced Data, Dept. of Computer Science & Technology, Xi'an Jiaotong University, China* 710049 *dept. of Computer Science & Technology, Nanjing University*, PII: S0031-3203(14)00484-doi:http://dx.doi.org/ 10.1016/ j.patcog.2014.11.014, Elsevier.

Path Planning of Autonomous Underwater Vehicle in Swimming Pool Under Communication Constraints

Bhaskar Jyoti Talukdar[1], Madhusmita Panda[1], Suvendu Narayan Mishra[1], Sushanta Kumar Biswal[2], Bikramaditya Das[3]

[1]Department of Electronics and Telecommunication Engineering, VSSUT, Burla, India
[2]Department of Pressure Vessel and Welding, Govt ITI Bhawanipatna, Bhawanipatna, India
[3]Department of Electronics and Telecommunication Engineering, BPUT, Rourkela, India
E-mails: Bjt10talukdar@gmail.com, mpanda_etc@vssut.ac.in, susoveny@ieee.org, sus6214@gmail.com, adibik09@gmail.com

Abstract

Path planning is an essential process of optimising the trajectories and achieving the desired mission objective efficiently. It has various applications such as search and rescue (SAR), underwater mapping and exploration, military and defence. Path planning of AUV is a serious concern, whenever a stationary obstacle is present in a real time environment. The successful deployment of an AUV in an assigned path encompasses different activities such as surveillance operations and monitoring the environment in presence of obstacles. In this study, the Dijkstra's algorithm has been proposed for solving the problem of path planning. Dijkstra's algorithm is a popular method of finding the nearest distance between the way-point along the desired path. It is often used in applications such as robotics, network routing, and transportation. The research mainly contributes in applying the Dijkstra's algorithm in resolving the path planning problem in presence of obstacles along with obtaining the shortest path. The work is carried out in MATLAB. The subsequent path is obtained in real-time and accordingly the outcomes of the simulation and the proposed method can resolve the issue of planning a path in a real time underwater environment.

Keywords: Autonomous underwater vehicle, dijkstra's algorithm, path planning, local minima problem

1. Introduction

AUVs are automated devices designed to run in underwater without the need of direct human intervention (Das B. a., 2014). They have sensors installed, propulsion systems, and navigation controls that allow them to navigate underwater autonomously (Das B. a., 2016). An AUV path planning involves choosing a path that avoids collisions between a starting point and a destination (Khan, 2020). There are various methods of deploying the AUV in underwater and navigating it to follow a path efficiently by avoiding collision.

Dijkstra's algorithm is very well known method and is used for determining the weighted graph's shortest path between a source node and every other optimised node. The algorithm guarantees the shortest path in graphs with non-negative edge weights. It was proposed by Edsger Dijkstra in 1959 (Dijkstra, 1959). The algorithm is extensively utilised in solving the problem on path planning of AUV (Sahoo, 2023). However, there are very less work that is found in the area of planning a path for AUV. Various path planning algorithms are applied as discussed in the literature. The research reviews of the Dijkstra's algorithm that optimises the best path for AUV is provided in this section. Das *et al.* focuses on controlling the leader-follower formation of several non-harmonic AUVs (Das B. a., 2016). Here in this work, a communication plan is developed for a good communication among the AUVs. Further, several algorithms are used for the optimisation of the path. Among all the optimisation techniques, the CLONAL selection algorithm provides better efficacy for cooperative motion of a group of AUV. The modified artificial potential field (APF) is utilised by Fan *et al.* to provide reliable path planning for the AUV (Fan, 2020). To enhance the local minima issues, the regular hexagon-guided approach is provided. The experimental outcome demonstrates that the suggested method can avoid obstacles and obtain an optimal path. Das *et al.* proposed an integrated approach combining Ant Colony Optimisation (ACO), APF, and the CLONAL selection algorithm for avoiding obstacles and path planning (Das B. a., 2016). Panda *et al.* employed the grey wolf optimisation technique which develops a global path planning solution for an AUV (Panda M. a., 2019).

This paper's major contribution is the practical usage of Dijkstra's algorithm to solve the challenges of path planning. Although Dijkstra's algorithm is widely recognised for its ability to optimise the shortest path, its implementation in real-time environments is rare. There are five sections in this study. In Section 2, the issue that has been raised and addressed. The Dijkstra's algorithm's theoretical framework is presented in Section 3. Section 4 shows the simulation outcomes of the AUV path planning methodology. The algorithm's performance and this work's contribution are summarised in Section 5.

2. The Problem Statement

The motion of the AUV along a horizontal plane is considered (Das B. a., 2020). Suppose that the obstacles are located on an N X M 2-D space. The AUV dimensions are thought to be significantly smaller with respect to the workspace dimensions. Figure 31.1 is the representation of the workspace (swimming pool) that is used throughout the experiment. The whole research work is carried out at swimming pool of the VSSUT, Burla.

Let x, y represents grid intervals for the corresponding x and y directions. A node X = (P, Q) is defined by any position in the grid, $0 \le P \le N, 0 \le Q \le M$. A path connecting the first node s_1 as well as a destination node d_1 is determined by a sequence of S = $s, ..., x_a$, $x_{a+1}, .., d_1$ and it is composed of segments of straight lines joining any two nearby nodes x_a, x_{a+1} (Das B. a., 2015). The amount of time needed to travel a specific course is calculated by summing the amount of time needed to travel each segment. Figure 31.2

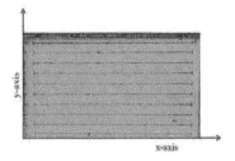

Figure 31.1: Workspace area-VSSUT swimming pool

Figure 31.2: Top view of the workspace

is the bird eye view of the workspace which is being used thorough the experiment. At any situation along the path at (x, y), the vehicle $\vec{v}_a$ (x, y) nominal velocity is determined by –

$$\vec{v}_a(x,y) = \vec{v}_r(x,y) + \vec{v}_c(x,y),$$

$$(x,y) \in x_{a-1}x_a \qquad (1)$$

$\vec{v}_r$ is the relative speed w.r.t. the current speed. The time taken to complete the path is calculated by,

$$t_a = \frac{|d_a \vec{e}_a|}{|\vec{v}_a|} \qquad (2)$$

The overall time needed to complete the path is the sum of all the travelling times for each sub-segment determines (Panda M. a., 2021).

3. Mathematical Modelling of Dijkstra's Algorithm

The Dijkstra's algorithm is being applied in this study to identify the shortest path. Initially, every neighbour node is inspected which is also known as relaxation and the distance between them is updated from the current node (Dijkstra, 1959). The mathematical expression for the relaxation is –

$$d(v) = d(u) + c(u,v), \text{if } d(u)$$
$$+ c(u,v) \le d(v) \qquad (3)$$

A new position is denoted by v and the previous position by u. The distance from the previous position is denoted as $d(v)$. In (3),

Figure 31.3: Starting point

Figure 31.4: Intermediate

Figure 31.5: Destination

$c(u,v)$ denotes the cost function between u and v.

For better understanding of the algorithm let us consider three different positions as shown in Figure 31.3. The destination is at position 3 and the direct distance or cost is unknown from the position 1 of AUV. There is no direct unobstructed route between the starting point and the destination, so at this moment, the movement will be from 1 → 2 and it is depicted in Figure 31.4. Further, to reach the destination it has to reach with the help of the intermediate node. Once, the intermediate node is reached after getting the cost function value $c(u,v)$ and satisfying (3), the destination can be reached and it is shown in Figure 31.5. Therefore, the Dijkstra's algorithm helps in optimising the shortest path using the above approach.

4. Simulation Result

This section testifies the Dijkstra's algorithm, and the results that are obtained while using the algorithm (Panda M. a., 2020). Figure 31.6 is the path planned for the AUV.

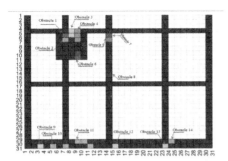

Figure 31.6: Modelled map for path planning

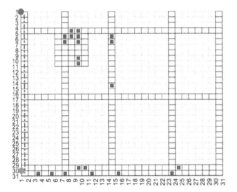

Figure 31.7: Movement of the AUV across the optimised path

The path is illustrated in Figure 31.7 (Das B. a., 2010). The map has varying numbers of static obstacles marked with different colours. In addition, the time is recorded for 20 rounds of completing a circle and reaching at the target. Figure 31.8 illustrates the amount of time the AUV took to travel from the start node to the destination (Rizwan, 2021).

5. Conclusion

The Dijkstra's algorithm is presented in this research, to address the path planning issue. The outcome of the research work proves that the algorithm is capable of resolving the issue of path planning and avoids static obstacles. The AUV reached its precise position in the shortest amount of time. There are number of static obstacles that are distributed across the map. Hence, in a 2D environment the Dijkstra's algorithm

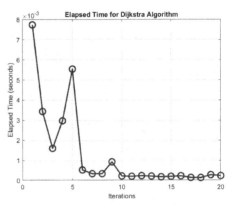

Figure 31.8: Number of iterations vs. Elapsed time graph

offers a viable optimal path based on computation time. For future work, hybrid algorithms can be developed by combining real-time path planning approaches with Dijkstra's algorithm.

Acknowledgement

This work is supported by IIT Guwahati Technology Innovation and Development Foundation (IITG TI&DF), which has been set up at IIT Guwahati as a part of the National Mission on Interdisciplinary Cyber Physical Systems (NMICPS), with the financial assistance from Department of Science and Technology, India through grant number DST/NMICPS/TIH12/ IITG/2020. Authors gratefully acknowledge the support provided for the present work.

References

Das, B. a. (2010). Efficacy of multiband OFDM approach in high data rate ultra wideband WPAN physical layer standard using realistic channel models. *International Journal of Computer Applications*, 2(2), 81-87.

Das, B. a. (2014). Adaptive sliding mode formation control of multiple underwater robots. *Archives of control Sciences*, 24(4), pp. 515-543.

Das, B. a. (2015). Employing nonlinear observer for formation control of AUVs under communication constraints. *International Journal of Intelligent Unmanned Systems*, 3(2/3), 122-155.

Das, B. a. (2016). Co-operative control coordination of a team of underwater vehicles with communication constraints. *Transactions of the Institute of Measurement and Control*, 38(4) pp. 463-481.

Das, B. a. (2016). Co-operative control of a team of autonomous underwater vehicles in an obstacle-rich environment. *Journal of Marine Engineering & Technology*, 15(3), 135-151.

Das, B. a. (2016). Cooperative formation control of autonomous underwater vehicles: An overview. *International Journal of Automation and computing*, 13, 199-225.

Das, B. a. (2020). Formation control of underwater vehicles using Multi Agent System. *Archives of Control Sciences*.

Dijkstra, E. W. (1959). *A note on two problems in connexion with graphs.*

Fan, X. a. (2020). Improved artificial potential field method applied for AUV path planning. *Mathematical Problems in Engineering*, 2020(1), 6523158. 1-21.

Huang, H. a. (2021). A novel particle swarm optimization algorithm based on reinforcement learning mechanism for AUV path planning. *Complexity*, 1-13.

Khan, M. R. (2020). Channel estimation strategies for underwater acoustic (UWA) communication: An overview. *Journal of the Franklin Institute*, 7229-7265.

Li, M. a. (2020). AUV 3D path planning based on A* algorithm. In *2020 Chinese Automation Congress (CAC)* (pp. 11-16). IEEE.

Panda, M. a. (2019). Grey wolf optimization for global path planning of autonomous underwater vehicle. *Proceedings of the Third International Conference on Advanced Informatics for Computing Research*, (pp. 1-6).

Panda, M. a. (2020). A hybrid approach for path planning of multiple AUVs. *Innovation in Electrical Power Engineering, Communication, and Computing Technology: Proceedings of IEPCCT 2019* (pp. 327-338). Springer.

Panda, M. a. (2021). Multi-agent system of autonomous underwater vehicles in octagon formation. In *Intelligent Systems: Proceedings of ICMIB 2020* (pp. 125-138). Springer.

Rizwan, K. M. (2021). A criterion based adaptive RSIC scheme in underwater communication. *Journal of Systems Engineering and Electronics*, 32(2), 408-416.

Sahoo, S. P. (2023). Hybrid Path Planning Using a Bionic-Inspired Optimization Algorithm for Autonomous Underwater Vehicles. *Journal of Marine Science and Engineering*, 11(4), 761.

Sun, B. a. (2018). An optimized fuzzy control algorithm for three-dimensional AUV path planning. *International Journal of Fuzzy Systems*, 20, 597-610.

An Intelligent WBAN to Predict Cardiovascular Diseases

Lomashradhha Parida[1,2], Tusharkanta Samal[2], Bivasa Ranjan Parida[3] , Suvendra Kumar Jayasingh[1], Mamata Ratha[2], Soumya Mohapatra[2], Durga Prasad Khanal[4]

[1]Department of Computer Science and Engineering, IMIT (BPUT), Cuttack, India
[2]Department of Computer Science and Engineering, DRIEMS University, Cuttack, India
[3]Department of Computer Science and Engineering, Silicon University, Bhubaneswar, India
[4]Saraswati Multiple Campus, Tribhuvan University, Nepal, India
E-mail: lomaparida@gmail.com, samaltushar1@gmail.com, bivasa.parida@silicon.ac.in, sjayasingh@gmail.com, mamata.rath200@gmail.com, soumyasnigdha.praharaj@gmail.com, durgapsdkhanal@gmail.com

Abstract

The accurate prediction is essential for the treatment of cardiovascular patients prior to the occurrence of a cardiac attack. This objective may be accomplished with the sophisticated machine-learning framework and a wealth of healthcare data on cardiac disorders. The existing systems use traditional approaches to choose characteristics from one dataset and assign an overall weight according to their relevance, which are ineffective in ameliorating the correct identification of heart disease. The proposed ensemble machine learning and feature fusion methodology using wireless body area network is used with an aim to develop an intelligent public healthcare system for the prediction of cardiac disease. The recommended system depicts supremacy over the existing approaches for predicting cardiac diseases with an accuracy of 89 %.

Keywords: Cardiovascular disease, machine learning, wireless body area network, internet of things

1. Introduction

The cardiovascular disease (CVD) manifests itself in the form of coronary heart illness, rheumatic heart disorder, cerebrovascular disorder, peripheral arterial disorder, and congenital heart illness (WHO, 2021). The wireless body area network (WBAN) remotely undertakes real-time observation of the health of a cardiac patient detected with indicators of cardiovascular ailment (Ahmed et al., 2020) by replacing traditional methods of hospital visits. Now, the CVD investigation uses a WBAN combined with the internet of things (IoT) and machine learning (ML) to improve the WBAN's effectiveness in detecting illnesses and monitoring patient health data generated by the sensors in a network (Latre et al., 2011). The IoT allows for detection, processing, and communication of physical and biological characteristics. The suggested ML model analyses data on important biological indications, such as respiration rate, body heat, oxygen levels, and heart rate from the cloud-based

Chapter 32 DOI: 10.1201/9781003581215

system. ML algorithms Extra Trees (ET), Ada Boost (AB), Gradient Boosting (GB), and Random Forest (RF) are used for this purpose.

Remaining portion of the paper portrayed as follows. Subsequent section reviews the existing literature, followed by the proposed approach and its performance evaluation. The last section summarises the paper.

2. Literature Review

This section outlines the contributions of numerous research that have employed WBAN in diagnostic purposes and public health applications using IoT and ML-based classification. Table 32.1 displays the different methodologies used in this area.

3. Proposed Methodology

The architecture of the recommended CVD prototype is shown in Figure 32.1. It is broken down into three sections: (1) Health-related smart devices, (2) Machine learning-based diagnosis, and (3) WBAN-IoT monitoring.

3.1 Health-Related Smart Devices

Using a software platform, the user makes an account and submits personal information such as name, phone number, date of birth, address, diet type, smoking habits, and so on. The person can choose to provide data about his or her complaints, such as blood pressure, chest pain, and muscular difficulties, after registration. The cloud system stores all personal and health-related data.

3.2 Machine Learning-based Diagnosis

To identify whether or not a person has heart disease, the CVD model uses cloud data. As shown in Figure 32.1, six IoT WBAN-based sensors have been mounted on the body of the user for uninterrupted monitoring.

3.3 WBAN-IoT based Monitoring

WBAN-based sensors, including a blood pressure sensor, pulse sensor, airflow breathing sensor, body temperature sensor, ECG sensor, and oxygen in the blood (SPO2) sensor are coupled to Arduino Uno and My Signals equipment kit using Zigbee technology. This data can be used to keep track of a patient's status in realtime or to obtain sensitive data that can be analysed for medical diagnosis.

4. Performance Evaluation

The suggested CVD-BAN approach is implemented by using the RF, ET, GB, and AB classifiers. Figure 32.2 presents the pervasiveness of the chest pain in

Table 32.1: List of different methodologies used in WBAN research area.

Authors	Technique used	Year
Kim et al.	WBAN with several hops and multi-channel TDMA	2015
Vogt et al.	For MRI scans, a Bluetooth low-energy sensor used	2016
Yang et al.	The long-range gadget is used to monitor the ECG	2016
Maity et al.	Wearable health monitoring	2017
Kumar et al.	IoT-based illness prediction system	2018
Jabeen et al.	A hybrid recommender system based on IoT	2019
Raj et al.	The opportunistic routing strategy saves energy	2020

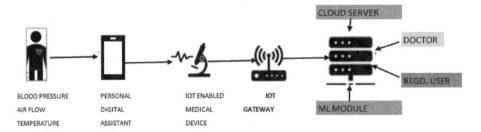

Figure 32.1: Proposed architecture

The Proposed Algorithm

Step 1:	Person registers using details via software applications; If a person not registered { Add a new record using person_id as primary key: }
Step 2:	Input user symptoms (Abnormal Blood Pressure (Yes/No), Pulse Rate, Chest Pain (Yes/No), Breathing Issues (Yes/No), Abnormal ECG (Yes/No), Headache (Yes/No));
Step 3:	Store these information to cloud;
Step 4:	Read data from the cloud, and apply the machine learning classifiers model to check Heart Patient or not;

Step 5:	If heart patient {
	Read and monitor parameters of the patient using CVD-BAN model;
	Store the reading in a cloud database and send to the patient;
	Continuous recording of data until the patient's symptoms are normal;}
	Else{
	Intimate to the patient about normal readings;}
	Set a reminder notification to check the symptoms periodically;

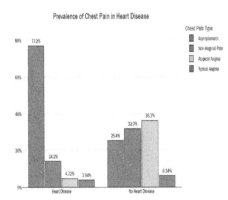

Figure 32.2: Prevalence of chest pain

cardiac disease. The chest discomfort (asymptomatic) is the utmost prevailing indicator among people with cardiac disease, accounting for about 77 percent. The feature's absolute Shapley value enumerates how much the characteristic modifies the anticipated chance of cardiac disease on average. There are 18 characteristics with an average absolute value of at least 0.1, which help in the model's forecasting. The final predictors have been chosen and their feature relevance are recorded in Table 32.2. The Table 32.3 exhibits the performance of the proposed model category

Table 32.2: Features importance in heart disease dataset.

Sl. No.	Features	Value	Sl. No.	Features	Value
1	Cholesterol Slope Flat	0.941	10	Age Sex M	0.153
2	Resting BP ST Slope Up	0.569	11	Resting BP Max HR	0.151
3	Sex M ST Slope Flat	0.512	12	Age Max HR	0.15
4	Chest Pain Type ATA	0.341	13	Chest Pain Type ATA ST Slope UP	0.146
5	Resting ECG Normal Cholesterol Sex M	0.319	14	Chest Pain Type NAP	0.144
6	Old peak Sex M	0.26	15	ST Slope UP Age	0.127
7	Cholesterol	0.252	16	Resting BP Old peak	0.124
8	Resting BP ST Slope Flat	0.16	17	Resting BP Cholesterol	0.118
9	Cholesterol Resting ECG Normal	0.159	18	Chest Pain Type ATA	0.115

Table 32.3: Model performance category wise (ET, RF, AB and GB) data models.

ML Model	AUC Value	Accuracy in %	F1 Score in %
Extra Trees	0.920	85.02	86.64
Random Forest	0.925	85.52	86.99
Ada Boost	0.897	84.06	85.35
Gradient Boosting	0.927	88.9	90.01

wise using different ML models like ET, RF, AB, and GB based on the parameters like area under the curve (AUC) value, accuracy and the F1 score. Where AUC value caters the measure of performance throughout all potential thresholds of classification, F1 score defines the harmonic mean of the accuracy and recall of the model of classification.

5. Conclusion

The WBAN monitoring system is linked to the body of the patient for constant monitoring if the user has been diagnosed with cardiac disease. The parameters, such as age, blood pressure, cholesterol, and kind of chest pain, are some of the most important predictors of heart disease in the model. Gradient Boosting has been found as the best model because it produces the greatest overall results and may be employed in the prompt diagnosis of cardiac disease, with an inclusive AUC value of 0.927, F1-Score of 90% and an accuracy of 89%. Other classification approaches might be employed in the future to increase accuracy and produce a long-term energy-efficient model for WBAN.

References

Ahmed, H., Younis, E. M., Hendawi, A., and Ali, A. A. (2020). "Heart disease identification from patients' social posts, machine learning solution on Spark." *Future Generation Computer Systems*, 111: 714-722.

Jabeen, F., Maqsood, M., Ghazanfar, M. A., Aadil, F., Khan, S., Khan, M. F., and Mehmood, I. (2019). "An IoT based efficient hybrid recommender system for cardiovascular disease." *Peer-to-Peer Networking and Applications*, 12: 1263-1276.

Kim, T. Y., Youm, S., Jung, J. J., and Kim, E. J. (2015, January). "Multi-hop WBAN construction for healthcare IoT systems." In *2015 International Conference on Platform Technology and Service* (pp. 27-28). IEEE.

Kumar, P. M., Lokesh, S., Varatharajan, R., Babu, G. C., and Parthasarathy, P. (2018). "Cloud and IoT based disease prediction and diagnosis system for healthcare using Fuzzy neural classifier." *Future Generation Computer Systems*, 86: 527-534.

Latre, B., Braem, B., Moerman, I., Blondia, C., and Demeester, P. (2011). "A survey on wireless body area networks." *Wireless Networks*, 17: 1-18.

Maity, S., Das, D., and Sen, S. (2017, July). "Wearable health monitoring using capacitive voltage-mode human body communication." In *2017 39th Annual International Conference of the IEEE Engineering in Medicine and Biology Society (EMBC)* (pp. 1-4). IEEE.

Raj, A. S. and Chinnadurai, M. (2020). "Energy efficient routing algorithm in wireless body area networks for smart wearable patches." *Computer Communications*, 153: 85-94.

Vogt, C., Reber, J., Waltisberg, D., Büthe, L., Münzenrieder, N., and Tröster, G. (2016, August). "A wearable bluetooth LE sensor for patient monitoring during MRI scans." In *2016 38th Annual International Conference of the IEEE Engineering in Medicine and Biology Society (EMBC)* (pp. 4975-4978). IEEE.

World Health Organization (2021, June 1). *Cardiovascular Diseases (CVDs)*.

Yang, Z., Zhou, Q., Lei, L., Zheng, K., and Xiang, W. (2016). "An IoT-cloud based wearable ECG monitoring system for smart healthcare." *Journal of medical systems*, 40: 1-11.

Online Auction Fraud Detection Using Deep Learning Network

Satyajit Panigrahi[1], Sharmila Subudhi[2]

[1]A. P. Moller, Maersk, Bengaluru, India, satyajit.panigrahi@maersk.com
[2]Department of Computer Science, Maharaja Sriram Chandra Bhanja Deo University,
Baripada, India, sharmilasubudhi@ieee.org

Abstract

This article introduces a new method for detecting fraud in online auctions and bids. This method uses two deep-learning models that apply kernel superposition. First, the data is preprocessed with careful attention to ethics and parameters to create two models that use differential reasoning to classify items as either "Fraud" or "Not Fraud". Two classification models are then generated using perceptron layers and an output layer. The models are optimised using the Adam optimiser and the mean squared error loss function. The callback object saves the best model with a patience value of 20 epochs. The generation of confusion matrices demonstrates the effectiveness of the proposed Deep Learning Models. Finally, a comparative analysis with two other approaches establishes the proposed schema's superiority.

Keywords: Online auction claims, fraud detection, deep learning, sequential model, functional model

1. Introduction

Online auctions or eAuctions held virtually over the internet are catching the bidders' attention in today's fast times (Abidi *et al.*, 2021). Fraud in online auctions is generally conducted on an auction website by creating a very lucrative auction, such as an adequate low initial threshold bid (Abidi *et al.*, 2021). Once a bidder wins an auction and finishes the payment, the forged seller will either not follow through with the delivery or send fake items. The presence of such malpractices could effectively curb online auction traffic in the forthcoming years.

Despite the availability of robust fraud detection measures like feedback ratings, counterfeit validation, and user authentication, they are hardly an appropriate deterrent against frauds. As a result, it has become indispensable for auction firms to monitor and overanalyze the proceedings of auctioneers and bidders.

This work presents an unconventional attempt at employing a deep neural network to break the fraud and genuine bidder scenarios. The highlights of the current work are mentioned as follows. (1) Two different Deep Neural Networks have been employed individually for behavioral analysis. (2) Perceptron filtering and Adam optimiser are used to create the best model callback to iterate through multiple epochs. (3) Facilitated automatic judging of the best model for itself without being restricted by user-limited standards and training cycles.

The structure of the paper is as follows. Section 2 discusses the related research, while Section 3 demonstrates the proposed model. A comparative and parallel functional model analysis is discussed in Section 4. Section 5 marks the conclusion with a summary of the contribution made.

2. Related Work

This section presents an extensive study on fraud detection models in electronic bidding and online auctions. Anowar and Sadaoui (2020) presented a computationally extensive clustering approach to label the biding records of iPhone7 available in eBay. Similarly, Gerritse and Wesenbeeck (2022) developed behavioral profiles of auctioneers through Local Outlier Factor (LOF) and Belief Propagation. using a small sample, thus making the model deprived of established ground truth values.

Abidi et al. (2021) suggested a fuzzy-infused machine learning models for detecting forged shill bidding records. A statistical-analysis based feature extraction technique along with the supervised classifiers was presented by Fire et al. (2023). However, these above models have not considered the model overfitting and data imbalance issues during operation.

Although researchers have applied various techniques to detect forged activities accurately, many false alerts are also raised, wasting time and resources. Besides, this affects the feasibility, usefulness, and reliability of a Fraud Detection System (FDS). Addressing this issue is one of the main motivating factors in this current investigation. Furthermore, the Deep Learning Models (DLMs) have already established their prowess in accurately capturing user profiles (Zhong et al., 2021; Zhang et al., 2021). Therefore, the application of DLMs has been explored in this context.

3. Proposed Model

In this section, we elaborate on the methodology of detecting intrusive activities in online auction databases using two Deep Learning Models, namely, the Deep Sequential Model (DSM) and the Deep Functional Model (DFM).

3.1. Deep Sequential Model

The proposed Deep Sequential Model (DSM) consists of three dense layers: one input layer, one hidden dense layer, and one output layer. The DSMs analyze the attribute correlation in a dataset to predict a future optimal solution (Zhong et al., 2021).

3.2. Deep Functional Model

The proposed Deep Functional Model (DFM) contains five layers: one input layer, two hidden dense layers, one concat layer, and one dense output layer.

3.3. Working Methodology

Figure 1 presents the suggested Online Auction Fraud Detection System (OAFDS) workflow. Once the dataset is preprocessed, 20% is kept as test data and the remaining as train data. The DSM and DFM techniques are used individually on the train data to build trained classifier models. The 10-fold cross-validation method is applied for training and validating the classifiers. Later, the test data is fed to the models to classify forged records from genuine ones.

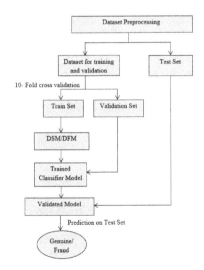

Figure 1: Workflow of Proposed OAFDS

4. Result Analysis

This section analyzes the experimental process and assesses the two proposed DSM and DFM models. The Python language in the Google Colab environment is used for implementation. Several tests were carried out on an anonymised Shill Bidding dataset (Alzahrani and Sadaoui, 2020).

4.1. Model Parameter

The hidden layers of the DSM and DFM are equipped with 60 kernels, each with the ReLU activation function. The output layer only consisted of one perceptron with the Sigmoid activation function. We have set the epochs number as 200 and the patience as 20. Figures 2 and 3 present the parameters used for building the DSM and DFM models, respectively.

Figures 2 and 3 present the parameters used for building the DSM and DFM models, respectively.

4.2. Performance Metrics

Five standard performance metrics - Sensitivity, Precision, Accuracy, F-score, and ROC curve are employed to measure the model's efficacy (Bradley, 1997).

Figure 4 illustrates the performance of our proposed models on the shill bidding dataset. It is evident from the table that the proposed DFM system outperforms the DSM model marginally in all parameters except in Sensitivity. Thus, it can be said that DFM helps in learning the random patterns and increases the robustness of the model in identifying fraudulent attacks. Figure 5 presents the ROC curve of both the proposed models.

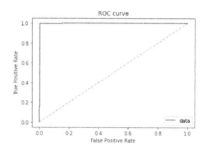

(a) Deep Sequential Model

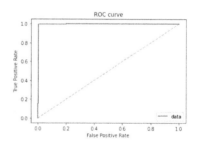

(b) Deep Functional Model

Figure 5: ROC of proposed models

```
Model: "sequential"

Layer (type)              Output Shape           Param #
=================================================================
dense (Dense)             (None, 60)             600

dense_1 (Dense)           (None, 60)             3660

dense_2 (Dense)           (None, 1)              61
=================================================================
Total params: 4,321
Trainable params: 4,321
Non-trainable params: 0
```

Figure 2: Parameters of Proposed DSM

```
Model: "functional"

Layer (type)          Output Shape      Param #   Connected to
=================================================================
input_1 (InputLayer)  [(None, 9)]       0

dense (Dense)         (None, 60)        600       input_1[0][0]

dense_1 (Dense)       (None, 60)        3660      dense[0][0]

concatenate (Concatenate) (None, 69)    0         input_1[0][0]
                                                  dense_1[0][0]

dense_2 (Dense)       (None, 1)         70        concatenate[0][0]
=================================================================
Total params: 4,330
Trainable params: 4,330
Non-trainable params: 0
```

Figure 3: Parameters of Proposed DFM

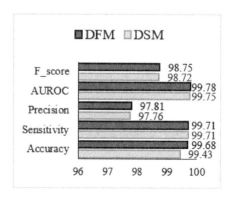

Figure 4: Performance of Proposed Systems on Shill Bidding Dataset

Table 1: Comparative analysis with other works

Method	Sensitivity (in %)	F-score (in %)
Proposed DFM-based OAFDS	99.71	98.75
Anowar and Sadaoui (2020)	98.15	98.27
Fire *et al.* (2023)	84.5	91.2

A comparative performance analysis of DFM-based OAFDS with two other contemporary works (Anowar and Sadaoui, 2020; Fire *et al.*, 2023) has been presented in Table 1 regarding Sensitivity and F-score metrics. The developed model identifies the forged activities effectively with better F-score and Sensitivity values.

5. Conclusion

Online Auction Fraud, though invariably rampant in the real world, has very soft measures employed against them. In this work, two deep learning systems, the Deep Sequential Model, and the Deep Functional Model, have been developed to tackle the fraud menace. The training and testing samples are extracted from the preprocessed dataset. The train set has been applied individually on these two systems for building individually trained supervised models. The test set is then used on the models, producing the result. Experiments were carried out to evaluate the effectiveness of the proposed system by using a real-world anonymised dataset. From the observations, we can infer that the Deep Functional Model produces the best outcome for all performance metrics with the highest F-score of 98.75% and accuracy of 99.68%. Besides, a comparative analysis with two contemporary works has proved the efficacy of the suggested model. Moreover, the future holds much room for improvement in model parameter tuning and emulation of bidders' profiles more accurately if the bidder is non-human.

References

Abidi, W. U. H., Daoud, M. S., Ihnaini, B., Khan, M. A., Alyas, T., Fatima, A., and Ahmad, M. (2021). "Real-time shill bidding fraud detection empowered with fussed machine learning." *IEEE Access*, 9: 113612-113621.

Alzahrani, A. and Sadaoui, S. (2020). "Clustering and labeling auction fraud data." In *Data Management, Analytics and Innovation: Proceedings of ICDMAI 2019*, (Vol. 1, pp. 269-283). Springer, Singapore.

Anowar, F. and Sadaoui, S. (2020). "Detection of auction fraud in commercial sites." *Journal of Theoretical and Applied Electronic Commerce Research*, 15 (1): 81-98.

Bradley, A. P. (1997). "The use of the area under the ROC curve in the evaluation of machine learning algorithms." *Pattern Recognition*, 30 (7): 1145-1159.

Gerritse, L. A., and van Wesenbeeck, C. F. A. (2022). Detecting Collusive Shill Bidding in Commercial Online Auctions. *Computational Economics*, 1-20.

Fire, M., Puzis, R., Kagan, D., and Elovici, Y. (2023). Large-Scale Shill Bidder Detection in E-commerce. In *Proceedings of the 27th International Database Engineered Applications Symposium* (pp. 79-86).

Zhong, M., Zhou, Y., and Chen, G. (2021). Sequential model based intrusion detection system for IoT servers using deep learning methods. *Sensors*, 21 (4): 1113.

Zhang, S., Liu, H., He, J., Han, S., and Du, X. (2021). Deep sequential model for anchor recommendation on live streaming platforms. *Big Data Mining and Analytics*, 4 (3): 173-182.

Advancing Medical Diagnosis

An In-Depth Exploration of Image Processing Techniques for Swift and Accurate Detection of Brain Tumors in Magnetic Resonance Imaging

Rojalin Mohapatra[1], Amaresh Sahu[2], Parimala Kumar Giri[1], Hemanta Kumar Bhuyan[3], Partha Sarathi Sahoo[4], Smitta Ranjan Dutta[2]

[1]CSE, GITA, Bhubaneswar, India, rosalinmohapatra000@gmail.com, parimala.6789@gmail.com
[2]MCA, ABIT, Cuttack, India, amaresh_sahu@yahoo.com, smittaranjandutta@gmail.com
[3]Department of IT and Computer Applications, Vignan's Foundation for Science, Technology & Research (Deemed to be University), Guntur, India, hmb.bhuyan@gmail.com
[4]Mentor Data Science, CSE, Ajay Binay Institute of Technology, Cuttack, India, partha.orcl@gmail.com

Abstract

Magnetic resonance imaging (MRI) represents a significant leap forward in medical technology, generating detailed, high-resolution images crucial for identifying and categorising diseases within the body's organs. One such condition detectable through MRI scans is the presence of brain tumors. The prolonged duration required for manually identifying and categorising brain tumors in MRI images could lead to delays in providing crucial medical interventions vital for a patient's path to recovery. The primary goal is to streamline the interpretation of medical images, especially those derived from MRI scans. The primary objective of this research is to comprehensively explore various techniques and methodologies, including K-Nearest Neighbors, Artificial Neural Networks, Convolutional Neural Networks, and Residual Neural Networks, utilised in the detection of brain tumors within MRI images.

Keywords: Machine learning, Artificial Neural Network (ANN), K-Nearest Neighbors (KNN), Convolution Neural Network (CNN), deep learning and Residual Neural Network (ResNet)

1. Introduction

In the realm of contemporary technological progress, numerous benefits have unfolded across various aspects of life, with the medical field notably witnessing substantial growth in magnetic resonance imaging (MRI) technology. This advanced technology plays a pivotal role in generating intricate images of internal organs, including the brain and tissues within the human body. The resulting high-resolution images serve as invaluable assets for identifying diseases that may affect different organs, with the presence of a brain tumor being one such condition readily discernible through MRI imaging. Despite these technological strides, the medical community still heavily relies on the manual examination conducted by doctors and radiologists to detect diseases (Sahu and Pattanaik, 2017). The dependence on human visual perception has proven inadequate for ensuring a swift and accurate diagnosis, even when carried out by seasoned medical professionals. This manual approach is susceptible to vulnerabilities and nuances, as there exists the potential for a brain tumor to seamlessly integrate with healthy brain tissue. Acknowledging these challenges, ongoing efforts are being directed towards addressing these issues and enhancing the efficiency and accuracy of brain tumor detection through the continued advancement of MRI technology suggested by (Josh and Shah, 2015), (Swapnil et al., 2016), (Himar et al., 2019). The emergence of digital image processing presents a compelling opportunity to accelerate the analysis of MRI images, providing valuable support to doctors and radiologists in the rapid identification of brain tumors presented by Swapnil et al. (2016). Diverse techniques and methods within the realm of image processing can be employed to extract crucial data and information relevant to the identification and subsequent classification of brain tumors proposed by (Sharma and Suji, 2016). These images can be subjected to further processing using various techniques or methods to aid interpretation by doctors and radiologists developed by (Akmalbek et al., 2023).

2. Literature Review

Digital image processing encompasses a vast array of methods dedicated to modifying digital images to enhance their quality, decrease their size, or extract meaningful data. This field of image processing is divided into four domains as image restoration and enhancement, image analysis, image coding with data compression, and image synthesis. In each domain, unique techniques and algorithms are utilised to fulfill specific goals, collectively driving advancements in the field and facilitating its extensive use across various industries and fields of study. Each of these distinct areas in image processing serves well-defined purposes, such as refining image quality, extracting meaningful insights, efficiently compressing data, and synthesising images for a multitude of applications. applications.

In the realm of medicine, addressing the processing of MRI images for the detection and classification of brain tumors poses a notable challenge. The swift and precise identification of brain tumors carries significant potential for streamlining the initial treatment procedures for patients, facilitating timely and focused therapeutic interventions as a pivotal first stride towards recuperation. Many image processing methods elevate the efficiency and effectiveness of managing medical practices in healthcare system (Shobana et al., 2018).

The domain of image processing can be subdivided into four principal domains: the restoration and enhancement of images, the analysis of images, encoding of images employing data compression methods, and the synthesis of images, explained by (Chithra and Santhanam, 2018).

To enhance the quality of digital image and to minimie the size many techniques were used. represented by (Chithra and Santhanam, 2018).

The process of image classification entails sorting pixels into specific sets based on their data values, discussed by (Wang et al., 2013). For a pixel to be assigned to a particular class, it must meet specific criteria established for that class, discussed by (Arman et al., 2021). The user might have predefined the classes using training data; otherwise, the classes are undetermined (Chithra et al., 1990). The crux of image classification hinges on the methodologies used for extracting and analyzing these image features, discussed by (Lorente et al., 2021)

2. Data and Variables

This paper Utilised the Kaggle dataset on brain tumors, comprising 98 files without tumors and 155 files with tumors. Verify the dataset's inclusivity to cover a wide range of brain tumor cases. Ensure strict

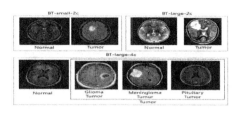

Figure 1: Brain Tumors images

adherence to ethical guidelines and regulations concerning patient privacy throughout the dataset utilisation process, dataset was represented by (JEAN_DJHONSON, 2019) in Kaggle.

Image resolution, orientation, and intensity levels were normalised to ensure uniformity. Noise reduction and contrast enhancement techniques were employed to improve the overall quality of the images. Advanced processing approaches, such as normalisation or augmentation, to augment model generalisation capabilities were investigated. Below images fetch from kaggle dataset as shown in Figure 1.

3. Methodology and Model Specifications

The research utilised a range of deep learning image classifiers, including KNN, ANN, CNN , ResNet. These models were chosen due to their compatibility with medical image analysis tasks and their ability to effectively manage the complexities associated with predicting brain tumor occurrences. Both technical and ethical aspects were considered in this research. Parameters were tuned for implementing classification methods. The dataset had been divided into training, validation, and test sets. To ensure robustness cross-validation technique was used, where twenty percent of data was taken for testing purpose. The final models were validated by using an independent test set to assess their generalisability to new and unseen brain MRI images. Also, the predictions against clinical outcomes were validated to ensure reliability. Validate predictions against known clinical outcomes to ensure reliability. To evaluate the performance of classification, statistical parameters like True Positives (TP), False Positives (FP), False Negatives (FN), and True Negatives (TN) were used with confusion matrix. From confusion matrix four statistical parameters had been derived, those are True Positives (TP), False Positives (FP), False Negatives (FN), and True Negatives (TN) used for evaluating the performance of a classification system. These indices are used to calculate various performance metrics, including accuracy, sensitivity, precision, and specificity. Certainly, accuracy is calculated as a percentage, representing the ratio of accurately predicted instances to the total number of instances. This ratio is then multiplied by 100 to express accuracy as a percentage. The F1 score serves as an evaluation metric in machine learning, amalgamating precision and recall into a singular measure, thereby offering a well-rounded assessment of a model's effectiveness. Lastly, the comparison bar graphs show the performance of all classifiers.

4. Empirical Results

This research Implemented four various image classifiers, including KNN, ANN, CNN, and ResNet. The rate with which a machine learning model accurately predicts the result is measured by its accuracy. We can determine the accuracy by dividing the total number of guesses by the number of right predictions. The Variation accuracy of four classifiers are depicted in Figures 2 (a), (b), (c) and (d). It indicates CNN method provide better accuracy results than others

A table used to describe how well a classification algorithm performs is called a confusion matrix. The performance of a classification algorithm is summarised and visualised using a confusion matrix.

The true Positive value is 10.False positive value is 1. The False Positive value is 2 and False Negative value is 25 as shown in Figure 3.

This bar graph show the accuracy of KNN,ANN,CNN and Res-Net is

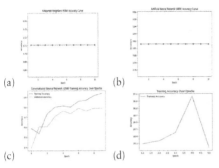

(a)   (b)

(c)   (d)

Figure 2: Represents CNN classifier has more predicting accuracy than other classifiers

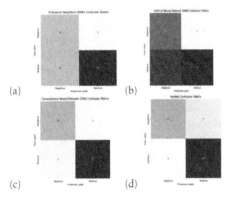

(a)   (b)

(c)   (d)

Figure 3: The confusion matrices of KNN, ANN, CNN and ResNet are given in Figure 3 (a), (b), (c) and (d)

70%,70%,85%,72% respectively as shown in Figure 4.

Below bar graph shows the F1 score of KNN, ANN, CNN and Res Net model are 78%,92%,95%,75%.

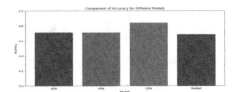

Figure 4: Accuracy of CNN is better than KNN, ANN and ResNet classifiers.

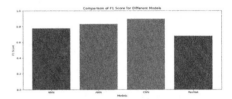

Figure 5: F1 score of CNN is better than KNN, ANN and ResNet classifiers

The utility of MRI technology is particularly pronounced in the early detection of brain tumor diseases within the medical domain. However, the manual interpretation of MRI images by healthcare practitioners remains a bottleneck in the diagnostic process, mainly due to the intricate structure of the human brain. The extended duration required for the manual detection and classification of brain tumors from MRI images can result in delays in administering vital medical treatments crucial for a patient's recovery. Acknowledging the urgent need for swift and accurate medical information to guide timely interventions, there has been a concerted effort to develop image processing techniques tailored for the analysis of MRI images. The primary goal is to streamline the interpretation of medical images, especially those derived from MRI scans. This study seeks to thoroughly investigate a variety of techniques and methodologies, encompassing KNN, ANN, CNN and ResNet utilised to identify and detect brain tumor by using MRI images.

5. Conclusion and Future Work

Magnetic resonance imaging (MRI) represents a significant leap forward in medical technology, generating detailed, high-resolution images crucial for identifying and categorising diseases within the body's organs. One such condition detectable through MRI scans is the presence of brain tumors.

References

Abdusalomov, A. B., Mukhiddinov, M., and Whangbo, T. K. (2023). "Brain tumor detection based on deep learning approaches and magnetic resonance imaging". *Cancers*, 16 (15): 4172.

Chithra, K. and Santhanam, T. (2018). "A novel denoising technique for mixed noise removal from grayscale and color images." *Journal of Theoretical & Applied Information Technology*, 96 (3).

Fabelo, H., Ortega, S., Szolna, A., Bulters, D., Piñeiro, J. F., Kabwama, S., and Sarmiento, R. (2019). "In-vivo perspectral human brain image database for brain cancer detection". *IEEE Access*, 7: 39098-39116.

Helms, R. M. (1990). "Introduction to image technology". *IBM Systems Journal*, 29 (3): 313-332.

Huang, Y., Wu, Z., Wang, L., and Tan, T. (2013). "Feature coding in image classification: A comprehensive study." *IEEE Transactions on Pattern Analysis and Machine Intelligence*, 36 (3): 493-506.

Joshi, M. A. and Shah, D. H. (2015). "Survey of brain tumor detection techniques through MRI images". *American International Journal of Research in Formal, Applied & Natural Sciences*, 10 (1): 2328-3785.

JEAN_DJHONSON, https://www.kaggle.com/datasets/jjprotube/brain-mri-images-for-brain-tumor-detection/data.

Keerthana, T. K. and Xavier, S. (2018). "An intelligent system for early assessment and classification of brain tumor." In *2018 Second International Conference on Inventive Communication and Computational Technologies (ICICCT)* (April):1265-1268.

Lorente, Ò., Riera, I., and Rana, A. (2021). "Image classification with classic and deep learning techniques". arXiv preprint arXiv:2105.04895.

Sahu, A. and Pattnaik, S. (2017). "Feature selection using evolutionary functional link neural network for classification." *International Journal of Advances in Applied Sciences*, 6 (4): 359-367.

Sarraf, A., Azhdari, M., and Sarraf, S. (2021). "A comprehensive review of deep learning architectures for computer vision applications". *American Scientific Research Journal for Engineering, Technology, and Sciences (ASRJETS)*, 77 (3): 1-29.

Shobana, G. and Balakrishnan, R. (2015). "Brain tumor diagnosis from MRI feature analysis-A comparative study." In *2015 International Conference on inovations in Information, Embedded and Communication Systems (ICIIECS)* IEEE. (March):1-4.

Suja, S., George, N., and George, A. (2018). "Classification of grades of Astrocytoma images from MRI using deep neural network." In *2018 2nd International Conference on Trends in Electronics and Informatics (ICOEI)* IEEE. (May): 1257-1262.

Telrandhe, S. R., Pimpalkar, A., and Kendhe, A. (2016). "Detection of brain tumor from MRI images by using segmentation & SVM." In *2016 World Conference on Futuristic Trends in Research and Innovation for Social Welfare*, (October): pp. 1-6.

Sharma, P., and Suji, J. (2016). "A review on image segmentation with its clustering techniques". *International Journal of Signal Processing, Image Processing and Pattern Recognition*, 9 (5): 209-218.

Comparative Analysis of Deep Learning Architectures for MRI Classification

Soumyajit Paul, Ira Nath, Souvik Kundu, Samadrita Kar

CSE, JIS College of Engineering, Kalyani, India
E-mail: soumyajitpaul2002@gmail.com, ira.nath@jiscollege.ac.in, souvik.kundu2403@
gmail.com, samadritakar02@gmail.com

Abstract

This research compares four Deep Learning models (VGG16, DenseNet121, ResNet50, and Google's Teachable Machine) for analyzing MRI scans. It includes a wide range of medical conditions to ensure accuracy across various diseases. The models are pre-trained on existing data and then fine-tuned for MRI analysis. Uniquely, the study uses attention mechanisms to see how the models "think" during diagnosis, fostering trust in their decisions. This comparison aims to identify the most effective AI for MRI analysis, ultimately leading to faster and more precise diagnoses, where the said tool provides better accuracy than the other models used (VGG16, DenseNet, and ResNet, given in order of decreasing accuracy).

Keywords: Deep learning, Google's Teachable Machine, MRI scans, ResNet50, VGG16

1. Introduction

Medical imaging, particularly MRI, plays a crucial role in diagnosis and therapy. Deep learning, especially Convolutional Neural Networks (CNNs), has revolutionised image analysis, offering superior accuracy.

This research delves into MRI classification using deep learning. We compare three established architectures (VGG16, ResNet50, DenseNet121) with Google's Teachable Machine. We focus on performance, interpretability, and clinical applicability.

The inclusion of Teachable Machine emphasises the importance of user-friendly tools. We aim to assess its potential for democratising medical image classification tasks, fostering wider adoption of these powerful technologies. Additionally, transfer learning is explored to understand its impact on model performance.

Ultimately, this research seeks to identify the optimal deep learning architecture for MRI classification, leveraging pre-trained models and interpretability metrics. This project aims to empower the medical imaging field to enhance patient care and diagnosis by bridging the gap between cutting-edge technology and practical application.

2. Literature Review

Recent advancements in brain tumor diagnosis leverage deep learning and image processing techniques for enhanced accuracy and efficiency. Srikanth *et al.* (2021) achieved 98% accuracy using a 16-layer VGG-16 neural network. Tandel *et al.* (2020) combined CNN with transfer learning, reaching up to 100% accuracy through hybrid approaches. Meanwhile, Konar *et al.* (2021) proposed

Chapter 35 DOI: 10.1201/9781003581215

a Quantum Fully Self Neural Network for segmenting cerebral lesions, introducing a novel quantum-based supervised learning technique. Pareek *et al.* (2020) employed supervised learning and principal component analysis for brain tumor categorisation, achieving 97% accuracy with KSVM. Khairandish (2022) introduced a hybrid CNN-SVM model with threshold-based segmentation, achieving an impressive 98.49% accuracy using various deep learning architectures and data augmentation techniques. Irmak *et al.* (2021) utilised a trio of CNN architectures, achieving high accuracies ranging from 92.66% to 99.33% for brain tumor detection and categorisation.

The paper comparison is given in Table 1.

3. Preliminaries

Brain tumors, arising in the cerebral region, affect neighbouring areas such as the pituitary gland and nerves. They are classified as malignant or benign, with symptoms including seizures, headaches, weakness, and speech and vision changes.

3.1. Dataset Description

The aim was to construct a Convolutional Neural Network (CNN) capable of discerning the presence of brain tumors from MRI scans. We utilised the Brain MRI Images for Brain Tumour Detection dataset, containing two classes: "YES" indicating images with tumors (encoded as 1), and "NO" indicating images without tumors (encoded as 0). The dataset comprised 155 tumor-positive images and 98 tumor-negative images for comparative study among various CNN architectures.

4. Description of Algorithm Used

Figure 1 outlines the project flow. To address limited data, data augmentation

Table 1: Papers comparison

Papers	Brief description of the proposition	Architectures and algorithms used	Accuracy rates
B. Srikanth *et al.*	VGG16 accepted pre-processed image input.	VGG-16	98%.
G.S.Tandel *et al.*	Improved performance using MRI images	AlexNet and cross-validation techniques.	87.14%, 93.74%, 95.97%, 96.65%, 100%.
M.Pareek *et al.*	Area and volume were employed for computation	KSVM	97%.
M.Khairandish *et al.*	Utilised pre-trained networks for feature extraction	GoogleNet, ResNet-18, AlexNet, and ShuffleNet.	98.49%.
Irmak *et al.*	Three different CNNs were proposed for distinct classifications	VGG-16, AlexNet, Inceptionv3, GoogleNet and ResNet-50.	98.14%,92.66%, 98.14%.
D. Konar *et al.*	Brain lesions segmentation using 3 states of qubits was introduced.	QFSNet supervised QINN networks.	100%.

(Source: Author's Composition)

was used, creating new data points by modifying existing images.

Data augmentation functions include random cropping, rotation, blur, contrast, etc.

4.2. Data Normalisation

Images with varying dimensions (as shown in Figures 2 and 3) were normalised to ensure accurate analysis. This involved identifying extreme points in each image to crop out the brain region, as illustrated in Figure 4.

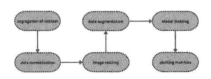

Figure 1: Flow diagram of the project
(Source: Author's Compilation)

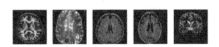

Figure 2: Images with no tumor
(Source: Author's Output)

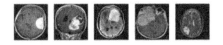

Figure 3: Images with tumor
(Source: Author's Output)

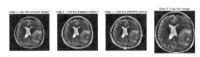

Figure 4: Normalisation Example
(Source: Author's Output)

Figure 5: Normalised image with no tumor
(Source: Author's Output)

4.1. Image Resizing

The cropped images were resized to (224, 224) for the ease of model training. These cropped and resized images were then stored in different directories namely, VAL_CROP, TEST_CROP, and TRAIN_CROP, each with their NO and YES directories (shown in Figures 5 and 6).

4.3. Data Augmentation

Data augmentation involves expanding the training dataset by introducing altered replicas of existing data. It enhances model classification capability, reduces overfitting, and addresses data scarcity. Various methods are used, as shown in Figures 7 and 8.

5. Result and Discussion

The models were trained and the accuracy and loss curves were found and have been

Figure 6: Normalised image with tumor
(Source: Author's Output)

Figure 7: Actual Image
(Source: Author's Output)

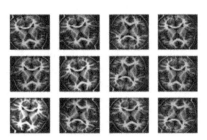

Figure 8: Augmented Images
(Source: Author's Output)

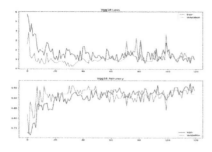

Figure 9: VGG16 evaluation
(Source: Author's Output)

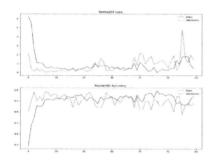

Figure 10: ResNet evaluation
(Source: Author's Output)

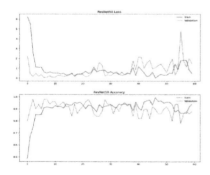

Figure 11: DenseNet evaluation
(Source: Author's Output)

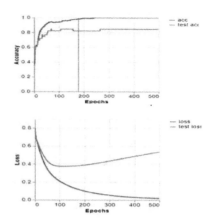

Figure 12: Teachable Machine evaluation
(Source: Author's Output)

6. Conclusion

This project advances brain tumor detection using diverse deep-learning models and techniques applied to MRI scans. Integrating data augmentation and state-of-the-art architectures like ResNet, DenseNet, Inception, and VGG16, it achieves exceptional precision and robustness. Comparative analysis reveals insights into model strengths and weaknesses, guiding further optimisation. The outcomes promise improved clinical applications in neuroimaging, enhancing patient care and advancing the capabilities of deep learning in healthcare.

References

Irmak, Emrah. (2021). "Multi-classification of brain tumor MRI images using deep convolutional neural network with fully optimized framework." *Iranian Journal of Science and Technology, Transactions of Electrical Engineering,* 45 (3): 1015-1036.

Konar, Debanjan, Siddhartha Bhattacharyya, Bijaya K. Panigrahi, and Elizabeth C. Behrman. "Qutrit-inspired fully self-supervised shallow quantum learning network for brain tumor segmentation." *IEEE Transactions on Neural Networks and Learning Systems* 33, no. 11 (2021): 6331-6345

illustrated in Figure 9, Figure 10, Figure 11, and Figure 12.

Limited dataset size and insufficient hardware resources hindered model training, causing instability in accuracy and loss curves. Adequate computing power and data are crucial for deep learning projects.

Khairandish, Mohammad Omid, Meenak-shi Sharma, Vishal Jain, Jyotir Moy Chatterjee, and N. Z. Jhanjhi. (2022). "A hybrid CNN-SVM threshold segmentation approach for tumor detection and classification of MRI brain images." *Irbm*, 43 (4): 290-299.

Navoneel Chakrabarty, Brain MRI Images for Brain Tumour Detection, https://www.kaggle.com/datasets/navoneel/brain-mri-images-for-brain-tumour-detection

Pareek, Meenakshi, C. K. Jha, and Saurabh Mukherjee. (2020). "Brain tumor classification from MRI images and calculation of tumor area." In *Soft Computing: Theories and Applications: Proceedings of SoCTA 2018*, pp. 73-83. Singapore: Springer Singapore.

Srikanth, B. and Venkata Suryanarayana, S. (2021). "WITHDRAWN: multi-class classification of brain tumor images using data augmentation with deep neural network."

Tandel, Gopal S., Antonella Balestrieri, Tanay Jujaray, Narender N. Khanna, Luca Saba, and Jasjit S. Suri. (2020). "Multiclass magnetic resonance imaging brain tumor classification using artificial intelligence paradigm." *Computers in Biology and Medicine*, 122: 103804.

Ranking Optimised Statistical Models for Time Series Forecasting of Crude Oil Price

Sourav Kumar Purohit[1], Sibarama Panigrahi[2]

[1]Department of Computer Science Engineering, Sambalpur University. Jyoti Vihar, Burla, India, sk27808@gmail.com, skpurohit@suiit.ac.in
[2]Department of Computer Science Engineering, National Institute of Technology, Rourkela, India, panigrahis@nitrkl.ac.in

Abstract

Crude oil prices (COP) have historically led the global energy market, affecting economic growth, inflation, and general financial stability. To make informed decisions and reduce possible risks, governments, and business stakeholders need reliable and precise forecasting of COP. In this study, Six foremost statistical models, including ARIMA, TBATs, ARFIMA, TBAT, ETS, and Naive, have been optimised by the Forecast package of R and used to predict the weekly, monthly, and daily COP time series. The simulation results indicate that the Naive, TBATS, and ARFIMA models are the best alternative to predict the weekly, monthly, and daily COP, respectively. Additionally, the TBATS model achieves the best overall rank in predicting monthly, weekly, and daily COP.

Keywords: Crude oil price forecasting, hybrid models, optimised statistical model

1. Introduction

Crude oil is the foundation and source of many sectors that are pivotal in the worldwide economy. It is a significant energy source, contributing as a necessary raw material and a key element in many production and transport activities (Abdollahi, 2020). Accurate COP forecasting can assist oil-exporting and oil-consuming nations in creating stabilisation strategies and budgets (Cabedo and Moya, 2003). Several studies have recently been published to explore the difficulty in predicting oil prices and accessing the best results. This research paper conducts comprehensive investigations to assess the effectiveness of six statistical models utilised for univariate forecasting of three different datasets of WTI COP (Daily, Weekly, Monthly). To ensure the findings' reliability, the study employs various train-test ratios, subjecting the results to rigorous statistical analysis.

2. Literature Review

Various time series modeling techniques have been proposed to forecast COP, which utilise historical price data and various predictor variables to estimate future price movements. These include, ARIMA (Gasper and Mbwambo, 2023), ARFIMA (Monge and Infante, 2022), ETS (Claveria, 2019), TBATS (Fajar and Nonalisa, 2021). The capacity to predict the actual COP using TVP-VAR models is discussed by Drachal, K. (2021). Gulen (1998) predicted the WTI COP using cointegration analysis. Allegret et al., (2015) use a global VAR method, which considers interdependencies in trade and finance across nations. In another study, the grey prediction model is suggested by Norouzi and Fani, (2020) and the findings demonstrate that applying the grey forecasting model considerably enhances the performance of COP forecasting. According to Miao. et al. (2017), the LASSO technique produces better forecasts to several standard models.

Despite the use of several statistical models for COP forecasting, there is a need for a comprehensive investigation to identify the best statistical model that can be used for precise COP forecasting. The existing literature has considered different performance measures like RMSE, SMAPE, MAE, MASE, etc. However, extensive statistical tests like Friedman Nemayi Hypothesis Test (FNHT) have not been utilised to draw firm conclusions. Additionally, none of the existing studies have incorporated daily, monthly, and weekly COP data to assess the efficiency of these statistical models.

3. Methodology

This study employs six different statistical models to predict the weekly, monthly, and daily COP. The methodology of the research is presented in Algorithm 1. To create the train and test sets, the dataset is divided into five distinct ratios ([90-10], [85-15], [80-20], [75-25], [70-30]), respectively. The model is built using the training set, and its performance is evaluated using the test set.

First, the COP time series data is taken as input. The data is scaled down using a method called min-max normalisation. Then, the statistical model is trained by using the training set. The normalised predictions are generated by applying the test set to the trained model. The data are transformed into a supervised learning problem during the model's training using the sliding window technique. The amount of the lag is defined as the number of steps taken in the data at a prior time. The Forecast package of R determines the value of lag (k)

for weekly, monthly, and daily COP datasets the time series s $=[y_1, y_2, \ldots\ldots\ldots\ldots, y_n]^T$ is converted to a pattern set of size $(n-k)$ by applying the sliding window technique. The real forecasts y' are then obtained by de-normalising the normalised forecasts. The model's performance measure is determined by taking into account the forecasts y' and the original time series y.

Algorithm 1: Methodology

Begin by inputting the COP time series data denoted as $[y_1, y_2, \ldots, y_n]^T$.

Normalise the COP time series data by applying the min-max normalisation algorithm.

Create the Training and Test sets by splitting the original dataset into two sets.

Employ the Training set to develop the optimised statistical model. Utilise the model to forecast the Test set by applying it to the respective test patterns.

Denormalise the predicted values to obtain actual forecasts, denoted as y'.

Determine the performance measure of the model by comparing the COP forecasts (y') with the original COP time series (y).

---

4. Result and Discussion

In this study, the WTI spot price is used to collect the daily, weekly, and monthly COP

Four distinct accuracy measures have been employed to evaluate the effectiveness of the statistical methods., namely MAE (Equation 1), MASE (Equation 2), RMSE (Equation 3) and SMAPE (Equation 4).

$$MAE = \frac{1}{n}\sum_{i=1}^{n}\left|y_i - y_i'\right| \qquad (1)$$

$$MASE = \frac{1}{n}\sum_{i=1}^{n}\frac{\left|y_i - y_i^{'}\right|}{\frac{1}{n-1}\sum_{j=2}^{n}\left|y_i - y_{i-1}\right|} \quad (2)$$

$$RMSE = \sqrt{\frac{1}{n}\sum_{i=1}^{n}\left(y_i - y_i^{'}\right)^2} \quad (3)$$

$$SMAPE = \frac{1}{n}\sum_{i=1}^{n}\frac{\left|y_i - y_i^{'}\right|}{\left(\left|y_i\right| + \left|y_i^{'}\right|\right)/2} \quad (4)$$

The models are ranked separately for MASE accuracy measures using FNHT on the results obtained from five different splits. The parameters of the most ungenerous model are found through the Forecast package of R (Hyndman and Khandakar, 2008; Hyndman et al., 2021). Once the model parameters of are optimised using the train COP data, the test COP data is predicted and the forecasting accuracies are obtained.

For the weekly COP, in (70-30) train-test split ratio TBATS offers the smallest RMSE of 2.79, MAE of 1.90, and MASE of 0.97 and Naive provides the least SMAPE of 3.59 among all the splits.

Similarly, for the monthly COP among all train-test splits, the (70-30) train-test

split ratio has the lowest RMSE of 5.60, SMAPE of 7.32, and MAE of 4.33 in the TBATS model.

Finally, across all daily COP data splits, the (70-30) split has a lower RMSE of 1.99 for the Theta model, while Naive has the lowest SMAPE of 1.89, MAE of 1.09, and MASE of 0.99

4.1. Overall Ranking of Statistical Forecasting Models

To rank the models, we have applied the FNHT regardless of different train and test set splits. The results of

FNHT on the MASE measure for daily COP are shown in Figure 1 shows that the Naïve model gets the top rank out of all the models taken into consideration, with a variance in mean rank bigger than the critical distance (3.4). Similarly, for weekly COP, TBATS, and monthly COP, the ARFIMA model outperforms all other models, as shown in Figures 2 and 3. Finally, for the daily, weekly and monthly COP, FNHT is carried out on the MASE measure to identify the top model for predicting COP and Figure 4 makes it clear that the TBATS model, with a rank variation greater than the critical distance (1.9), is the unsurpassed model for overall COP prediction among various models that are examined in this research.

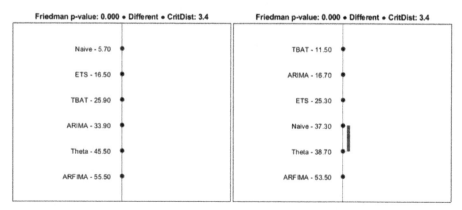

Friedman p-value: 0.000 • Different • CritDist: 3.4

Naive - 5.70
ETS - 16.50
TBAT - 25.90
ARIMA - 33.90
Theta - 45.50
ARFIMA - 55.50

Friedman p-value: 0.000 • Different • CritDist: 3.4

TBAT - 11.50
ARIMA - 16.70
ETS - 25.30
Naive - 37.30
Theta - 38.70
ARFIMA - 53.50

Figure 1: Rank of Forecasting Models for Daily COP using FNHT employing MASE measure

Figure 2: Rank of Forecasting Models for Weekly COP using FNHT employing MASE measure

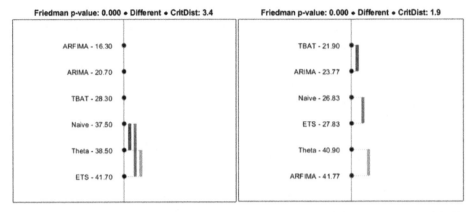

Figure 3: Rank of Forecasting Models for Monthly COP using FNHT employing MASE measure.

(Source: Author's Compilation)

Figure 4: Rank of Forecasting Models for Daily, Weekly, and Monthly COP using FNHT employing MASE measure.

5. Conclusion

This study proposes a cutting-edge learning framework for the effective assessment of statistical models on daily, weekly, and monthly COPs. Further, a thorough statistical analysis is carried out to determine the optimum statistical model for COP prediction. According to the study, the Naive, TBATS, and ARFIMA models are the best models for forecasting daily, weekly, and monthly COP, respectively. Additionally, the study indicated that, the TBATS model is the most suitable out of the six models considered in this study for forecasting any COP while simultaneously considering Daily, Monthly, and Weekly COP data.

To increase the accuracy of the prediction, shallow ML (Purohit and Panigrahi, 2024) and DL models can be employed. The performance of these models may be further improved by applying evolutionary and swarm algorithms.

References

Abdollahi, H. (2020). "A novel hybrid model for forecasting crude oil price based on time series decomposition. *Applied energy*, 267: 115035.

Allegret, J. P., Mignon, V., &Sallenave, A. (2015). "Oil price shocks and global imbalances: lessons from a model with trade and financial interdependencies." *Economic Modelling*, 49: 232-247.

Claveria, O. (2019). "Forecasting the unemployment rate using the degree of agreement in consumer unemployment expectations." *Journal for Labour Market Research*, 53(1): 3.

Drachal, K. (2021). "Forecasting crude oil real prices with averaging time-varying VAR models." *Resources Policy*, 74: 102244.

Fajar, M. and Nonalisa, S. (2021). "Forecasting chili prices using TBATS." *International Journal of Science and Research in Multidisciplinary Studies*, 7(2).

Gasper, L. and Mbwambo, H. (2023). "Forecasting crude oil prices by using ARIMA model: evidence from Tanzania."

Gülen, S. G. (1998). "Efficiency in the crude oil futures market." *Journal of Energy Finance & Development*, 3 (1): 13-21.

Hyndman, R., Athanasopoulos, G., Bergmeir, C., Caceres, G., Chhay, L., O'Hara-Wild, M., Petropoulos, F., Razbash, S., Wang, E., and Yasmeen F. (2021). _"Forecast: forecasting functions for time series and linear models_. R package version 8.15." https://pkg.robjhyndman.com/forecast/.

Hyndman, R. J., Khandakar, Y. (2008). "Automatic time series forecasting: the forecast package for R." *Journal of Statistical Software*, 26 (3): 1-22.

Miao, H., Ramchander, S., Wang, T., and Yang, D. (2017). "Influential factors in crude oil price forecasting." *Energy Economics*, 68: 77-88.

Monge, M. and Infante, J. (2022). "A fractional ARIMA (ARFIMA) model in the analysis of historical crude oil prices." *Energy Research Letters*, 3(Early View).

Norouzi, N. and Fani, M. (2020). "Black gold falls, black plague arise-An Opec crude oil price forecast using a gray prediction model." *Upstream Oil and Gas Technology*, 5: 100015.

Purohit, S. K. and Panigrahi, S. (2024). "Forecasting crude oil prices: a machine learning perspective." In *International Conference on Computing, Communication and Learning* (pp. 15-26). Cham, Springer Nature Switzerland.

Arduino Uno-based Women Safety Device with IoT

Arpita Sahoo, Sartaj Singh, Madhusmita Panda

Department of ECE, Siksha 'O' Anusandhan Deemed to be University, Bhubaneswar, India, madhusmitapanda@soa.ac.in

Abstract

This paper introduces a smart security system called the women's safety device system is to increase women's safety. It serves as an alarm system as well as a security precaution. The gadget utilises the Global Positioning System (GPS) to detect the victim's location and transmits it through the Global System for Mobile Communications (GSM). Additionally, Internet of Things (IoT)-based features simultaneously use the Wi-Fi ESP-32 module to transmit messages and an emergency email. This system focuses on such concerns where victims cannot discreetly reach their smartphones, offering an efficient security tool designed to empower women in awful circumstances. Unlike traditional approaches such as cumbersome belts, different clothing, or terrible smartphone applications our goal focuses to develop a smart, comfortable and adaptable gadget, prioritizing ease of use without compromising on functionality.

Keywords: Women Safety Device, GPS, GSM, IoT, Arduino Uno

1. Introduction

The Georgetown Institute for Women, Peace and Security (GIWPS) has highlighted India's ranking as the 148th safest country out of 170, underlining the demanding concern for women's safety intensified by the enhancing crime rates. A biggest challenge lies in the delayed response of police, often leaving victims vulnerable due to their inability to discreetly seek help.

To address these issues, we present an Arduino Uno-based women's safety device, offering an additional layer of security for women. Triggered a panic button, the system quickly utilises GPS and GSM module to pinpoint the victim's location and notify emergency contacts and police. Additionally, an email system can be activated with another button press, ensuring immediate assistance with location details sent to multiple phones. This initiative aims to empower women, providing them with a reliable means to seek help and stay secure in potentially dangerous situations.

2. Related Work

Various researchers have proposed an innovative solution to address women's safety concerns in today's society. Bhilare et al. (2015) suggest the installation of GPS and GSM-based vehicle tracking systems for female employees, allowing for real-time tracking and emergency alerts. Sogi et al. (2018) introduce SMARISA, a wearable device containing a Raspberry Pi, a camera, and emergency button to capture an attacker's image and notify emergency contacts. Punjabi et al. (2018) and Ahir et *al.* (2018) have developed a small security gadget triggered by a weight switch, which sends location alerts and initiates calls to authorities in an emergency. Harikiran *et al.* (2016) and Monisha *et al.* (2016) proposed an IoT-based solution combining a smart band and a smart phone to transmit distress signals and coordinates to authorities and family members. Lastly *et al.* (2023), present a Raspberry Pi-based women's safety device that activates upon pressing a panic button by sending location details via SMS to predefined contacts. These initiatives aim to empower women with affordable and adequate security measures against potential threats such as harassment and assault, promoting gender equality and safety in the workplace and beyond.

3. Design Proposal for a women's Safety Device

Figure 1, shows the block diagram of the device, detailing all the necessary components for the device. Our safety equipment has a button that, when pressed, sounds an alert. This will start looking for the victim using the GPS module, and the GSM will transmit the message to the specified mobile numbers. The message will contain an emergency message with the location of the victim. We will use Arduino Uno as the microcontroller to send commands throughout the circuit. Additionally, we have also incorporated an emergency emailing system. The Wi-Fi module will connect to the victim's hotspot and send the emergency mail with the victim's location to the given mail. Lastly, we added a feature of sending messages to four different numbers. When the victim presses the button the first time, the message will be sent to a first mobile number, if it is pressed a second time, then to second number; if pressed the third time, then to third mobile number; and if pressed the fourth time, then to fourth selected number. This provides an extra layer of security by ensuring that someone will come to the rescue. Our proposed device would be wearable as a band, or it can also be carried as it would be compact and minimalistic.

3.1. Components

The prototype utilises the following components.

3.1.1 Arduino Uno

The Arduino Uno microcontroller, shown in Figure 2, serves as the central decision-making unit within the system.

It interfaces with various sensors, switches, and modules, collecting data from them and making decisions based on the received signals.

3.1.2 Wi-Fi ESP-32 Module

ESP32 (See Figure 3) supports wired and wireless communication methods, with wireless connectivity enabled through Wi-Fi and Bluetooth technologies.

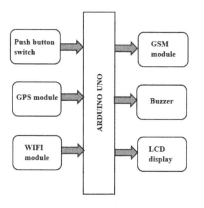

Figure 1: Block diagram of women's safety system

Figure 2: Arduino module

3.1.3. GSM SIM 900A

The GSM/GPRS module (See Figure 4) SIM 900A enhances communication capabilities for various electrical devices, offering voice and data transmission capabilities. It uses AT instructions to create communication with external devices, giving users control over its features and connection to the desired network.

3.1.4. GPS Module

A GPS module (See Figure 5) is a piece of equipment used to gather exact location, velocity, and time data by receiving signals from GPS satellites.

3.1.5 Buzzer

Buzzers (See Figure 6) are used to produce auditory alerts, notifications, or warnings in various electronic devices and systems.

Figure 3: WI-FI ESP-32 module.

Figure 4: GSM Sim 900A.

Figure 5: GPS module

Figure 6: Buzzer

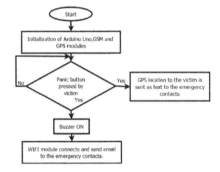

Figure 7: Flowchart showing the sequence of operations in the smart safety system

4. Methodology

Figure 7, shows the workflow of the proposed device in effectively addressing the sensitive issue of women's safety. In this case, when the alert mechanism is triggered, GPS and GSM are used to send the message containing the location of the victim to relatives and officials. The area is sent as a Google Maps link for easy access.

5. Results

We used the Arduino Integrated Development Environment (IDE) to create and run our code. The C programming language is used to create our code. Figure 8 displays the live location of the victim on the map. Additionally, Figure 9 indicates the women's safety device created using Arduino Uno, GPS, GSM and Wi-Fi modules.

6. Conclusion

The proposed Arduino UNO-based women's safety device with IoT provides personal security, creates emergency alerts,

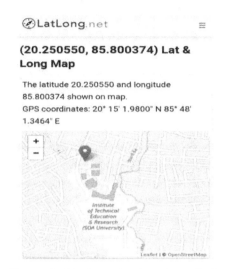

Figure 8: Displays the location of the victim.

Figure 9: The proposed hardware design for enhancing women's safety.

location tracking of the victim, shocks to the attacker and sends messages to the police or family and friends, with a user-friendly design.

Our design provides live location tracking through the GPS and messages are sent to the respective numbers through the GSM. The device also creates a sound that provides a shock for the attacker. Also, it sends an email to the mail id with the press of a button. In addition, it is compact and easy to use and understand. The use of this tool will help women get over their fear of taking risks for pursuing their professions and occupations.

References

Ahir, S., Kapadia, S., Chauhan, J., and Sanghavi, N. (2018). "The personal stun-a smart device for women's safety." In *2018 International Conference on Smart City and Emerging Technology (ICSCET)* (pp. 1-3). IEEE.

Bhilare, P., Mohite, A., Kamble, D., Makode, S., and Kahane, R. (2015). "Women employee security system using GPS and GSM based vehicle tracking." *International Journal for Research in Emerging Science and Technology*, 2(1): 65-71.

Harikiran, G. C., Menasinkai, K., and Shirol, S. (2016). "Smart security solution for women based on Internet of Things (IOT)." In *2016 International Conference on Electrical, Electronics, and Optimization Techniques (ICEE-OT)* (pp. 3551-3554). IEEE.

Monisha, D. G., Monisha, M., Pavithra, G., and Subhashini, R. (2016). "Women safety device and application-FEMME." *Indian Journal of Science and Technology*, 9(10): 1-6.

Punjabi, S. K., Chaure, S., Ravale, U., and Reddy, D. (2018). "Smart intelligent system for women and child security." In *2018 IEEE 9th Annual Information Technology, Electronics and Mobile Communication Conference (IEM-CON)* (pp. 451-454). IEEE.

Sogi, N. R., Chatterjee, P., Nethra, U., and Suma, V. (2018). "SMARISA: a raspberry pi based smart ring for women safety using IoT." In *2018 International Conference on Inventive Research in Computing Applications (ICIR-CA)* (pp. 451-454). IEEE.

Sawant, S. A., Gurakhe, S., Shaikh, T. S., Bagmare, S., Rathad, C., and Sobale, S. (2023). "Safety with technology: a smart SOS device." In *2023 7th International Conference on Computing, Communication, Control and Automation (ICCUBEA)* (pp. 1-6). IEEE.

River Pollution Detection and Quantification with Aerial Imagery using YOLOv8

Yash Vyavahare, Tanmayee Mali, Lilavati Mhaske, Sayali Patil, Meenakshi Thalor

Department of Information Technology, All India Shri Shivaji Memorial Society's Institute of Information Technology, Pune, India,
E-mail.vyavahare.yash@gmail.com

Abstract

River pollution is a pressing environmental concern with far-reaching implications for ecosystems, biodiversity, and human health. As pollutants, including plastic waste and chemical runoff, continue to infiltrate the water bodies, the need for effective monitoring and remediation approaches becomes paramount. The escalating issue of river pollution necessity innovative methods for detection and remediation. In response, this research introduces a methodology for river pollution detection utilising drone imagery, YOLOv8 segmentation, and Canny edge detection. By leveraging drones equipped with high-resolution cameras, this research achieves accurate river segmentation and identifies plastic pollutants within the river environment. This approach holds immense promise for environmental monitoring, offering the ability to pinpoint polluted regions, trace pollution sources, and deploy drones for cleaning operations. The ongoing monitoring and maintenance facilitated by this methodology contribute to the conservation of river ecosystems and the preservation of aquatic life. As pollution challenges persist, this research paves the way for future interdisciplinary collaborations and global implementations, aiming to secure cleaner and healthier waterways.

Keywords: River-pollution, drone, YOLOv8, segmentation, edge-detection

1. Introduction

River pollution is a pressing environmental concern with far-reaching implications for ecosystems, biodiversity, and human health. As pollutants, including plastic waste and chemical runoff, continue to infiltrate the water bodies, the need for effective monitoring and remediation approaches becomes paramount. In this context, this research presents a novel methodology that harnesses the capabilities of drone technology, sophisticated image analysis techniques, and machine learning to tackle the complex issue of river pollution. By combining the agility and perspective of drones with YOLOv8 (You Only Look Once version 8) segmentation and Canny edge detection, this research aim to achieve a comprehensive solution for river pollution detection and management.

The methodology commences with the collection of aerial drone imagery, where drones equipped with high- resolution cameras capture detailed snapshots of river environments. These images are

Chapter 38 DOI: 10.1201/9781003581215

then subjected to YOLOv8 segmentation, enabling precise river delineation and the identification of plastic pollutants within the river. Furthermore, the Canny edge detection algorithm is applied to highlight pollutant contours, providing a visual representation of the pollution extent.

The applications of this methodology are vast, encompassing environmental monitoring, pollution source identification, and efficient deployment of drones for cleaning operations. It offers the ability to not only locate polluted regions but also trace the pollution back to its source, facilitating legal actions against polluters. The ongoing monitoring ensures the maintenance of clean river segments, thereby contributing to the long-term preservation of aquatic ecosystems.

As river pollution challenges persist on a global scale, this research holds the promise of shaping the future of environmental conservation. It underscores the importance of interdisciplinary collaborations, automation improvements, and worldwide implementations, as collective efforts are made towards cleaner and healthier river ecosystems.

2. Literature Review

Previous research has made significant strides in utilising remote sensing techniques for water body segmentation and pollution detection. For instance, García, M., Alcayaga, H., Pizarro, A. Automatic Segmentation of Water Bodies Using RGB Data: A Physically Based Approach (García et al., 2023) developed a physically based model for water body segmentation using RGB data, achieving high accuracy in estimating water surface area and perimeter without requiring a data training process. While their method showed robust segmentation results, it lacked real-time capabilities and was limited to RGB data, potentially limiting its applicability in dynamic environments. M. Rizk, T. N. Al-Deen, H. Diab, A. M. Ah- mad and Z. El-Bazzal, "Proposition of Online UAV- Based Pollutant Detector and Tracker for Narrow- Basin Rivers: A Case Study on Litani River," (Rizk et al., 2018) focused on using UAVs for real-time pollutant detection and tracking in narrow-basin rivers. While the approach addressed the limitations of traditional methods and satellite-based techniques, it primarily focused on pollutant tracking rather than segmentation. Additionally, the scalability and cost-effectiveness of their method for large-scale applications were not thoroughly discussed. L. Schreyers, T. van Emmerik, L. Biermann and M. van der Ploeg, "Direct and Indirect River Plastic Detection from Space," (Schreyers et al., 2022) explored the use of satellite imagery for direct and indirect detection of riverine plastics. While their indirect approach showed promise in detecting a high number of plastic items, the direct detection method was limited to large objects. Additionally, the study did not extensively discuss the methodology's

Figure 1: Flight path and parameters
Source: Maps Pilot Pro

scalability and real-time monitoring capabilities. Satellite images have also been used for river pollution detection by performing frequency domain analysis (S. Manickam, 2018). Sentenel-2 images are also used for measuring coastal water quality indicators (S. Hafeez, 2019).

3. Data and Variables

3.1. Study Area

Authors gathered drone images from the Mula Mutha river, located at coordinates 18.5212° N and 73.8514° E in Pune, India. This choice of study area was made in consideration of its significance in the context of pollution management in the region. The selection of the study area took into account the river's contribution to pollution downstream, as well as the practical aspects of data collection using unmanned aerial vehicles (UAVs). This region had not been previously studied for individual plastic object detection, making it an ideal location for assessing and evaluating various plastic monitoring methods. By collecting data in this area, authors aimed to contribute to the understanding and mitigation of plastic pollution in the Mula Mutha river basin.

3.2. Dataset

3.2.1. Data Collection

Aerial data acquisition was performed using a DJI Mini 2 drone, flying at an altitude of 67 meters above the terrain with a resolution of 3 centimeters. The drone was equipped with a 4K resolution camera, which provided high-quality imagery for the analysis. To ensure systematic and organised data collection during the survey, authors employed the use of the Maps Pilot Pro software as shown in Figure 1. This software played a crucial role in creating precise and efficient flight plans for the drone missions. By utilising Maps Pilot Pro, authors were able to design flight paths that optimised coverage of the study area as shown in Figure 2, taking into account the specific requirements of the research objectives. Arepository of 985 raw images was

obtained through aerial surveillance of the river. To create a comprehensive dataset for training and evaluation, various pre-processing methods were applied, resulting in a dataset comprising 2,361 images. This dataset was divided into subsets, with 2,064 images designated for training, 200 for validation, and 97 for testing.

3.3. Performance Indicators

3.3.1. Mean Average Precision (mAP)

The Mean Average Precision (mAP) is an essential performance metric used to evaluate the effectiveness of object detection and segmentation models. Unlike traditional object detection, instance segmentation not only identifies objects but also distinguishes individual instances of the same class, assigning unique labels to each. To calculate mAP for instance segmentation, a precision-recall curve is generated for each instance.

$$IoU = \frac{Area\,of\,Overlap}{Area\,of\,Union} \tag{1}$$

The mAP for instance segmentation assesses the accuracy of both object detection and instance labeling, taking into account the intersection over union (IoU) between predicted segmentation masks and ground truth masks. The IoU measures the degree of overlap between these masks, and it's used to determine whether a predicted instance corresponds to a true positive (TP) or a false positive (FP). Similar to traditional object detection mAP, the mAP for instance segmentation calculates the average precision across all instances and classes while considering multiple IoU thresholds. This comprehensive metric provides insights into the model's performance in accurately delineating individual instances within a class, making it a vital tool for assessing the quality of instance segmentation algorithms.

3.3.2. F1 Score

The F1-Score is a valuable performance metric in in- stance segmentation, measuring the model's accuracy in distinguishing and segmenting individual instances within an image. Unlike traditional object detection, in- stance segmentation assigns unique labels to each object instance of the same class and delineates their precise boundaries. The F1-Score for instance segmentation is a crucial indicator of the model's ability to balance both precision and recall effectively.

In the context of instance segmentation, the F1-Score is computed as the harmonic mean of precision and recall, ranging from 0 to 1. Precision is the ratio of correctly identified instances to the total number of predicted instances, and recall represents the ratio of correctly detected instances to the total number of ground truth instances. A high F1-Score indicates a model that achieves both high precision and recall in segmenting object instances accurately.

$$Precision = \frac{TP}{TP + FN} \qquad (2)$$

$$Recall = \frac{TP}{TP + FP} \qquad (3)$$

The F1-Score is particularly useful in instance segmentation tasks where the model is required to not only identify objects but also provide precise delineations for each instance. Achieving a high F1-Score signifies that the model is adept at accurately identifying, labelling, and segmenting individual instances within the same class, making it a valuable metric for evaluating the quality of instance segmentation algorithms.

$$F1\,Score = 2\,\frac{Precision * Recall}{Precision + Recall} \qquad (4)$$

4. Methodology and Model specifications

The input image as shown in Figure 2 is the 4k image captured by the drone over the study. It is passed to the river segmentation model of YOLOv8. The output of model is the river segment mask highlighted in the river segment image. This result marks the first crucial step towards identifying and managing river pollution. The next step involves converting the input image to grayscale image, it converts the RGB image to single band image for applying Edge

detection algorithms. Using Canny edge detection, the edges in the image are identified, outlining edges and contours within the entire image enhances the ability to identify features of the pollutants. To focus specifically on the river segment, river segment mask is used to effectively isolate regions of interest within the water body. The edges detected in the river segment are used to draw the contours over the pollutants. The output image as shown in Figure 2 demonstrates the detected plastic pollutants within the river with a highlighted segments for both river and pollutants within it.

4.1. YOLOv8 Instance Segmentation

YOLOv8 excels in instance segmentation, a process that involves the identification and outlining of individual objects within an image. It provides precise bounding boxes and accurate masks, making it an excellent choice for tasks requiring pixel-level analysis. Instance segmentation goes beyond object detection by outlining the precise shapes of individual objects in an image. The output includes masks or contours outlining each object along with corresponding class labels and confidence scores. This method is beneficial when the exact shapes of objects are required, not just their presence within an image (Maharjan *et al.*, 2022).

To distinguish the river from the captured images and facilitate the identification of plastic pollutants within the river segment, the YOLOv8 instance segmentation model was employed. This deep learning model, known for its object detection capabilities, was adapted to segment the river area in the images. By training the model on labelled data, this study aims to obtain a precise river mask, effectively separating it from the surrounding landscape.

4.2. Edge Detection using Canny Algorithm

Edge detection using the Canny algorithm is a well-regarded technique in the domains of computer vision and image processing, allowing for precise edge identification within images. The Canny method

Figure 2: Input image (left), output image (right)

effectively detects a wide range of edges while reducing image noise. The process involves noise reduction using a Gaussian filter to smoothen the image, followed by gradient calculation to highlight regions with rapid intensity changes. Further steps involve non-maximum suppression to maintain consistent one-pixel wide edges, thresholding to distinguish strong and weak edges, and edge tracking by hysteresis to link weak edges to strong ones. The Canny algorithm is esteemed for its ability to produce precise and comprehensive edges, making it widely applicable in various image processing tasks and computer vision applications.

To enhance the identification of pollutants within the river segment, the Canny edge detection algorithm was applied to the river mask. This algorithm effectively detected the edges and boundaries of objects within the river, including plastic pollutants. Subsequently, contours were drawn around these detected edges, providing a visual representation of the pollutants within the river mask. This step facilitated further analysis and quantification of the extent and distribution of plastic pollutants in the river (Schreyers *et al.*, 2022).

5. Empirical Results

During the training of YOLOv8 for river pollution detection, the observed trends in the metrics revealed notable insights into the model's learning process (Figure 3). While the segmentation loss (seg_loss) displayed a consistent decrease over training iterations, indicating that the model was progressively improving its ability to align

predicted segmentation masks with the ground truth river boundaries, other metrics exhibited distinctive trends. Notably, precision and recall, which assess the accuracy and completeness of plastic pollutant detection, demonstrated an increase over time. This indicated that the model was becoming more adept at accurately identifying pollutants within the river segment, balancing the trade-off between true positives and false positives. Simultaneously, the mean average precision (mAP) witnessed an upward trajectory, signifying the model's enhanced object detection capabilities, considering various Intersection over Union (IoU) thresholds. These dynamic trends in the metrics underscore the model's evolving proficiency in river pollution detection as it undergoes training and optimisation.

6. Performance Evaluation

The model's performance was rigorously assessed using standard metrics. Figure 6 shows one sample input and output image of the system where the use of YOLOv8 and instance segmentation led to exceptional precision (0.957), recall (0.913), mAP50 (0.944), and mAP50-95 (0.897).

This is a notable improvement compared to previous models, signifying a substantial reduction in false positives and higher accuracy in identifying pollutants within the river. Edge detection further enhanced precision in pollutant quantification.

7. Conclusion

This methodology for river pollution detection using drone imagery, along with

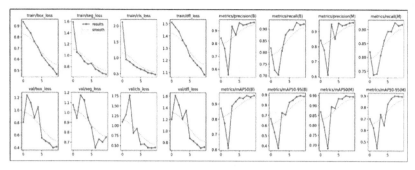

Figure 3: Training metrics graphs

YOLOv8 segmentation and Canny edge detection, represents a promising approach in the field of environmental monitoring and conservation. The YOLOv8 river segmentation model demonstrated exceptional performance ensuring accurate and comprehensive delineation of river segments. By efficiently identifying polluted areas, tracing pollution sources, and deploying drones for cleaning, this methodology equips researchers, environmental agencies, and policymakers with valuable tools to combat river pollution effectively. The ongoing monitoring and maintenance facilitated by this technology further contribute to the long-term well-being of river ecosystems, providing a foundation for sustainable environmental conservation efforts. As pollution challenges persist on a global scale, this research, with its remarkable segmentation precision and recall, serves as a beacon for future interdisciplinary collaborations and worldwide implementations, ultimately striving for a cleaner and healthier environment.

References

García, M., Alcayaga, H., Pizarro, A. (2023). "Automatic segmentation of water bodies using RGB data: a physically based approach." *Remote Sensing*, 15: 1170. https://doi.org/10.3390/rs15051170.

Hafeez, S. and Wong, M. S. (2019). "Measurement of coastal water quality indicators using Sentinel-2, an evaluation over Hong Kong and the Pearl River Estuary." *IGARSS 2019 - 2019 IEEE International Geoscience and Remote Sensing Symposium*, Yokohama, Japan, pp. 8249-8252, doi: 10.1109/IGARSS.2019.8899342.

Manickam, S., Priyadharshini, V. M., and Anitha Rajathi, V. M. (2018). "Digital monitoring of river pollution using satellite images in frequency domain." *2018 2nd International Conference on Trends in Electronics and Informatics (ICOEI)*, Tirunelveli, India, pp. 142- 146, doi: 10.1109/ICOEI.2018.8553809.

Maharjan, Nisha, Hiroyuki Miyazaki, Bipun Man Pati, Matthew N. Dailey, Sangam Shrestha, and Tai Nakamura. (2022). "Detection of river plastic using UAV sensor data and deep learning." *Remote Sensing*, 14 (13): 3049. https://doi.org/10.3390/rs14133049.

Rizk, M., Al-Deen, T. N., Diab, H., Ahmad, A. M. and El-Bazzal, Z. (2018). "Proposition of online UAV- based pollutant detector and tracker for narrow- basin rivers: a case study on Litani River." *2018 International Conference on Computer and Applications (ICCA)*, Beirut, Lebanon, pp. 189-195, doi: 10.1109/COMAPP.2018.8460374.

Schreyers, L., van Emmerik, T., Biermann, L., and van der Ploeg, M. "Direct and indirect river plastic detection from space." *IGARSS 2022 - 2022 IEEE International Geoscience and Remote Sensing Symposium*, Kuala Lumpur, Malaysia, 2022, pp. 5539-5542, doi: 10.1109/IGARSS46834.2022.9883

Survey on Diverse Access Control Techniques in Cloud Computing

Amiya Sahoo[1], A. Aparna Rajesh[1], G. M. Kiran[2]

[1]Aryan Institute of Engineering and Technology, Bhubaneswar, India
[2]Shridevi Institute of Engineering and Technology, Tumkur, India
E-mail: dramiya79@gmail.com, aparna.subhadra@gmail.com, kirangm900@gmail.com

Abstract

Cloud computing is one of the most alternative paradigms for providing computer resources and services online As more and more companies move their activities to the cloud, it is critical to make sure cloud resources are accessible securely. Protecting sensitive data, controlling access to cloud resources, and preventing illegal activity are all made possible by access control techniques. This study provides an extensive overview of cloud computing access control approaches, emphasising their salient characteristics, advantages, and drawbacks. We examine each model's features, such as its adaptability, scalability, and fit for different cloud deployment scenarios. We also go about the difficulties and future trends in access control for cloud computing, including dynamic policy enforcement, fine-grained access control, and compliance management. This research provides useful information to practitioners, academics, and businesses aiming to strengthen the security of their cloud installations through an examination of the evolution and current status of access control models in cloud computing.

Keywords: Cloud, scalability, survey

1. Introduction

The Internet-based applications and services that are available now, thanks to the cloud computing technology of today, offer users some very amazing benefits. For cloud users, managing resource access is essential since uncontrolled access often leads to security issues. Errors in the system might also result from this. Access control systems need to include a number of features in order to implement access control concepts. By providing user features or functions that are appropriate for the processes followed, methodology, and administrative features, these systems eliminate the need to execute the activities independently in different domains. As such, determining the priorities of the systems is straightforward. Furthermore, it is probably going to execute the procedures for accepting or rejecting the jobs based on priority according to a set of standard criteria. This study classifies cloud computing access control strategies into four categories: Fine Grained Based Approaches, Encryption Based Approaches, Extended Approaches, and Traditional Approaches.

1.1. Traditional Approaches

There are 3 types of models in this category: "*Role-Based Access Control (RBAC), Discretionary Access Control (DAC) and Mandatory Access Control (MAC).*"

Chapter 39 DOI: 10.1201/9781003581215

1.1.1. Role Based Access Control

The term "Role Based Access Control (RBAC)" was first used in the 1970s, Users are allocated specific functions in accordance with RBAC, and these roles facilitate the exercise of the necessary rights, restrictions, and authorisations (Sandhu *et al.*, 1996). The overall structure of RBAC states that permission assignment (PA) determines each role's rights inside the system, and user assignment (UA) determines which roles are allocated administration permissions, permissions and restrictions of the client from the function of the user (Oh *et al.*, 2006).

1.1.2 Discretionary Access Control

Graham and Dennis created "Discretionary Access Control (DAC)," whose organisational framework forms the basis of security systems. The one who is in possession of the items is referred to as the Owner, and as such, they assign power; hence, they also choose the safety rules. According to his own preferences, the system owner limits access, imposes limitations, and gives users the necessary authorisation (Khan and Sakamura, 2015). Control over users or groups is the basis of this access control. (Msahli *et al.*, 2014).

1.1.3 Mandatory Access Control

Experts concluded that because DAC cannot manage all data, it lacks security, which is why Mandatory Access Control (MAC) was created. A central system, as opposed to an owner, makes choices in MAC (Servos *et al.*, 2017, Shen *et al.*, 2006). As a result, security model activities are made more dependable (Auxilia and Raja, 2012). The most crucial aspects of the system are its completeness and privacy, followed by security (Zou *et al.*, 2009). It may therefore maximise system security to the greatest extent feasible (Fan *et al.*, 2009). By imposing restrictions, it enables even the system administrator to function in accordance with the recommended norms (Blanc and Lalande, 2013).

1.2 Extended Approaches

Cloud applications are massive, shared, virtualised, complex information systems that are difficult to access using standard technology and access control methods.

Nonetheless, new methods have been employed as CC has progressed. Enhancing the defining, giving, and/or restriction of access to cloud resources was the aim of these new models.

1.2.1 Attribute Based Access Control

In Attribute-Based Access Control, user attributes are essential (ABAC). These parameters may be modified based on the user's general features, which include age, height, and personality factors (Tawosi, 2016). According to the subjects, the characteristics are established. Since the qualities are related to the data input, they may be categorised into three groups: topic attributes, resource attributes, and environment attributes. The systems that can use ABAC in the most useful manner are the distributed and open ones. It demonstrates how much faster these systems are progressing toward dependable and adaptable access control solutions (Pussewalage and Oleshchuk, 2016).

1.2.2 Task Based Access Control

Thomas *et al.* introduced a task-based access control paradigm in 1997 after adopting the idea of a task. It may quickly use different methods for different work-related activities. The system functions, the role, and the task may all be combined into a single task by incorporating the idea of task into the RBAC architecture.

The task and the authority are linked in order to achieve dynamic management of access rights. Because some of the authorisation job would be shared by the cloud server, the user's burden would be decreased.

1.2.3 Action Based Access control (AcBAC)

The idea of an action-based access control model (AcBAC) looked into the link between the roles, tenses, and environment and offered a formal definition of the action state management function in order to handle the problem of information system resource authorisation decision-making. AcBAC is a versatile technique that takes into account and carefully analyzes the role, the circumstance, and the environment in order to handle access control

1.3 Encryption Based Access Control

Traditional access control approaches' main drawback is their high implementation costs. Moreover, cloud computing service providers and data owners do not belong to the same trusted domain. As a result, it is crucial to safeguard the information by encrypting it in a way that is incomprehensible to outside parties. Thus, we'll talk about a number of access control model encryption techniques in this part.

1.3.1 *Identity Based Encryption*

The IBE paradigm connects and validates the signatures fast, without revealing the public or secret key. The IBE paradigm is based on the public key cryptography technique, wherein the client selects a random permutation and gives the other party an identifying code. The user has the option to select any string. Nevertheless, the right private key (PKG) is generated using a Personal Key Generation Center.

1.3.2 *Hierarchical Identity Based Encryption*

Security is made clear by the HIBE scheme notion. A single private key generator (PKG) in a conventional IBE (1-HIBE) system assigns a private key to each user and gives them with an arbitrary primitive ID (PID) as their public key. Three elements make up a two-level HIBE (2-HIBE) system: users, domain PKGs, and a root PKG. All three are connected to PIDs. The public key for each user is composed of their PIDs and domain. If a user has previously

requested their domain secret key (SK) from the root PKG, domain PKGs are able to calculate the private key PK of any user within their domain.

1.3.3 *Timed Release Encryption*

Data owners can encrypt their data and make it available to authorised users by using the timed-release encryption (TRE) approach. After the predetermined future time has elapsed, the user with the necessary credentials can obtain the key to decrypt the encrypted content. Chen *et al.* proposed a non-centralised TRE mechanism and an efficient coding scheme with privacy protection features, which are suitable for application scenarios needing just a somewhat dependable time server. Unruh *et al.* suggested a reversible time-based encryption method based on quantum cryptography, along with an encryption procedure for an anonymous recipient.

1.3.4 *Role Based Encryption*

Based on the layered RBAC architecture, the role-based encryption (RBE) idea is used to construct the data access control technique. Here, the owner stored the data on a cloud server after encrypting it. Only authorised users are able to decode the data and reveal the plain content. With this system, senders can encrypt data using given responsibilities. Encrypted material can only be successfully decoded by higher role users.

1.4 Fine Grained Access Control

While the aforementioned encryption methods can be employed to protect data privacy, they are not suitable for meeting requirements for flexible data sharing. The administrators should be able to set up and prepare the proper degree of exposure and specificity so that the system may assign different levels of access control to users who belong to the same group. While defining discrete user access entitlements, the Fine-Grained methods offer flexibility (Table 1).

Table 1:

Slno.	Approach	Types	Feature				
			F1	F2	F3	F4	F5
1	Traditional	Dac	Yes	Low	No	N/a	Owner
		Mac	No	Low	No	N/a	Csp
		Rbac	No	Low	No	N/a	Admin
2	Extended	Abac	Yes	Low	No	Low	Admin
		Tbac	No	High	No	Low	Admin
3	Encryption	Ibe	No	Low	Low	Low	Csp
		Hibe	No	High	Low	Low	Csp
		Pre	No	High	Medium	No	Owner
4	Fine-grained	Abe	No	High	High	No	Admin
		Cp-abe	Yes	High	High	Yes	Admin
		Kp-abe	No	High	High	No	Admin

1.4.1 *Attribute Based Encryption*

The bulk of encrypted data may be subject to fine-grained access control using ABE with the assistance of users and attribute permission. Users can either create an access structure inside the ciphertext or label it with a descriptive encryptor and assign it a key to access datas. The main versions of the model are described in the next subsections.

1.4.2 *Key-policy Attribute based Encryption*

KP-ABE is a technique for one-to-many communications in which cipher texts are annotated with attribute sets and users' keys are linked to policies. On KP-ABE, the user's characteristics are supplied in the leaves of an access tree structure that specifies the user's private key. When the ciphertext's properties align with the data access structure, a user can only decrypt it with a certain key. One important aspect of the KP-ABE approach is the secrecy of outsourced data, as mentioned in. There are a number of issues with this paradigm, many of which are related to its inefficiency and absence of an attribute revocation method. Among the limitations are key escrow, key cooperation, and key revocation.

1.4.3 *Cipher Text Attribute based Encryption*

In the second generation of ABE, called CP-ABE, which connects ciphertexts to rules, the user's key is defined as a string with several descriptive attributes. Using the access tree architecture that private keys match, a cryptor creates the rule to decode the message. A client can decode a cypher code with a given password if only the data access structure has the qualities that the private key to the tree nodes provides. While using a somewhat different method, this model shares efficiency difficulties with KP-ABE when it comes to attribute revocation.

1.5 Results and Analysis

Access control models should be selected in real-world situations based on a variety of applications and scenarios. Models of access control are assessed based on a set of metrics that include: distributed (F1), processing overhead (F2), fine-grained (F3), security (F4), forward or backward, and controller (F5).

2. Conclusion

To safeguard the information and resources of information systems, businesses are finding that access control is an increasingly important information security strategy. Research on access control models has advanced significantly after many years of development. Due to the unique nature of cloud computing, access control is crucial for efficiently safeguarding its resources. Thus, one of the most crucial subjects in the current study on cloud security is authentication and permission in cloud computing. The examination of several approaches is evaluated according to the following standards: distributed, fine-grained, controller, security, and overhead processing.

References

Auxilia, M. and Raja, K. (2012). "A semantic-based access control for ensuring data security in cloud computing." In *International Conference on Radar, Communication and Computing (ICRCC)*, (pp. 171-175).

Blanc, M. and Lalande, J. F. (2013). "Improving mandatory access control for HPC clusters." *Future Generation Computer Systems*, 29 (3): 876-885.

Briffaut, J., Lalande, J. F., and Smari, W. W. (2008). "Team-based MAC policy over security-Enhanced Linux." In *Second International Conference on Emerging Security Information, Systems and Technologies, SECURWARE'08.* (pp. 41-46).

Chatterjee, S., Gupta, A. K., Mahor, V. K., and Sarmah, T. (2014). "An efficient fine grained access control scheme based on attributes for enterprise class applications." In *International Conference on Signal Propagation and Computer Technology (ICSPCT)*, 2014 (pp. 273-278).

Ed-Daibouni, M., Lebbat, A., Tallal, S., and Medromi, H. (2016). "A formal specification approach of privacy-aware attribute based access control (pa-abac) model for cloud computing." In *International Conference on Systems of Collaboration (SysCo)*, (pp. 1-5).

Fan, Y., Han, Z., Liu, J., and Zhao, Y. (2009). "A mandatory access control model with enhanced flexibility." In *International Conference on Multimedia Information Networking and Security, MINES'09.* (Vol. 1, pp. 120-124). IEEE.

Ferraiolo, D. F., Sandhu, R., Gavrila, S., et al. (2001). "Proposed NIST standard for role-based access control." *ACM Transactions on Privacy and Security (TISSEC)*, 4 (3): 224-274.

Jin, P. and Fang-Chun, Y. (2006). "Description logic modeling of temporal attribute-based access control." In *First International Conference on Communications and Electronics, 2006. ICCE'06.* (pp. 414-418).

Khan, M. F. F. and Sakamura, K. (2015). "Fine-grained access control to medical records in digital healthcare enterprises." In *International Symposium on, Networks, Computers and Communications (ISNCC)*, 2015 (pp. 1-6). IEEE.

Lei, Z., Hongli, Z., Lihua, Y., and Xiajiong, S. (2011). "A mandatory access control model based on concept lattice." In *International Conference on Network Computing and Information Security (NCIS)*, (Vol. 1, pp. 8-12).

Msahli, M., Chen, X., and Serrhrouchni, A. (2014). "Towards a fine-grained access control for cloud." In *IEEE 11th International Conference on, e-Business Engineering (ICEBE)*, 2014 (pp. 286-291).

Oh, S., Sandhu, R., Zhang, X. (2006). "An effective role administration model using organization structure." *ACM Transactions on Privacy and Security (TISSEC)* 9 (2): 113-137.

Pussewalage, H. S. G., and Oleshchuk, V. A. (2016). "An attribute based access control scheme for secure sharing of electronic health records." In *IEEE 18th International Conference on e-Health Networking, Applications and Services (Healthcom)*, (pp. 1-6).

Sandhu, R., Coyne, E. J., Feinstein, H. L., et al. (1996). "Role-based access control models." *Computer*, 29 (2): 38-47.

Sandhu, R., Bhamidipati, V., Munawer, Q. (1999). "The ARBAC97 mode for role-based administration of roles. ACMTrans." *Information Systems Security (TISSEC)*, 2 (1): 105-135.

Sandhu, R., Munawer, Q. (1999). "The ARBAC99 model for administration of roles." In: *Proceedings of 15th Annual Computer Security Applications Conference*, pp. 229-238. IEEE, New York, NY.

Servos, D. and Osborn, S. L. (2017). "Current research and open problems in attribute-based access control." *ACM Computing Surveys (CSUR)*, 49 (4): 65.

Shen, H. B. and Hong, F. (2006). "An attribute-based access control model for web services." In *Seventh International Conference on Parallel and Distributed Computing, Applications and Technologies*, 2006. PDCAT'06. (pp. 74-79).

Tawosi, V. (2016). "A light weight dynamic attribute based access control module integrated with business rules." In *IEEE 10th International Conference on Application of Information and Communication Technologies (AICT)*, (pp. 1-5).

Zou, D., Shi, L., and Jin, H. (2009). "DVM-MAC: a mandatory access control system in distributed virtual computing environment." In *15th International Conference on Parallel and Distributed Systems (ICPADS)*, (pp. 556-563). IEEE.

A Mathematical Modelling Approach for Customised Robot in Smart Manufacturing

Kali Charan Rath[1], Alex Khang[2], Gopal Krushna Mohanta[1], Sasadhara Sahu[1]

[1]Department of Mechanical Engineering; GIET University, Gunupur, India, rathkalicharan1980@gmail.com, gmohanty@giet.edu, sasadharasahu2002@gmail.com
[2]Universities of Science & Technology, Vietnam and United States; AI & Data Scientist, Global Research Institute of Technology & Engineering, NC, United States, alex.khang@outlook.com

Abstract

This paper presents a novel mathematical modelling approach for tailored robots in smart manufacturing, focusing on kinematic analysis. It addresses challenges in integrating robots into dynamic processes by formulating a comprehensive model considering various degrees of freedom and operational constraints for adaptability and precision. Objectives include developing a robust model, evaluating performance through simulation and experimentation, and optimising customised robotic systems for enhanced efficiency. Novelty lies in customised kinematic analysis, adaptability integration, and validation via simulation, advancing customised robots in Industry 4.0.

Keywords: Mathematical modelling, customised robots, smart manufacturing, kinematic analysis, adaptability and precision, industry 4.0 advancements

1. Introduction

In Smart Manufacturing, custom robotics are vital for precision and adaptability. Ding et al. (2020), Qu et al. (2019), Tao et al. (2018), Nguyen et al. (2023). Mathematical modelling key for customising robots in smart manufacturing, optimising performance through tech, focusing on end-effector, payload, and kinematics Mittal et al. (2019), Sarcinelli-Filho and Carelli (2023).

The core of this research is the creation of a customised mathematical model using advanced computational algorithms tailored for specific robots. Xu et al. (2019) Engineers will use numerical methods to capture the robot's manipulator system kinematics Hameed et al. (2023), Nguyen et al. (2023), Rajendran et al. (2023) and mathematical optimization applied Lin et al. (2017), Namjoshi and Rawat (2022), Zhao et al. (2023). Geometric optimization of the end-effector precision provided Maloisel et al. (2019), Meng et al. (2023), Sahoo and Lo (2022).

This paper presents a math model for RRP robots with kinematic analysis, using precise equations and simulations to understand behavior. It's a significant advancement for robotics design and motion planning, aiming to revolutionise smart manufacturing.

2. Robot in Smart Industry

The robot epitomises Industry 4.0, integrating cutting-edge tech to transform smart manufacturing with its agile 6-DOF arm, enhanced sensors, and seamless IoT compatibility for efficient production and collaboration.

| (a) | (b) | (c) |

Figure 1: (a) Robot structure, (b) Robot and Trajectory, (c) End-effector and spline

Using robots (Figure 1) in industry offers numerous benefits, revolutionising manufacturing processes. Firstly, robots enhance productivity by executing repetitive tasks with high precision and speed, leading to increased output and reduced production time. Secondly, they improve workplace safety by taking on hazardous or strenuous tasks, thus minimising the risk of accidents and injuries to human workers. Additionally, robots contribute to consistent product quality by ensuring uniformity in manufacturing processes, reducing errors and defects. Moreover, they enable flexibility in production, allowing for quick reconfiguration and adaptation to changing production demands without significant downtime. Lastly, robots can work in environments unsuitable for humans, such as extreme temperatures or confined spaces, expanding the scope of industrial applications while optimising resource utilisation.

3. Geometry and Kinematic Analysis of Robot

Robot geometry defines limits; kinematics plans; computational tools model. Robot uses DH parameters for kinematic analysis, employing transformation matrices to determine end-effector frame.

T = A1 × A2 ×A3 ×A4 ×A5 ×A6

Homogeneous Transformation Matrix T is:

$$T = \begin{bmatrix} R & P \\ 0 & 1 \end{bmatrix}$$

The forward kinematics equation relates the joint angles (θ) to the end-effector position and orientation:

T = F ($\theta1, \theta2, ... \theta6$)

4. RRP customised robot

Explore DH parameterisation (Figure 2) and kinematics analysis for the versatile RRP robot with two revolute joints and one prismatic joint. let's define the DH parameters:

Link	a_i	α_i	d_i	θ_i
1	a_1	α_1	d_1	θ_1
2	a_2	α_2	d_2	θ_2
3	a_3	α_3	d_3	θ_3

Figure 2: RRP Robot

Joint Sl. No	Type of the Joint	Joint offset value (b) m	Joint angle value (θ) deg	Link Length value (a) m	Twist angle value (α) deg	Initial joint value (JV) deg	Final joint Value (JV) deg
1	R	0.05	Variable value as desired	0.15	70	0	180
2	R	0.05	Variable value as desired	0.15	90	90	180
3	P	variable	90	0.075	90	0	0.1

Link-1 configuration matrix is

1	0	0	0.15
0	0.34202	-0.939693	0
0	-0.939693	0.34202	0.05

Link-2 configuration matrix is

0	0	1	0
1	0	0	0.1
0	1	0	0.05

Link-3 configuration matrix is

0	0	1	0
1	0	0	0.075
0	1	0	0

End-effector configuration :

0	1	0	0.15
-0.939693	0	0.34202	-0.08326
0.34202	0	0.939693	0.186722
0	0	0	1

Forward Kinematics Result Output:

-0.25882	-0.96593	0	0.18945
0.96593	-0.25882	0	2.63896
0	0	1	1
0	0	0	1

Link-1 configuration matrix is

1	0	0	0.15
0	0.34202	-0.939693	0
0	-0.939693	0.34202	0.05

Input initial joint angle and ranges, loop through combinations to calculate end effector positions, and output the list for various joint configurations (Figure 3 & 4).

Link-2 configuration matrix is

0	0	1	0
1	0	0	0.1
0	1	0	0.05

Link-3 configuration matrix is

0	0	1	0
1	0	0	0.075
0	1	0	0

Figure 3: RRP- Robot end-effector trajectory

End-effector configuration:

0	1	0	0.15
-0.939693	0	0.34202	-0.08326
0.34202	0	0.939693	0.186722
0	0	0	1

Figure 4: End-effector trajectory

4.3. Result Discussion

The 2D plot visualises how joint angles and prismatic displacement affect the end-effector's path. It helps researchers and engineers understand the robot's workspace and motion capabilities, aiding in path planning and performance optimisation.

Kinematics analysis via DH parameters yields forward kinematics, deriving transformation matrices for individual joints.

$T_i = Rot(Z,\theta i) \times Trans(Z,di) \times Trans(X, ai) \times Rot(X, \alpha i)$ DH parameters :

$a1 = 1, \alpha1 = 0 , d1 = 0, \theta1 = \dfrac{\pi}{4}$

$a2 = 1, \alpha2 = 0 , d2 = 0, \theta2 = \dfrac{\pi}{3}$

$a3 = 1, \alpha3 = 0 , d3 = 0, \theta3 = 0$

4.4. End-effector Position and its Geometric Point

A 3-DOF RRP robot (Figure 5) offers precise motion in 3D space, optimised for tasks requiring both rotational and translational movements.

Figure 5: Trajectory of RRP end-effector

Table 1: End-effector geometry

x	y	z
0.15	-0.08326	0.186722
0.150949	-0.08103	0.186722
0.163732	-0.01678	0.187294
0.127529	0.08058	0.190256
0.059177	0.121865	0.193195
-0.0281	0.105122	0.192325
-0.04434	0.090207	0.189063
-0.04965	0.083719	0.186899
-0.05	0.08326	0.186722

Quintic equation is a polynomial equation of degree 5, and it generally has the form:

ax^5+bx^4+cx^3+dx^2+ex+f =0

Three variables (x, y, z) have corresponding sets of values. Three separate quintic equations, one for each variable, are required to fit the data points. Here's a generalised form of a quintic equation for each variable (x, y, z):

ax^5+bx^4+cx^3+dx^2+ex+f =0
gx^5+hx^4+ix^3+jx^2+kx+l =0
mx^5+nx^4+ox^3+px^2+qx+r =0

Now it is required to determine the coefficients (a, b, c, d, e, f, g, h, i, j, k, l, m, n, o, p, q, r) based on specific data points.

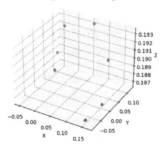

Figure 6: Scatter plot in the work volume

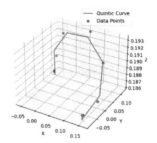

Figure 7: 3D line plot with quintic equation for real tool movement

4.5. Error Correction Algorithm for RRP Customised Robot

To correct errors in a 3D line plot using a quintic equation for real tool movement in an RRP robot (Figure 6, 7 & 8) , follow these steps: a) Generate Quintic Polynomial Trajectory; b) Apply Inverse Kinematics; c) Perform Forward Kinematics ; d) Integrate Sensors and Feedback; e) Calculate Error; f) Apply PID Control; g) Continuously Iterate Trajectory; h) Integrate Collision Detection; i) Test Error Correction; j) Fine-tune Algorithm Parameters

Error free RRP model:

Figure 8: Simulation model of error free RRP Robot with three D.O.F and C^0 continuity

Floating and mobile base robots use virtual linkage for motion like fixed-base ones. "Workspace" in robotics refers to end effector's range, influencing optimal design within spatial constraints.

5. Conclusion

This work pioneers tailored mathematical modeling for custom robots in smart manufacturing, focusing on kinematic analysis to integrate robots into dynamic processes. It aims for a robust model, performance assessment through simulation/experimentation, and insights for optimising customised robotic systems, advancing Industry 4.0.

5.1. Future Scope of the Work

Future work in robot kinematic modeling aims to improve robotic systems' accuracy, versatility, and efficiency. This involves developing advanced algorithms for dynamic environments, integrating sensors like lidar for better perception, exploring AI-driven motion planning, and creating modular models for easy deployment across different robots and industries.

References

Ding, H., Gao, R. X., Isaksson, A. J., Landers, R. G., Parisini, T., and Yuan, Y. (2020). "State of AI-based monitoring in smart manufacturing and introduction to focused section." *IEEE/ASME Transactions on Mechatronics*, 25 (5): 2143-2154.

Hameed, A., Ordys, A., Możaryn, J., and Sibilska-Mroziewicz, A. (2023). "Control system design and methods for collaborative robots." *Applied Sciences*, 13 (1): 675.

Lin, Y. C., Hung, M. H., Huang, H. C., Chen, C. C., Yang, H. C., Hsieh, Y. S., and Cheng, F. T. (2017). "Development of advanced manufacturing cloud of things (AMCoT)—a smart manufacturing platform." *IEEE Robotics and Automation Letters*, 2 (3): 1809-1816.

Mittal, S., Khan, M. A., Romero, D., and Wuest, T. (2019). "Smart manufacturing: characteristics, technologies and enabling factors." *Proceedings of the Institution of Mechanical Engineers, Part B: Journal of Engineering Manufacture*, 233 (5): 1342-1361.

Maloisel, G., Schumacher, C., Knoop, E., Grandia, R., and Bächer, M. (2023). "Optimal design of robotic character kinematics." *ACM Transactions on Graphics (TOG)*, 42 (6): 1-15.

Meng, Q., Li, J., Shen, H., Deng, J., and Wu, G. (2023). "Kinetostatic design and development of a non-fully symmetric parallel Delta robot with one structural

simplified kinematic linkage." *Mechanics Based Design of Structures and Machines*, 51 (7): 3717-3737.

Namjoshi, J. and Rawat, M. (2022). "Role of smart manufacturing in industry 4.0." *Materials Today: Proceedings*, 63: 475-478.

Nguyen, T. Q., Pham, K. B., and Chi, D. T. K. (2023). "Modeling robotic arm with six-degree-of-freedom through forward kinematics calculation based on deep learning." *International Journal of Intelligent Systems and Applications in Engineering*, 11 (2): 293-300.

Qu, Y. J., Ming, X. G., Liu, Z. W., Zhang, X. Y., and Hou, Z. T. (2019). "Smart manufacturing systems: state of the art and future trends." *The International Journal of Advanced Manufacturing Technology*, 103: 3751-3768.

Rajendran, V., Debnath, B., Mghames, S., Mandil, W., Parsa, S., Parsons, S., and Ghalamzan-E, A. (2023). "Towards autonomous selective harvesting: a review of robot perception, robot design, motion planning and control." *Journal of Field Robotics*.

Sahoo, S., and Lo, C. Y. (2022). "Smart manufacturing powered by recent technological advancements: a review." *Journal of Manufacturing Systems*, 64: 236-250.

Sarcinelli-Filho, M. and Carelli, R. (2023). "Kinematic Models." In *Control of Ground and Aerial Robots* (pp. 5-22). Cham: Springer International Publishing.

Tao, F., Qi, Q., Liu, A., and Kusiak, A. (2018). "Data-driven smart manufacturing." *Journal of Manufacturing Systems*, 48: 157-169.

Xu, Z., Zhou, Q., and Yan, Z. (2019). "Special section on recent advances in artificial intelligence for smart manufacturing–part I intelligent automation and soft computing." *Intelligent Automation and Soft Computing*, 25 (4): 693-694.

Zhao, J., Wang, X., Xie, B., and Zhang, Z. (2023). "Human-robot kinematics mapping method based on dynamic equivalent points." *Industrial Robot: The International Journal of Robotics Research and Application*, 50 (2): 219-233.

Precision Cardio Care

Unveiling the Power of Machine Learning Models for Cardiovascular Risk Prediction

Satyaprakash Swain[1], Kartick Swain[1], Swagatika Sahoo[1], Suvendra Kumar Jayasing[1], Kumar Janardan Patra[1], Soumya Ranjan Prusty[2]

1Department of Computer Science and Engineering, Institute of Management and Information Technology, Cuttack, BPUT, India
2Vihave.ai, Hyderabad, India
E-mail: satyaimit@gmail.com, kartickswain18@gmail.com, swagatika.sahoo1609@gmail.com, sjayasingh@gmail.com, Janardanpatra1997@gmail.com, soumya777prusty@gmail.com

Abstract

Cardiovascular diseases have emerged as a leading cause of mortality worldwide, necessitating advanced predictive models for timely identification and intervention. This research delves into the efficacy of three prominent machine learning algorithms named Naive Bayes (NB) second one is Random Forest (RF) and third one is Sequential Minimal Optimisation (SMO) in predicting heart attack risk. Using a dataset of 1320 records from Kaggle, we employed a ML ensemble approach to combine their predictive power. RF emerged as the most accurate, with 98.71% overall accuracy, highlighting its ability to discern complex patterns. Our findings endorse RF as the preferred model for heart disease prediction due to its exceptional accuracy. This study provides valuable insights into the comparative performance of popular algorithms, aiding practitioners in making informed choices for heart disease prediction models.

Keywords: Machine learning, SMO, NB, RF, heart disease detection

1. Introduction

Cardiovascular diseases present a global health challenge, prompting the quest for predictive models to identify and intervene in such ailments promptly. This study examines heart attack detection, utilizing a dataset of 1320 records sourced from Kaggle. NB, RF and SMO were evaluated, employing a ML ensemble approach. RF stood out with 98.71% accuracy, showcasing its ability to decipher complex data patterns effectively. However, NB and SMO demonstrated more modest accuracies at 68.23% and 71.34%, respectively, indicating limitations in capturing intricate data relationships. These findings underscore RF's potential as a robust contender in cardiovascular health predictive modeling. Yet, they also highlight the need for nuanced understanding and consideration of algorithm performance. This research contributes valuable insights for proactive healthcare management and underscores the importance of informed algorithm selection in addressing cardiovascular health challenges.

This understanding of varying accuracies and capabilities among these models paves the way forward in our pursuit of

Chapter 41 DOI: 10.1201/9781003581215

effective heart disease predictive models. Amidst these insights, the RF stands out as a beacon, hinting at its potential as the preferred model for heart disease prediction. Its exceptional accuracy not only signifies its prowess in decoding intricate data but also charts a path toward precision healthcare, potentially pivotal in identifying individuals at risk of heart attacks for proactive preventive measures. The implications of this research transcend disciplinary boundaries, echoing across the medical landscape. As healthcare evolves, our study lays a groundwork for informed decisions in designing predictive models, catalyzing a shift from reactive to anticipatory healthcare, potentially reshaping the trajectory of global heart disease outcomes.

2. Literature Review

A plethora of studies showcase the versatile applications and promising outcomes of machine learning across diverse domains. Galleta *et al.* (2019) utilise Data Envelopment Analysis and Naïve Bayes to assess city competitiveness, achieving a 90.64% accuracy rate vital for decision-making in regional development. Roshini *et al.* (2019) delves into social media's impact, noting reduced stress levels but highlighting concerns about overdependence. Azinet *et al.* (2021) focus on stroke diagnosis, finding Random Forest outperforms with 96% accuracy, crucial for early treatment. Fouedjio *et al.* (2020) propose a spatial prediction method for geochemical datasets using regression random forest. Wei *et al.* (2023) optimises concrete frame structures with random forest regression, offering wide-ranging applications. Usha *et al.* (2022) forecast diabetes using Random Forest, emphasizing Hyperparameter tuning methods' effectiveness. Elias *et al.* (2022) predicts stroke risk with a stacking method, boasting an impressive AUC of 98.9%. Shafiul *et al.* (2020) explore stroke risk prediction using various algorithms, achieving 99.98% accuracy. Mishra *et al.* (2023) detects heart attacks early with SVM,

emphasizing its crucial role in survival forecasts. Hung *et al.* (2020) predict stroke outcomes, with Naïve Bayes aiding early risk assessment. Swain *et al.* (2023) explore IoT and machine learning for heart disease prognosis, advocating for advanced technologies. These studies collectively underline machine learning's potential in advancing research and addressing critical challenges across industries.

3. Research Design

3.1. Ensemble Learning

Ensemble learning involves the fusion of multiple models to heighten predictive accuracy and resilience. Through the amalgamation of diverse algorithms or akin models trained on varied data subsets, ensemble methods strive to alleviate individual model limitations. This approach substantially enhances overall performance and dependability in generating predictions or classifications. Here we have only used the Voting Model for risk prediction.

3.2. Naive Bayes (NB)

NB stands as a probabilistic classification algorithm rooted in Bayes' theorem, assuming feature independence. Despite its "naive" assumption, it calculates the probabilities of a data point's class association, opting for the class exhibiting the highest probability. It stated mathematically as following:

$$p\left(\frac{A}{B}\right) = \{P\,P(A)\}\{P(B)\}$$

where A and B are events and $P(B) \neq 0$

- P(A) represents the prior probability of A
- P(B) denotes the marginal probability
- P(A|B) signifies the posterior probability of A given B

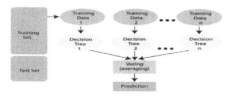

Figure 1: Random Forest Architecture

3.3. Random Forest (RF)

RF stands as an ensemble learning technique recognised for its proficiency in both classification and regression tasks. Consisting of multiple decision trees as shown in below Figure 1, RF functions by creating numerous trees and consolidating their predictions to generate a conclusive result.

3.4. Sequential Minimal Optimisation (SMO)

SMO emerges as a renowned algorithm tailored for SVMs training. Its strategy involves optimizing SVMs by fragmenting extensive optimisation challenges into more manageable, smaller sub-problems. Recognised for its efficacy, particularly in managing substantial datasets, SMO has garnered popularity for its prowess in training SVMs, especially in contexts featuring high-dimensional data and the graphical classification present in Figure 2.

4. Methodology

4.1. Flow of Work

The detail work structure of our research is presented bellow in Figure 3.

4.2. Voting Model

Voting, a core ensemble technique, aggregates predictions from several models to formulate a conclusive decision. Within this method, each model contributes its prediction and the ensemble output is determined by the most frequent prediction or through aggregation, such as averaging probabilities or outputs.

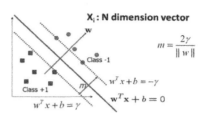

Figure 2: Sequential Minimal Optimisation graphical classification

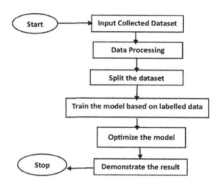

Figure 3: Flow work of proposed model

Figure 4: Graphical presentation of Voting Model Accuracy

For this research we have chosen three different machine learning models like NB, RF and SMO and train these models and found the accuracy. Here NB gives accuracy of 68.23%, RF gives 71.34%, SMO shows the accuracy of 98.71%.

The Graphical representation of accuracy shown in below Figure 4.

4.3. Data preparation

We've processed a dataset comprising 1320 records, featuring vital parameters such as age, chest pain type, gender, blood pressure, cholesterol levels, and more. Employing meticulous cleaning methods, including mean, mode, and median techniques, we ensured data accuracy. Subsequently, we partitioned the dataset into three segments: 70% for training, 20% for testing, and 10% for validation. This stratification enables robust model development, evaluation, and validation, essential for reliable predictions and insights.

5. Comparison Analysis

After successfully clean the data set, we split our whole dataset into 3 equal sets and train our taken models and store the Accuracy, MAE, RMSE and RAE value in below table I, II, III and IV, the graphical representation of tables is shown in Figures 5, 6, 7 and 8 which are given below.

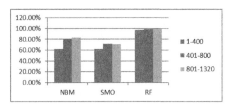

Figure 5: Graphical presentation of accuracy

Table 2: MAE Value for Voting Model Of NBM, SVM, SMO

Dataset	NBM	SMO	RF
1-400	0.3722	0.381	0.0641
401-800	0.1866	0.2882	0.0308
801-1320	0.1801	0.2948	0.025

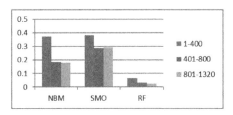

Figure 6: Graphical presented of MAE values

Table 3: RMSE value For Voting Model of NBM, SVM, SMO

Dataset	NBM	SMO	RF
1-400	0.5854	0.6172	0.1657
401-800	0.4074	0.5369	0.099
801-1320	0.3981	0.543	0.0822

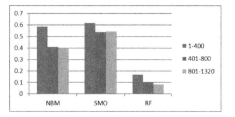

Figure 7: graphical presentation of RMSE vale

Table 4: RAE value of Voting Model NBM, SVM and SMO

Dataset	NBM	SMO	RF
1-400	77.3008%	79.1185%	13.3148%
401-800	39.7557%	61.4061%	6.5514%
801-1320	38.1514%	62.4511%	5.3022%

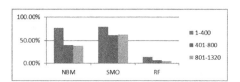

Figure 8: Visual representation of RAE values

6. Conclusion

In summary, our study on ML Models for heart disease classification reveals RF's exceptional accuracy of 98.71%, signalling its potential for robust risk identification. Despite lower accuracies, Naive Bayes (68.23%) and Sequential Minimal Optimisation (71.34%) offer contextual value. This research underscores the importance of precise predictive models in preventive healthcare, advocating for Random Forest adoption.

Future directions include feature enrichment, algorithmic fine-tuning, and real-world validation to enhance predictive accuracy and applicability, promising advancements in proactive healthcare interventions.

References

Galleta, D. T. and Carpio, J. T. (2019). "Predicting regional development competitiveness index using naive bayes: basis for recommender system." 2019 IEEE 4th International Conference on Computer and Communication Systems (ICCCS), Singapore, pp. 115-119, doi: 10.1109/CCOMS.2019.8821666.

Roshini, T., Sireesha, P. V., Parasa, D., and Bano, S.. (2019) "Social media survey using decision tree and naive bayes classification," 2019 2nd International Conference on Intelligent Communication and Computational Techniques (ICCT), Jaipur, India, pp. 265-270, doi: 10.1109/ICCT46177.2019.8969058.

Tahia Tazin, et al. (2021). "Stroke disease detection and prediction using robust learning approaches". Journal of Health Care Engineering 2021 p. 7633381. ISSN:2040-2295. DOI: 10.1155/2021/7633381.

Francky Fouedjio. (2020). "Exact conditioning of regression random forest for spatial prediction, artificial intelligence in geosciences, volume 1, pp. 11-23, ISSN 2666-5441, https://doi.org/10.1016/j.aiig.2021.01.001.

Wei, L. (2023). "Genetic algorithm optimization of concrete frame structure based on improved random forest." 2023 International Conference on Electronics and Devices, Computational Science (ICEDCS), Marseille, France, pp. 249-253, doi: 10.1109/ICEDCS60513.2023.00051.

U. N. A and Dharmarajan, K. (2022). "Diabetes prediction using random forest classifier with different wrapper methods." 2022 International Conference on Edge Computing and Applications (ICECAA), Tamilnadu, India, pp. 1705-1710, doi: 10.1109/ICECAA55415.2022.9936172.

Dritsas, Elias, and Maria Trigka. (2022). "Stroke risk prediction with machine learning techniques." Sensors, 22 (13): 4670. https://doi.org/10.3390/s22134670

Azam, Md Shafiul, Md Habibullah, and Humayan Kabir Rana. (2020). "Performance analysis of various machine learning approaches in stroke prediction." International Journal of Computer Applications, 175 (21): 11-15.

Mishra, I., Mohapatra, S. (2023). "An enhanced approach for analyzing the performance of heart stroke prediction with machine learning techniques." International Journal of Information Technology, 15: 3257–3270. https://doi.org/10.1007/s41870-023-01321-8.

Hung, Ling-Chien, Sheng-Feng Sung, and Ya-Han Hu. (2020) "A machine learning approach to predicting readmission or mortality in patients hospitalized for stroke or transient ischemic attack." Applied Sciences, 10 (18): 6337. https://doi.org/10.3390/app10186337

Swain, S., Behera, N., Swain, A. K., Jayasingh, S. K., Patra, K.J., Pattanayak, B. K., Mohanty, M. N., Naik, K. D., Rout, S. S. (2023). "Application of IoT framework for prediction of heart disease using machine learning." International Journal of Computers, Communications & Control, 11 (10s): 168–176. https://ijritcc.org/index.php/ijritcc/article/view/7616.

Design and Implementation of Viterbi Decoder Using Verilog HDL

Khirod Kumar Ghadai[1], Subrat Kumar Nayak[1], Biswa Ranjan Senapati[1], Kumar Janardan Patra[2], Satyaprakash Swain[2], Pintu Naik[3]

[1]Department of CSE, SOA (Deemed to be University) Bhubaneswar, India
[2]Department of CSE, Institute of Management and Information Technology, Cuttack, BPUT, India
[3]Department of Computer Science and Engineering, IIT, Jammu, India
E-mail: khirod1990@gmail.com,subratnayak@soa.ac.in,biswaranjansenapati@soa. ac.in,Janardanpatra1997@gmail.com, satyaimit@gmail.com, pintunaik300@gmail.com

Abstract

For reliable data transfer in noisy situations, Viterbi decoding is essential. We concentrate on putting into practice a Spartan-6 FPGA-based Viterbi Decoder with a constraint length of 3 and a coding rate of ½. Using Verilog HDL, we design the decoder and execute it on the FPGA board. This practical implementation demonstrates the effectiveness of Convolutional codes in error detection and correction. We utilise Xilinx ISE 14.7 and Model Sim for synthesis and simulation, respectively, ensuring accurate performance evaluation. The hardware-centric approach suits real-time applications, optimizing efficiency in noisy communication channels. This work underscores the significance of FPGA technology in deploying error-correcting codes, particularly in scenarios influenced by additive white Gaussian noise.

Keywords: Convolution encoder, HDL, viterbi decoder, AWGN, FPGA

1. Introduction

In modern digital communication systems, noise in channels can lead to errors in input-output sequences, necessitating a low error probability, ideally $\leq 10(-6)$, for reliable communication. Channel coding adds redundancy to enhance reliability, while source coding reduces redundancy to boost efficiency. Channel coding involves mapping incoming data sequences to channel input sequences and inverse mapping them back to output data sequences to minimise the impact of noise. Convolutional Encoding, a common technique, addresses issues like quantisation distortions, channel fading, and AWGN in wireless digital communication systems.

The transmitter, aided by an encoder, handles mapping, while the decoder in the receiver performs inverse mapping. Convolutional codes require decoding to retrieve original information, with options including the Fano algorithm or the Viterbi Algorithm. The Viterbi Algorithm excels in forward error detection and correction, despite being resource-intensive for shorter constraint lengths (k < 10). The presented digital communication system employs Convolutional Encoder and Viterbi Decoder at the transmitting and receiving ends respectively.

Chapter 42 DOI: 10.1201/9781003581215

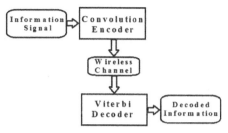

Figure 1: Digital communiqué structure with Convolution Encoder and Viterbi Decoder

The digital communication system comprises encoding the input signal and decoding the received signal. This paper outlines the plan & execution of a Viterbi Decoder, offering insights into its architecture and FPGA board implementation. Figure 1 delivers a pictorial demonstration of the system's structure, illustrating the role of the encoder and decoder in ensuring reliable communication amidst channel noise.

2. Literature Review

Schuh F. et al. (2013) propose an extension of matched decoding for PAM-transmission of convolutionally encoded signals over ISI-channels. Their optimised super-trellis representation reduces states while maintaining Euclidean distance, enhancing decoding efficiency. The method is further extended to punctured convolutional codes, introducing a time-variant, non-linear trellis description to improve decoding efficiency.Li X. et al. (2019) tackle beam alignment issues in millimetre-wave (mm-Wave) communications by leveraging the sparse scattering nature of channels. They introduce the GF-Code technique, framing beam alignment as a sparse encoding and phase-less decoding problem. Simulations confirm the efficiency of the proposed method in retrieving sparse signals from compressive phaseless data.Zhou Q. et al. (2020) present a novel cross-modal dual deep generative model (CDDG) for neural encoding and decoding of retinal ganglion cells (RGCs)

in brain-machine interfaces. Tested on salamander RGC spike datasets, the model outperforms other encoding models by effectively addressing nonlinear circuit problems and seamlessly integrating encoding and decoding processes. Rowshon M. et al. (2021) introduce polarisation-adjusted convolutional (PAC) codes, combining a one-to-one convolutional transform with the polar transform to enhance error correction. Integration of list and sequential decoding techniques, along with complexity-reducing strategies, improves performance and reduces decoding complexity compared to conventional methods.In decoding, Hu X. et al. (2021) propose three improved bit-flipping techniques for polar codes. These techniques enhance decoding precision without significantly increasing complexity, achieving better performance over conventional methods.Zhang W. et al. (2023) present a low-latency approach for bit flipping in successive cancellation list decoding of polar codes, demonstrating significant improvement over existing SCL bit-flipping techniques in simulations.Lastly, Yang D. et al. (2022) analyze error propagation in successive cancellation decoding of polar codes, shedding light on the impact of initial incorrect information bits on subsequent ones and how channel polarisation affects bit-channels in polar code decoding. Wieckowski A. et al. (2021) compare the Fraunhofer VVenC decoder to official test models, showing substantial bitrate savings with shorter runtimes, offering quality and speed trade-offs in encoding.

3. Research Design
3.1. Convolutional Encoder

In a digital communication system, Convolutional Codes play a crucial role as error-correcting codes, distinguishing themselves from block codes. These factors denoted as (n, k, m):

n: output bits k: input bits

m: memory registers

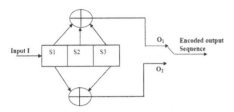

Figure 2: Convolutional Encoder with Code rate ½ and Constraint Length 3

Code rate, represented by k/n, serves as a metric for code efficiency, ensuring that there are either more or equal to the number of input bits in the output. The parameters k and n vary between 1 and 8, while m ranges from 2 to 10.A convolutional code's constraint length (K) establishes how many shifts a single message bit can affect the encoder output and is stated as K = (m+1). Convolutional Codes, also known as Trellis codes, typically employ shift registers comprising flip-flops and Modulo-2 adders.

The block diagram depicts a ½ rate Convolutional Encoder, as illustrated in Figure 2.

3.2. Basic Architecture of Viterbi Decoder

When a bitstream has been encoded using a Convolution Encoder, a Viterbi Decoder uses the Viterbi Algorithm to decode it. Below the essential modules within a Viterbi Decoder.

3.3. Branch Metric Unit

The Branch Metric Unit is the first module of the Viterbi decoder, as seen in Figure 3.

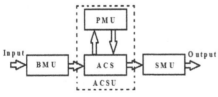

Figure 3: Sectional Layout of the Viterbi Decoder

Here, the ideal encoder outputs from the transmitter are compared to the received data bits to calculate the branch metric. When calculating branch metrics, the Hamming distance is utilised. An analog signal is received at the decoder side.

4. Methodology

4.1. ACS Unit

The ACS Unit is the second module of the Viterbi Decoder. This ACSU consist of two sub-modules, one is a Path Metric Unit (PMU) and the other one is an Add Compare Select (ACS).

4.2. Flow of Work

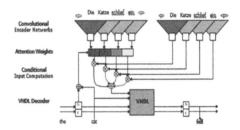

Figure 4. Work flow of proposed Encoder and Decoder models

4.3. Path Metric Unit

The PMU determines the path metrics of a given trellis stage.

4.4. Add Compare Select

By supplementing partial path metrics with branch metrics, the Add Compare Select approach computes state metrics. It compares two incoming states based on their metrics, selecting the state with the highest value. Branch metrics, derived from XORing expected and received code symbols "00" and "11," contribute to state metric updates. This process determines optimal state selection in Viterbi decoding, facilitating efficient error detection and correction in Convolutional Codes.

4.5. Survivor Memory Management Unit

As the last part, the SM unit is in charge of keeping track of the choices the ACS unit made and using those choices to calculate the output that has been decoded. Path history management in the SMU employs two primary techniques: the Trace-Back and Register-Exchange approaches.

5. Empirical results

5.1. Implementation of the Viterbi Decoder using Spartan-6 FPGA board

The design of the Viterbi Decoder is performed using Verilog Code. It is

Table 1: Testing value of the proposed model

ViterbiDecoder	
BELS	99
LUT2	5
LUT3	8
LUT4	21
LUT5	19
LUT6	41
MUXF7	5
Flipflops/Latches	3
FDR	3
ClockBuffers	1
BUFGP	1
I/O Buffers	13
I Buff	10
O Buff	3

synthesised and simulated using Xilinx 14.7 and implemented in Spartan-6 FPGA board. The synthesis report shows that 94 slice LUTs have been used in this design. Also, 94 slice registers and 13 I/O buffers used.Modelsim software is used for the simulation operations, while ISE foundation from Xilinx is used for all synthesis and routing. After implementation we test our model using primitive and black box usages and store the output value in below

When we apply VHDL model at time of decoding it takes 0.002ms / step to decode the encoded inputs. The RTL schematic

Figure 5: RTL schematic of Viterbi Decoder

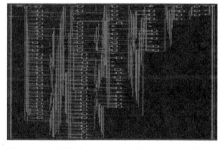

Figure 6: Technology schematic of the Viterbi Decoder

Table 2: Utilisation structural Model of Viterbi decoder Data

Inputs	Totalinputsise	Usedsise	Usedin%
SliceLUTs	9112	94	1
LUT FlipFlop	94	94	100
BondedIOB	232	14	6
BUFG/BUFGCTRs	16	1	6

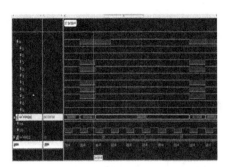

Figure 7: Simulation Results of Convolutional Encoder Input to the encoder is 010. Output of the encoder is 00 11 10 11 00

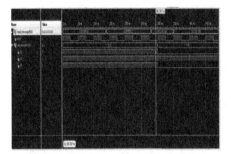

Figure 8: Simulation results of Viterbi Decoder with error detection & Correction, Input to the decoder is 00 11 10 11 00/10 11 10 11 00 Output of the decoder is 010

and the Technology schematic are shown in Figures 5 and 6 respectively. Similarly Figures 7 and 8 shows the output of the encoder and decoder.

Here, it has been observed that the Viterbi Decoder decodes all the encoded received signals correctly. Even, with a one-bit error in the received encoded signal, the Viterbi Decoder provides correct information signal (decoded output).

6. Conclusion

This paper presents a Viterbi decoder tailored for a Convolutional Code with a ½ rate and constraint length of 3. Key components include the BMU, ACSU, & SMU. Designed in Verilog HDL, synthesis, and simulation with Xilinx 14.7 tools precede successful implementation on a Spartan-6 FPGA Board. Utilizing the (111,101) generator polynomial, the ½ rate Convolutional Encoder encodes the information sequence, transmitted over AWGN channel. Upon reception, the Viterbi Decoder effectively operates with low device utilisation, ensuring error detection and correction. Future work may explore heightened circuit complexity with longer constraint lengths.

References

Fazeli, A., Vardy, A., and Yao, H. (2021). "List decoding of polar codes: how large should the list be to achieve ML decoding?" *2021 IEEE International Symposium on Information Theory (ISIT)*, Melbourne, Australia, pp. 1594-1599, doi: 10.1109/ISIT45174.2021.9517940.

Li, X., Fang, J., Duan, H., and Li, H. (2019). "A sparse encoding and phaseless decoding approach for fast mmwave beam alignment." *ICASSP 2019 - 2019 IEEE International Conference on Acoustics, Speech and Signal Processing (ICASSP)*, Brighton, UK, pp. 4714-4718, doi: 10.1109/ICASSP.2019.8682482.

Rowshan, M., Burg, A., and Viterbo, E., (2021). "Polarization-adjusted convolutional (PAC) codes: sequential decoding vs list decoding." In *IEEE Transactions on Vehicular Technology*, 70 (2): 1434- 1447. doi: 10.1109/TVT.2021.3052550.

Schuh, F., Schenk, A., and Huber, J. B. (2013). "Matched decoding for punctured convolutional encoded transmission over ISI-Channels," SCC 2013; *9th International ITG Conference on Systems, Communication and Coding, Munich, Germany*, pp. 1-5.

Zhou, Q., Du, C., Li, D., Wang, H., Liu, J. K., and He, H. (2020). "Simultaneous neural spike encoding and decoding based on cross-modal dual deep generative model." *2020 International Joint Conference on Neural Networks (IJCNN)*, Glasgow, UK, pp. 1-8, doi: 10.1109/IJCNN48605.2020.9207466.

Enhancing Environmental Impact Assessments for Sustainable Development

A Machine Learning Approach

Suprava Ranjan Laha[1], Debasish Swapnesh Kumar Nayak[1], Chinmayee Senapati[2], Satyaprakash Swain[1], Anshuman Samal[3], Binod Kumar Pattanayak[1], Saumendra Pattnaik[1]

[1]Department of Computer Science and Engineering, FET-ITER, Siksha 'O' Anusandhan (Deemed to be University), Bhubaneswar, India
[2]Department of Civil Engineering, FET-ITER, Siksha 'O' Anusandhan (Deemed to be University), Bhubaneswar, India
[3]Department of Computer Science and Engineering, Institute of Management and Information Technology, Cuttack, BPUT, India
E-mail: supravalaha@gmail.com, swapnesh.nayak@gmail.com, satyaimit@gmail.com, anshumansamal3@gmail.com, chinmayees3752@gmail.com, binodpattanayak@soa.ac.in, saumendrapattnaik@soa.ac.in

Abstract

The need for sustainable development and growing concerns about environmental degradation has led to a significant change in the approaches taken when conducting Environmental Impact Assessments (EIAs). The objective of this research paper is to improve the accuracy, efficiency, and overall effectiveness of Environmental Impact Assessments (EIAs) in promoting sustainable development by applying Machine Learning (ML) methodologies. This study aims to justify the advantages and difficulties of incorporating machine learning into the Environmental Impact Assessment (EIA) process by means of an extensive analysis of pertinent scholarly literature, and developing trends. The subtopics that have been identified consist of data pre-processing, decision support systems, predictive modelling, and ethical issues that arise when technology and environmental conservation come together. To sum up, this research proposes that environmental science and sophisticated machine learning techniques work together to create a future that is more resilient and sustainable.

Keywords: Sustainable development, ML, DSS, predictive modelling

1. Introduction

In the face of unprecedented environmental challenges, conventional methods for assessing impacts undergo scrutiny due to biodiversity loss and climate change (M. et al. (2023)). While Environmental Impact Assessments (EIAS) traditionally evaluate proposed policies and actions, their reliance on outdated knowledge and simplistic designs is increasingly conspicuous in dynamic, interconnected environmental systems (T. Hundloe (2021)). Machine learning, which can detect complicated patterns from massive datasets and create comprehensive projections, might

Chapter 43 DOI: 10.1201/9781003581215

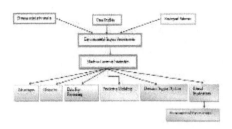

Figure 1: ML for environmental impact

revolutionise environmental impact assessments. Objective of this study is:

- Examine methods for handling environmental data pre-processing.
- Investigate methods for environmental impact assessment that use predictive modelling.
- Assess decision support tools designed to promote sustainable development.
- Examine moral issues that are relevant to environmental impact assessments using machine learning.
- Evaluate stakeholder engagement strategies and interdisciplinary collaboration tactics.

This research examines ML's role in improving EIAs' accuracy, efficacy, and depth (Figure 1), merging ML techniques with environmental science for complex EIA enhancement. It explores the intricate balance between tech progress and ecological preservation (T. Hundloe (2021), L. Goparaj (2015)).

2. Methodology

This work improves Environmental Impact Assessments (EIAs) for sustainable development using Machine Learning (ML). We identify trends by examining relevant scientific literature and case studies, emphasising the benefits and drawbacks of incorporating ML models into Environmental Impact Assessments (EIAs). The study highlights how, in order to promote a resilient and sustainable future, environmental science and cutting-edge machine learning algorithms must work together.

3. Preparing the Data

Environmental data, which comes from many sources and fluctuates spatially and temporally, must be pre-processed to ensure reliability and relevance. This section discusses the problematic processes of preparing environmental data for machine learning applications and the unique challenges. Environmental datasets often have outliers, missing values, and regional and temporal variations (A. M. Denton and A. Roy(2017)). By preparing environmental data, ML can precisely decipher complex environmental patterns, enabling more accurate forecasts and sustainable development decisions. Figure 2 shows a time series graph of environmental data for Bhubaneswar, Cuttack, and Puri, India. The graph shows that environmental data values for all three cities fluctuate with time, but they are not constant. Bhubaneswar's environmental data rose from 2010 to 2012, whereas Cuttack's fell.

4. Predictive Modelling for Environmental Impact Assessment

Predictive modelling uses statistical representations of past and current data to predict future environmental patterns. Regression analysis is a crucial EIAS predictive modeling tool. The relationship between environmental consequences and independent variables like pollution, land use, and climate is analysed via regression equations. These analytical equations estimate impact magnitude and trajectory by

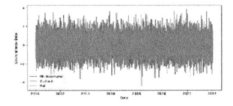

Figure 2: Temporal Fluctuations of Odisha Environmental Data

modifying independent parameters (Laha et al., S. (2022)). Equation 1 shows the regression equation.

Where:
- Y is the dependent variable.
- β_0 is the intercept.
- β_1, β_2,.., βn are the coefficients representing the impact of independent variables $X_1, X_2,.., Xn$
- € is the error term.

Predictive modelling increases impact evaluations and lets decision-makers evaluate scenarios to measure the prospective consequences of different actions. Here, the predictive modelling paradigm becomes essential for understanding complex environmental connections and promoting equitable growth through informed decision-making.

5. Decision Support Systems (DSS) for Sustainable Development

Using ML analytics, DSS becomes a dynamic framework that processes complicated environmental data to provide sustainable development knowledge (T. et al., Decision Support Systems for Sustainable Development). A key component of probabilistic modelling, Bayesian networks provides a probabilistic inference method for decision-making by efficiently capturing correlations between different elements. Equation 2 is a statistical method that helps decision-makers assess environmental risks and build predictive models.

Where
- P (Environmental Impact | Variables) : posterior probability
- P(Variables | Environmental Impact)" indicates how likely it is that the variables will be as observed considering the impact.
- The term "P (Environmental Impact)" denotes the previous algorithmic intervention that was created to reduce bias in the impact's probability.

- The probability of the variables that were observed is indicated by "P(Variables)".

Random Forest ensemble learning improves DSS by consolidating ML model predictions. To improve prediction accuracy and control environmental system complexity, this group approach employs several views.

6. Ethical Considerations for Environmental Impact Assessments

Integrating ML with Environmental Impact Assessments (EIAs) poses moral concerns that must be addressed appropriately to guarantee the implementation of ethical and just environmental science technology. ML algorithms that exacerbate environmental injustice are ethical concerns (F. Yilmaz-2021). DIR reduces bias by correcting model forecasts for integrity.

$$P(Y=1|S=s)=P(Y=1) \tag{3}$$

Here in equation 3:
- $P(Y=1| S=s)$ Positive outcome given a sensitive attribute s.
- $P(Y= 1)$: Positive outcome.

Black-box deep neural network models may confuse users about decision-making. Solution: employ interpretable models like decision trees and model-independent interpretability methods.

Environmental ML applications raise data privacy problems since sensitive environmental data may identify groups or persons. Model training with decentralised data is possible while preserving data privacy thanks to federated learning and other privacy-preserving technologies.

A thorough investigation of the moral implications of ML-assisted EIAs necessitates the creation of moral standards, continuing discussion, and cooperation amongst multidisciplinary teams.

7. Result & Discussion

Environmental Impact Assessments (EIAs) can benefit greatly from machine learning (ML), which can help with pre-processing data, predictive modelling, decision support systems, and ethical issues. Advanced data preparation techniques refine environmental datasets, handling complexities like missing values and outliers. ML-driven data preparation enhances model building by providing cleaner datasets. Predictive modeling, employing regression and classification methods, offers more accurate forecasts of environmental impacts. Probabilistic modeling, such as Bayesian networks, improves prediction in decision support systems. Ethical machine learning promotes responsible technology and stakeholder involvement, prioritising openness, privacy, and inclusivity. Collaborative efforts and stakeholder participation enrich ML-supported EIAs, integrating local insights and human-centric design. Continuous dialogue, education, and adaptive approaches are vital for unleashing Machine Learning's potential in environmental assessments and sustainable development.

8. Conclusion

ML in Environmental Impact Assessments (EIAs) is revolutionary for sustainable development. This study delves into the intricate facets of environmental research and technology, including data pre-processing, predictive modeling, and ethical considerations. Leveraging machine learning enhances EIAs, fostering inclusive decision-making and stakeholder engagement while upholding privacy and impartiality. Ethical technology alignment and interdisciplinary collaboration are pivotal for sustainable progress.

References

Fattahi, M. and Sharbatdar, M. (2023). "Machine- learning-based personal thermal comfort modeling heat recovery using environmental parameters." *Sustainable Energy Technologies and Assessments*, 57: 103294.

Hundloe, T. (2021), "Linking eia to the principles of sustainable development." *Environmental Impact Assessment*, pp. 31-48, 2021.

Goparaju, L. (2015), "Geospatial Technology in environmental impact assessments retrospective." *Present Environment Sustainable Development*, 9 (2): 139-148.

Denton, A. M. and Roy, A. (2017). "Cluster-overlap algorithm for assessing pre-processing choices in Environmental Sustainability." *2017 IEEE International Conference on Big Data (Big Data)*.

International Research Journal of Modernization in Engineering Technology and Science. (2023). "Optimizing data pre-processing: the data pre-processing interface." T. X. Bui. Decision Support Systems for Sustainable Development, pp. 1-10.doi: 10.1007/0-306-47542-1_1

Yilmaz, F. (2021). "Performance and environmental impact assessment of a geo-thermal-assisted combined plant for multi-generation products." *Sustainable Energy Technologies Assessments*, 46: 101291.

Laha, S. R., Pattanayak, B. K., & Pattnaik, S. (2022). "Advancement of environmental monitoring system using IoT and sensor: a comprehensive analysis. *AIMS Environmental Science*, 9 (6): 771-800.

Laha, S. R., Pattanayak, B. K.., Pattnaik, S., and Kumar, S. (2023). "A smart waste management system framework using IoT and LoRa for Green City Project." *IJRITCC*, 11 (9): 342–357.

Unveiling Deception

Safeguarding Supply Chain IoT from Data Fraudulence with Context-Aware Deep Learning

Susovan Chanda[1], Surjya Ghosh[2], Indranil Sengupta[3], D. Sinha Roy[1]

[1]Department of Computer Science & Engineering, NIT Meghalaya, Shillong, India
[2]Department of Computer Science & Information Systems, BITS Pilani, Goa, India
[3]Department of Computer Science & Engineering, IIT Kharagpur, India

Abstract

The increasing integration of Internet of Things (IoT) devices in the supply chain has enhanced visibility and efficiency for businesses. However, the rise of false alarms and data falsification presents significant security risks, particularly in critical sectors like the healthcare vaccine supply chain, where such issues can result in catastrophic consequences, including loss of human lives. To address this problem, this research introduces an innovative context-based approach to accurately and efficiently detect data falsification. The approach employs a hierarchical clustering algorithm to define the context of an object, converts this context information into representational vectors using an auto-encoder, and utilises an LSTM-based multi-layered deep learning technique for detection. The proposed method has been validated using two large datasets.

Keywords: IoT, anomaly, neural network, data falsification, supply chain, big data analytics

1. Introduction

In today's interconnected world, the Internet of Things (IoT) has transformed the supply chain industry, improving operational efficiency and enabling real-time tracking of goods. However, this digital shift has also introduced significant security challenges, making it essential to safeguard IoT devices to ensure the integrity of goods, data, and customer trust (Shashank Kumar *et al.*, 2022). With the IoT adoption rate in the supply chain industry soaring, it is projected that there will be 30 billion connected devices by 2025. This rapid growth has expanded the attack surface for cybercriminals.

Data falsification has become a serious threat in the IoT landscape of the supply chain, leading to various severe consequences such as false alarm generation, loss of human lives, economic loss, denial of services, and false data injection attacks (Lei Cui, 2022). For instance, in the healthcare sector, falsified data has resulted in a $3 billion loss (Madhav Chinta *et al.*), and the unavailability of COVID-19 vaccines has caused 2 million deaths worldwide (Bowen Wang *et al.*, 2020). In the utility sector, falsified data has led to approximately $6 billion in losses (S. Bhattacharjee and S. K. Das, 2018).

IoT devices generate a vast volume of data, resulting in numerous data communications and transmissions (Honghao Gao *et al.*, 2022). The primary challenge lies in distinguishing between genuine and falsified

Chaper 44 DOI: 10.1201/9781003581215

readings (Jun Liu *et al.*, 2022). For example, differentiating between a legitimate fire alarm triggered by an actual fire and one manipulated by an adversary is complex. However, every incident is somehow related to others, and using context can help detect falsified data (Baibhab Chatterjee, 2021).

This paper proposes a context-based solution for detecting data falsification. The approach involves two steps. First, it defines the context for each event by considering all IoT devices in the environment and extracting contextual information. A hierarchical clustering algorithm is used to define the context. Second, the context information is fed into a sequence encoder created using an autoencoder, which automatically learns the representation from the context (Marco Maggipinto *et al.*, 2022). Finally, a multilayered LSTM-based deep neural network is used to detect data falsification. Experimental results indicate that this proposed solution achieves at least 93% accuracy in detecting data falsification.

2. Literature Review

In statistical based method solely rely on the statistical analysis of the data to detect the fraud. Most of the proposed solutions build a model using the available sample data. And then comparative analysis is done between the current data and the model to detect the fraud. Basically these solutions decides the anomaly based on the threshold limit and they assume that data generated by the devices are normally distributed (Woojin Cho *et al.*, 2020). Some well known statistical model is Gaussian distribution (Woojin Cho *et al.*, 2020), Markov chain model (Elise Epaillard, et.al. 2019) and Naive Bayes algorithm (Stephen McLaughlin *et al.*, 2013). In Naive Bayes scheme event profile is created for individual device and then this event profile is used in future to detect if the data is falsified or not. Statistical methods may be effective when the data volume is really low. It performs poorly for the large volume dataset.

In Classifier based model method, classifier is constructed from from the labeled

sample data 28. It is suitable when it is easy to get the falsified data samples. Clusters based model groups the unlabelled data into different clusters and label them. These labelled data is used for anomaly prediction. Needless to mention that clustering of unlabelled data mostly rely on the parameters such as threshold similarity, cluster size, minimum sample size etc used to define cluster(Xiaoxue Ma *et al.*, 2023). Moreover cluster based algorithm merely label the unlabelled data but they don't consider the context. Due to this reason, cluster based algorithm could not detect anomaly properly (Liu Yang *et al.*, 2021).

Local density-based method used the contexts and densities of their sample distribution in the neighborhood from a local perspective to detect the data falsification (Wei Wang *et al.*, 2023). The categorisation of local density-based techniques encompasses various approaches: DBSCAN (Density-Based Spatial Clustering of Applications with Noise) oriented, neighborhood radius-driven, local anomaly factor-centered, local correlation integral-focused, and local peculiarity factor-oriented methods.

3. Proposed Method

Within the scope of this paper, we present an innovative contextual model tailored to identify the data falsification for Internet of Things (IoT). Our novel model unfolds across a two-tiered structure, where each tier serves a distinct purpose. In the initial phase, we undertake the pivotal task of elucidating the contextual parameters encompassing the data that necessitates validation. Subsequently, the second phase orchestrates a synergistic fusion of context and data to effectively ascertain the presence of falsification. Intricacies abound in our model's architectural composition. The first phase concentrates on the delineation of data context. This endeavor entails a meticulous segmentation of data streams originating from diverse IoT devices. To achieve this, we employ the hierarchical agglomerative clustering method, enabling us to categorise data points into discrete events. These events stand as distinct manifestations of data generation, each contributing unique attributes to the overall contextual landscape. The second phase takes a cohesive stride by amalgamating events from assorted devices to collectively define the context of a particular event. This harmonious fusion substantiates the overarching contextual framework against which data authenticity is gauged.

3.1. Detection of Data Falsification Framework

The structure introduced is illustrated in Figure 1. This structure comprises two distinct phases: firstly, the data produced by IoT devices is merged with their contextual information to create an input, wherein a representation is acquired; subsequently, this derived representation is fed into a Multi-layered LSTM-based Neural Network to detect instances of data falsification. The proposed Multi-layered Neural Network has been employed for the purpose of detecting data falsification, as illustrated in Figure 2. In the implementation of Multi-task Learning, a configuration of 3 shared layers and 1 event-specific layer has been utilised. These layers are initiated by taking the derived representation vectors as input. Initially, the shared layers aim to enhance learning by capitalising on commonalities across events within various historical data.

In this model, separate tasks are created for each event, effectively treating each event as an individual task. During the training process, batches of data are composed of specific events. This approach facilitates the identification of data falsification within the context of a given event, allowing prediction errors to be

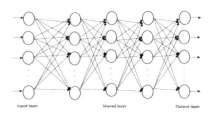

Figure 2: LSTM based multi-layered neural network

backpropagated for weight adjustments across both shared and task-specific layers. This iterative process is repeated for all events, iteratively updating event-specific weights. This methodology allows the model to develop a generalised representation applicable to all events. The incorporation of Multi-task Learning serves as a regularisation technique to mitigate the risk of overfitting.

Furthermore, to enhance robustness, dropout has been applied following the final shared layer. The activation function used is Softmax.g

4. Evaluation

In this section, we begin by outlining our experimental setup. Following this, we embark on a two-step evaluation of our proposed framework.

4.1. Experimental Setup

Each warehouse's data is considered individually and divided into a 60:40 ratio, with the initial 60% designated for model training and the remaining 40% for testing. The acquisition of representation is exclusively accomplished through the training data.

4.2. Performance of the Proposed Model

Success of the proposed model depends on two factors. First, how accurately it detects the events (context) using the data produced by the IoT devices. Second, how efficiently it detects the manipulation done on the the data produced by the IoT devices.

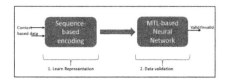

Figure 1: Proposed Framework

4.2.1. Detection of Events

Figures 3, 4 and Table 2 captured to evaluate the event detection performance of the proposed model. Fig. 3 show that any changes on the event related information made by the adversaries are successfully detected by the proposed model and it can remove them efficiently. We also evaluated the cosine similarity of the events to determine how efficiently model and learn the representation. It shows that different event pair has less similarity than the same event pair. In Figure 4 we show that event classification performance of the model. It shows that different event pair has less similarity than the same event pair. In Figure 4 we show that event classification performance of the model. It depicts all the events are identified with more than 80% AUCROC.

4.2.2. Detection of Data Manipulation

The assessment of data falsification detection performance is presented in Figure 4. Specifically, the warehouse-wise classification accuracy (aucwt) is illustrated in the same figure. Notably, it is observed that for 75% of the warehouses, the aucwt value exceeds 80%, contributing to an average AUCROC of 84% (standard deviation: 8%). The calculated 93.6% confidence interval for aucwt values falls within the range of [81.1%, 87.7%].

When comparing with baselines, the effectiveness of the proposed MTL-NN model's performance has been substantiated. Our findings demonstrate that the MTL-NN model consistently outperforms the baseline models (STL). The baseline model achieves an average aucwt of 74% with a standard deviation of 12%.

Events	e1	e2	e3	e4
e1	0.846	0.811	0.803	0.752
e2	0.811	0.921	0.844	0.868
e3	0.803	0.844	0.939	0.877
e4	0.752	0.868	0.877	0.957

Table 1: Average cosine similarity of the representation vectors obtained for every pair events.

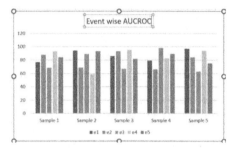

Figure 4: Event-wise AUCROC

5. Conclusion

In this research, we introduced a novel and sophisticated approach that leverages context-aware deep learning to effectively identify instances of data fraudulence within the realm of IoT devices. The salient feature of this proposed protocol resides in its exceptional accuracy, marking a significant step forward in the arena of data integrity assurance. In conclusion, our research underscores the significance of context-awareness in fortifying data security within the dynamic landscape of IoT.

References

Bhattacharjee, S. and Das, S. K. (2018). "Detection and forensics against stealthy data falsification in smart metering infrastructure." *IEEE Transactions on Dependable and Secure Computing*, pp. 1–1.

Baibhab Chatterjee, Dong-Hyun Seo, Shramana Chakraborty, Shitij Avlani,

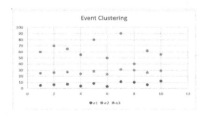

Figure 3: Event Clustering

Xiaofan Jiang, Heng Zhang, Mustafa Abdallah, Nithin Raghunathan, Charilaos Mousoulis, Ali Shakouri, Saurabh Bagchi, Dimitrios Peroulis, and Shreyas Sen. (2021). "Context-aware collaborative intelligence with spatio-temporal in-sensor-analytics for efficient communication in a large-area iot testbed." *IEEE Internet of Things Journal*, 8 (8): 6800–681.

Madhav Chinta, Abdelsalam Helal, and Choonhwa Lee. (2003). "ILC-TCP: an interlayer collaboration protocol for TCP performance improvement in mobile and wireless environments." In *2003 IEEE Wireless Communications and Networking*, WCNC 2003, New Orleans, LA, USA, 16-20 March, 2003, pages 1004–1010. IEEE.

Woojin Cho, Youngrae Kim, and Jinkyoo Park. (2020). "Hierarchical anomaly detection using a multioutput gaussian process." *IEEE Transactions on Automation Science and Engineering*, 17 (1): 261–272.

Lei Cui, Youyang Qu, Gang Xie, Deze Zeng, Ruidong Li, Shigen Shen, and Shui Yu.(2022). "Security and privacy-enhanced federated learning for anomaly detection in iot infrastructures." *IEEE Transactions on Industrial Informatics*, 18 (5): 3492–3500.

Elise Epaillard and Nizar Bouguila. (2019). "Variational bayesian learning of generalized dirichletbased hidden markov models applied to unusual events detection." *IEEE Transactions on Neural Networks and Learning Systems*, 30 (4): 1034–1047.

Honghao Gao, Binyang Qiu, Ramon J. Duran Barroso, Walayat Hussain, Yueshen Xu, and Xinheng Wang. (2022). "Tsmae: a novel anomaly detection approach for internet of things time series data using memory-augmented autoencoder." *IEEE Transactions on Network Science and Engineering*, pp. 1–1.

Shashank Kumar, Rakesh D. Raut, Pragati Priyadarshinee, Sachin Kumar Mangla, Usama Awan, and Balkrishna E. Narkhede. (2022). "The impact of iot on the performance of vaccine supply chain distribution in the covid-19 context." *IEEE Transactions on Engineering Management*, pp. 1–11.

Jun Liu, Jingpan Bai, Huahua Li, and Bo Sun. (2022). "Improved lstm-based abnormal stream data detection and correction system for internet of things." *IEEE Transactions on Industrial Informatics*, 18(2):1282–1290.

Xiaoxue Ma, Jacky Keung, Pinjia He, Yan Xiao, Xiao Yu, and Yishu Li. (2023). "A semi-supervised approach for industrial anomaly detection via self-adaptive clustering." *IEEE Transactions on Industrial Informatics*, pp. 1–12.

Marco Maggipinto, Alessandro Beghi, and Gian Antonio Susto. (2022). "A deep convolutional autoencoder-based approach for anomaly detection with industrial, non-images, 2-dimensional data: A semiconductor manufacturing case study.:" *IEEE Transactions on Automation Science and Engineering*, 19 (3): 1477–1490.

Stephen McLaughlin, Brett Holbert, Ahmed Fawaz, Robin Berthier, and Saman Zonouz. (2013). "A multi-sensor energy theft detection framework for advanced metering infrastructures." *IEEE Journal on Selected Areas in Communications*, 31 (7): 1319–1330.

Bowen Wang, Yanjing Sun, Trung Q. Duong, Long D. Nguyen, and Lajos Hanzo. (2020). "Riskaware identification of highly suspected covid-19 cases in social iot: a joint graph theory and reinforcement learning approach." *IEEE Access*, 8: 115655–115661.

Wei Wang, Jianbo Li, Ying Li, and Xueshi Dong. (2023) "Predicting activities of daily living for the coming time period in smart homes." *IEEE Transactions on Human-Machine Systems*, 53 (1): 228–238.

Liu Yang, Yinzhi Lu, Simon X. Yang, Tan Guo, and Zhifang Liang. (2021). "A secure clustering protocol with fuzzy trust evaluation and outlier detection for industrial wireless sensor networks." *IEEE Transactions on Industrial Informatics*, 17 (7): 4837–4847.

Ultrasonic Communication Using ASK Modulation and Demodulation Technique for Implanted Sensor in Wireless Body Area Network

Annapurna Sahoo[1], Leena Samantaray[2]

[1]Research scholar, BPUT, Odisha, Assistant Professor of ABIT, Cuttack, India, manini.
n07iitkgp@gmail.com
[2]ABIT, Cuttack, India, leena_sam@rediffmail.com

Abstract

Radiofrequency is the main choice to transmit data for wireless body area network devices due to its wide range of spectrum, distance, and penetration capability in different obstacles or mediums. It is versatile in supporting different communication protocols. It is also reliable, scalable, and cost-effective. Ultrasonic communication has very less adverse effects on the skin and body and whole systems are also suitable for transmission of information with short and medium-range security. Though this transducer transmits data with a short range and narrow bandwidth, different digital modulation and demodulation techniques can be implemented at both the transmitter and receiver side to achieve a good BER and signal-to-noise ratio to enhance the system's performance. In this paper, an ultrasonic transducer with amplitude shift keying (ASK) modulation and demodulation techniques for wireless communication is simulated. The sensing data is taken from a Galvanic skin response sensor implanted on a human body to the ultrasonic trans-receiver system followed by ASK modulation and demodulation process.

Keywords: Ultrasonic communication, ASK modulation and demodulation, GSR sensor, ultrasonic transducer, WBAN, NFC

1. Introduction

Nowadays most medical devices are digital and equipped with wireless transmission facilities, and the patient's health status monitoring data can be sent wirelessly to the doctor or any medical staff for further examination and necessary action Ultrasonic communication involves the transmission and reception of high-frequency sound waves beyond the human auditory range, typically above 20 kHz.

Utilised in various fields like animal communication, medical imaging, and technology, ultrasonic signals carry information through modulation, enabling data transfer in forms such as text, images, or commands. Ultrasonic communication refers to the transfer of information through high-frequency sound waves that are beyond the range of human hearing, typically above 20 kHz. for various purposes, including data transmission, object detection, medical imaging, and animal communication.

Chapter 45 DOI: 10.1201/9781003581215

Wireless modules play a significant role in enabling connectivity and data transfer in various medical sensors. In most medical sensor devices, Bluetooth Low Energy wireless technology is used in wearable health trackers for ideal short-range communication between smartphones or tablets, other WiFi for high-speed wireless communication over long distances, Zigbee, NFC (near field communication) for short distances, and also cellular communication technology for long-distance real-time data transmission to health care provider are used.

In ultrasonic communication for Wireless Body Area Networks (WBANs), various modulation techniques are employed to transmit data effectively and reliably through the human body. ASK modulation and demodulation techniques are simple and also offer the benefits of being more immune to interference, cost-effective, and easy to implement.

2. Background

2.1. Ultrasonic Communication

Ultrasonic communication in the medical sector utilises sound waves beyond the range of human hearing (above 20 kHz).

Ultrasonic communication facilitates wireless data transmission in implantable medical devices with continuous monitoring of physiological parameters, such as blood pressure, temperature, and glucose levels, transmitting vital data wirelessly to external monitoring systems.

2.3. Galvanic Skin Response (GSR) Sensor

Galvanic Skin Response (GSR) measures the changes in the electrical conductance of the skin in response to emotional arousal or stress. These sensors detect fluctuations in sweat gland activity, providing insights into an individual's psychological state. GSR sensors are commonly used in various fields like psychology, medicine, and human-computer interaction to assess emotional reactions, stress levels, and arousal. Most of the GSR sensor uses Ag/AgCl electrodes which are often placed on the surface of the skin, commonly on the fingers or palm, to detect changes in sweat gland activity. The amount of sweat glands varies across the human body but is the highest in hand and foot regions (200–600 sweat glands per (Salimpoor *et al.*, 2009). The electrodes pick up the subtle variations in the electrical properties of the skin in response to emotional arousal or stress, allowing the GSR sensor to record these physiological reactions accurately.

2.2. General Specification of GSR Sensor

Appropriate cables or connectors to link the output of the GSR sensor to the input of the modulator circuit are essential for further processing. Implantable devices must be made from those materials that can be tolerated by the human body. The transfer of energy from external to internal part with high data rate and power consumption

TABLE 1: GSR sensor specification (https://community.element14.com/members-area/b/blog/posts/ouch-sensing-galvanic-skin-response-gsr)

Size	25mm
Power interface pin	yes
Working current	1mA
Power voltage	3.3V-5V
Weight	4g
Shipment Weight	0.01kg

Figure 1: GSR sensor (https://community.element14.com/members-area/b/blog/posts/ouch-sensing-galvanic-skin-response-gsr)

TABLE 2: General Specification of Ultrasonic
Transducer EC4012 (https://www.electronicscomp.com/ultrasonic-transnitter-receiver-40khz-transducer-india)

Sensor type	Ultrasonic
Frequency	40kHz
Sound pressure level	112Db min
Operating Temperature Range	

Figure 2: EC4012 Transducer (https://www.electronicscomp.com/ultrasonic-transnitter-receiver-40khz-transducer-india)

plays a vital role in designing biomedical implantable devices. The digital modulation and demodulation techniques are important sections of the whole system design.

2.4. ASK Modulation and Demodulation

The amplitude shift keying (ASK) is the simplest digital modulation used in wireless telemetry bio-devices and is suitable for biomedical implanted devices. Here the information is digital having two states i.e. logic '0' and 'and shift carrier signal amplitude corresponds to the digital signal that will be modulated. ASK modulation technique uses analog signal as carrier signal (Pratama *et al.*, 2016).

2.6. EC4012 Ultrasonic Transducer (UT)

We can use an ultrasonic transducer as a transmitter and receiver whose center frequency is about 40kHz. We can also use an operational amplifier in transmitter UT and receiver UT circuits to amplify weak signals from sensors or other sources before transmitting them, helping boost the signal strength for better transmission over long distances or through noisy environments (Pratama *et al.*, 2016).

2.5. System Architecture of Ultrasonic WBAN Communication

The proposed method can be divided into the following sections 1. GSR sensor 2. ASK modulator and demodulator. 3. Ultrasonic trans-receiver, and a display unit as shown in Figure 3.

Data acquisition and processing section where the data can be taken from the GSR sensor attached to the patient's finger. The output of this sensor ASK modulation circuit can be connected to the input of an ultrasonic transducer as transmitter and another ultrasonic transducer as receiver and again to the ASK Demodulator circuit. We can use a display device to visualise and measure the received signal.

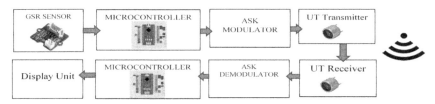

Figure 3 Block diagram of a proposed ultrasonic communication system for WBAN

2.7. ASK Modulator and Demodulator

The process of ASK modulation involves encoding digital data onto a carrier signal by varying its amplitude. A binary 0 is represented by one amplitude level (e.g., low amplitude), while a binary 1 is represented by another level (e.g., high amplitude).

Ultrasonic transducers are operated in the frequency range of 30-500kHz and use air as a transmitting medium. The low-frequency sensors operating between 30-80 kHz are suitable and effective for long-range communication in meters whereas high-frequency sensors are more effective for short-range communication.

In the proposed method ultrasonic transducer EC4012 whose center frequency is 40kHz can be used as a transmitter and receiver reciprocally (Pratama *et al.*, 2016).

The distance covered by the signal from UT Tx to UT Rx can be tested by applying a signal of 40 kHz with an amplitude 2 to 5-volt UT transmitter and receiving the signal at the UT Rx side (Pratama *et al.*, 2016).

2.9. Experimental Procedures for Stress Detection in Patients

Attach the sensor to the finger of the patient. b) GSR sensor's output connected to the microcontroller's input. c) UT is connected to the output of the MCU(1) via the ASK modulator circuit. d) Another UT is connected to collect the data to the input of MCU(2) on the receiver side via the ASK demodulator circuit. e) Finally, the receiver output can be sent to the laptop screen and the stressed or relaxed conditions of the patient can be decided (Gogate *et al.*, 2019).

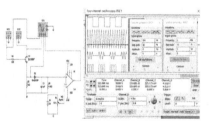

Figure 4: Circuit Diagram of ASK Modulator and Demodulator at carrier frequency 1kHz with simulation result.

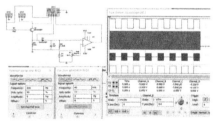

Figure 5: Circuit Diagram of ASK Modulator and Demodulator at carrier frequency 40kHz with simulation result.

2.8. Duty Cycle

The duty cycle represents the total active period as compared to the total period. It has also implications for bandwidth Utilisation, power consumption, and data transmission rate in digital communication. A higher duty cycle might allow a faster data transmission rate but is susceptible to issues like signal distortion, noise, and interference due to the longer duration of signal activity.

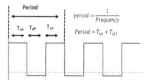

Figure 6: Duty cycle from the simulation result

3. Result Analysis and Discussion

The figure below shows both the circuit and simulation results of ASK modulation using transistor BC548BP with signal frequency 200Hz and amplitude 5V which is modulated with a carrier with frequency 1 kHz and 40 kHz with amplitude 2V using signal generator and 4 channel oscilloscope. The ASK Demodulation is carried out using OPAMP LS741-IC. The demodulated output duty cycle is less than the original baseband signal.

In the above simulation, the duty cycle of the demodulated signal compared to the original modulating signal is slightly lower indicating that the transmission path has introduced distortion. The characteristics of the communication channel might affect the signal fidelity. There may be attenuation, reflections, or any other path loss that alters the signal shape and affects the duty cycle.

4. Conclusion

In this paper, to transmit data from the implant body sensor wirelessly to the nearest receiver which is a very simple ASK modulation and Demodulation circuit designed and simulated at two different frequencies of the carrier signal. For ASK modulation both the modulating signal and carrier signal are taken from the signal generator and the output was observed in 4 channel oscilloscope using circuit simulator software.

References

Pratama, Muhammad Harry Bintang., Arif, M., Khusnil, M. and Zahra, A. (2016). "Department of Electrical Engineering Diponegoro University Semarang, Indonesia." Implementation of Ultrasonic Communication for Wireless Body Area Network Using Amplitude Shift Keying Modulation" *IEEE Conference*, pp. 3790-3793.

Salimpoor, V.N., Benovoy, M., Longo, G., Cooperstock, J. R., and Zatorre, R. J. (2009). "The rewarding aspects of music listening are related to the degree of emotional arousal." *PLoS ONE*, 4: e7487.

Mahammad A. Hannan, Saad, M., Abbas, Salina A., Samad and Aini Hussain. (2012). "Modulation techniques for biomedical implanted devices and their challenges." *Sensors*, 12: 297-319.

Gogate, Utara., Bakal, Jagadish. (2019), "Hunger and stress monitoring system using galvanic skin Response." *Indonesian Journal of Electrical Engineering and Computer Science*, 13: 81-85.

Escudero, J. and Rai, G. S. (2011). "An Investigation into Ultrasonic Communication for Near-Body Networks".

Jiang, Wentao and William, M. D. Wright. (2015). "Multi-channel ultrasonic data communications in air using range-dependent modulation schemes". *IEEE Transaction*.

Harmony in Harvest: Machine Learning's Symphony for Sustainable Agriculture

Srutipragyan Swain, Dibyalaxmi Priyadarshini, Laxmipriya Mallik, Suvendra Kumar Jayasing, Sujata Ray, Kumar Janardan Patra

Department of Computer Science and Engineering, Institute of Management and Information Technology, Cuttack, BPUT, India
E-mail: sruti56@gmail.com, dibyalaxmi1990@gmail.com, laxmipriya.mallik94@gmail.com, sjayasingh@gmail.com, sujata1.ray@gmail.com, Janardanpatra1997@gmail.com

Abstract

Global population increase and food scarcity are the two biggest obstacles to sustainable development. machine learning revolutionises agriculture by predicting optimal crop types. Utilizing a dataset of 2200 observations on agricultural parameters, including soil nutrients and environmental factors, four machine learning models were evaluated. Results show Random Forest (91.23%), Light Gradient Boosting Machine (92%), and Naive Bayes (96.2%) as top performers, highlighting their ability to comprehend complex relationships. In contrast, Decision Stump exhibited lower accuracy (36.36%). This underscores the efficacy of ensemble methods in predicting crop suitability. By leveraging machine learning, farmers can optimise productivity and resource management, ensuring sustainable agricultural practices for enhanced profitability and food security.

Keywords: Smart farming, machine learning, crop prediction, soil analysis

1. Introduction

In the face of global challenges like swelling populations and food scarcity, machine learning emerges as a transformative force in agriculture. With its ability to distill insights from vast datasets, it promises to redefine how we cultivate sustenance, ushering in a future where abundant harvests thrive in harmony with nature's rhythms. Equipped with predictive models, farmers become active participants in orchestrating nature's design, sowing seeds with confidence in optimal conditions.

At the heart of this transformation lies a meticulous dataset of 2200 observations, capturing the intricate dance between soil, environment, and crop growth. Machine learning approaches such as first one is Random Forest, second one is Light Gradient Boosting Machine, third one is Naive Bayes, and last is Decision Stump unravel this cryptic code, revealing the harmony between agricultural variables. Results demonstrate the power of complex methods, with Naive Bayes shining at 96.2% accuracy, while simpler models lag behind.

Beyond statistical achievements, this endeavor holds transformative potential for agriculture. Empowered by predictive prowess, farmers can tailor crop selection to their specific landscapes, maximizing yields while conserving resources. This integration of machine learning into farming practices promises not only improved crop yields but also fosters environmental sustainability, paving the way for a future of bountiful harvests and agricultural harmony.

2. Existing Work

Mohamed et al. (2021) introduces a smart farming model integrating IoT sensors for real-time monitoring of agricultural parameters, facilitating functions like harvesting and pest control. Javaid et al. (2022) advocate Agriculture 4.0, stressing the importance of rural communication infrastructure for effective IoT utilisation in agriculture. Verdouw et al. (2021) propose Digital Twins for remote farm management, outlining their integration with AI/ML for enhanced decision-making across farming sectors. Jerhamre et al. (2022) explore AI implementation challenges and opportunities in agriculture, emphasizing factors like data ownership and cybersecurity. Durai et al. (2022) propose a farmer-centric model aiding crop selection and cost estimation, enhancing profitability. Alfred et al. (2021) highlight Big Data, ML, and IoT's transformative role in rice production, proposing a framework for precision agriculture. AlZubi et al. (2023) advocate for AI-IoT integration in agriculture for improved farming practices, addressing challenges like IoT deployment and data sharing. Jayasingh et al. (2022) discuss innovative approaches to data classification and weather prediction using diverse ML techniques, with RF outperforming other models. Swain et al. (2023) explore machine learning for heart disease prognosis, advocating for advanced technologies and IoT.

3. Research Design

3.1 Ensemble Learning

Ensemble learning serves as a collaborative approach where multiple models unite to tackle a problem, akin to a team of experts

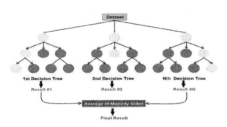

Figure 1: RF Architecture

pooling their insights. By amalgamating predictions from various models, often simpler ones, ensemble learning enhances accuracy and resilience. It's akin to crafting a dish using different cooking methods—each model contributes its unique aspect to the final prediction.

3.2 Random Forest

Instead of relying on one expert, imagine a team of them, each with their own perspective and expertise. That's the essence of a Random Forest. It builds multiple decision trees, like individual analysts, each trained on a unique slice of data and features. By pooling their predictions, the forest cancels out individual biases and noise, yielding a more accurate, as shown in Figure 1.

3.3 LMT (Light Gradient Boosting Machine)

LMT is a standout machine learning algorithm recognised for its efficiency and accuracy, especially in managing large datasets. Created by Microsoft, this method falls under the gradient boosting ensemble approach, showcasing prowess in regression and classification tasks.

3.4 Decision Stump

A decision stump is a fundamental machine learning model employing a solo decision node, dividing data based on a selected feature. This uncomplicated decision tree, featuring just one split, is commonly utilised in boosting methods like AdaBoost. While decision stumps provide clear-cut decisions, their simplicity leads to low complexity,

3.5 Naïve Bayes

Naive Bayes (NB) operates on Bayes' theorem, a probabilistic classification approach assuming feature independence, suitable for small datasets. It computes label probabilities from observed features, selecting the most likely class.

4. Methodology

4.1 Flow of Works

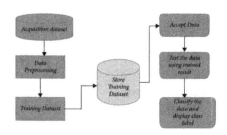

Figure 2: Flow work of our proposed model

4.2 Voting Model

The voting method in machine learning amalgamates multiple models for predictive or classification tasks. It includes Hard Voting, where models vote for a class, determining the majority-voted class as the final prediction, and Soft Voting, which averages predicted probabilities from

various models to select the class with the highest average probability. By utilising diverse models, voting methods heighten predictive accuracy, mitigating individual model biases and encompassing diverse data aspects

In our research we taken RF, LMT, NB and DS machine learning models to build the proposed model.

4.3 Data Preprocessing

We collected the dataset for our research from Kaggle dataset. The dataset has 2200 numbers of data that correspond to characteristics like temperature, humidity, rainfall, pH potential of hydrogen, K-potassium, P-nitrogen, and P-phosphorus.

Using mean, mode and median we clean our dataset. After successfully clean we tokenise the dataset and split the dataset into 4 equal number of dataset.

5. Comparison and Analysis

We split our collected data set into three categories 70% data for training, 20% data for testing and 10% for validation. We compare between taken RF, LMT, NB and DS models and store the accuracy, MAE, RMSE, RAE and RRSE value in Tables 1, 2, 3, 4 and 5. The graphical presentation shows in Figures 3, 4, 5, 6 and 7 respectively.

Table 1: Accuracy of Taken Models.

Data sets	RF	LMT	NB	DS
dataset 1	91.23	92	96.2	36.363
dataset 2	91.24	93.55	96.88	36.3636
dataset 3	91.23	93.32	96.23	36.363
dataset 4	95	96	97	36.3663

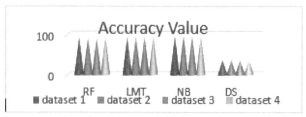

Fig. 3 graphical presentation of accuracy

Table 2: MAE value of RF, LMT, NB and DS.

Datasets	RF	LMT	NB	DS
dataset 1	0.0045	0.0119	0.0004	0.2165
dataset 2	0.009	0.0039	0.0019	0.2182
dataset 3	0.0012	0.0323	0.0011	0.2184
dataset 4	0.0036	0.0099	0.0004	0.2182

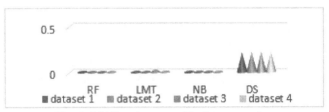

Fig. 4. MAE values graphical presentation

Table 3: RMSE value of ML model

Dataset	RF	LMT	NB	DS
dataset 1	0.0202	0.0374	0.0073	0.329
dataset 2	0.0437	0.0414	0.024	0.3303
dataset 3	0.0165	0.061	0.0045	0.3306
dataset 4	0.0206	0.0387	0.0108	0.3303

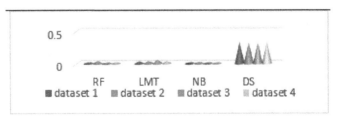

Fig. 5. Graphical presentation of RMSE values

Table 4: RAE value of RF, LMT, NB and DS.

Dataset	RF	LMT	NB	DS
dataset 1	1.641	4.3306	0.1633	78.5636
dataset 2	3.2755	1.4268	0.7058	79.1921
dataset 3	0.4444	11.7269	0.5024	79.2654
dataset 4	1.2935	3.5873	0.1441	79.1921

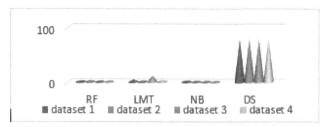

Fig. 6. Histogram for RAE values

Table 5: RRSE value for taken models

Dataset	RF	LMT	NB	DS
dataset 1	5.4417	10.077	1.9586	88.6405
dataset 2	11.7824	11.1668	6.4569	88.9943
dataset 3	4.4448	16.4424	0.0003	89.08
dataset 4	5.5406	10.4181	2.8975	88.9943

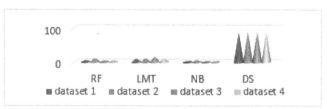

Fig. 7. Graphical presentation of RRSE values

6. Conclusion

The role of machine learning in optimising agriculture is paramount, particularly in predicting crop suitability, showcasing its transformative potential. Employing diverse models on a comprehensive dataset revealed nuanced accuracies, notably Random Forest achieving 91.23% and Light Gradient Boosting Machine at 92%, illustrating their superiority in discerning intricate environmental-crop relationships. This signals a pivotal shift in farming decisions, empowering farmers with precise insights for crop selection based on environmental nuances. Future directions may involve refining models with finer environmental data, expanding datasets for heightened accuracy, and integrating real-time data streams and satellite imagery to bolster predictive capabilities for proactive farming strategies.

References

Elsayed Said Mohamed, AA. Belal, Sameh Kotb Abd-Elmabod, Mohammed A El-Shirbeny, A. Gad, Mohamed B Zahran. (2021). "Smart farming for improving agricultural management." *The Egyptian Journal of Remote Sensing and Space Science*, 24 (3, Part 2): 971-981.

Mohd Javaid, Abid Haleem, Ravi Pratap Singh, Rajiv Suman. (2022). "Enhancing smart farming through the applications of Agriculture 4.0 technologies." *International Journal of Intelligent Networks*, 3: 150-16.4.

Cor Verdouw, Bedir Tekinerdogan, Adrie Beulens, Sjaak Wolfert. (2021). "Digital twins in smart farming." *Agricultural Systems*, 189.

Elsa Jerhamre, Carl Johan Casten Carlberg, Vera van Zoest. (2022). "Exploring the susceptibility of smart farming: Identified opportunities and challenges." *Smart Agricultural Technology*, 2: 100026..

Senthil Kumar Swami Durai, Mary Divya Shamili. (2022). "Smart farming using machine learning and deep learning techniques." *Decision Analytics Journal*, 3: 100041.

R. Alfred, J. H. Obit, C. P. -Y. Chin, H. Haviluddin and Y. Lim. (2021). "Towards paddy rice smart farming: a review on big data, machine learning, and rice production tasks." In *IEEE Access*, 9.

A. A. AlZubi and K. Galyna. (2023) "Artificial intelligence and internet of things for sustainable farming and smart agriculture." In *IEEE Access*, 11.

Jayasingh, S.K., Mantri, J.K., Pradhan, S. (2022). "Smart weather prediction using machine learning. In: Udgata, S.K., Sethi, S., Gao, XZ. (eds) *Intelligent Systems. Lecture Notes in Networks and Systems*, vol 431. Springer, Singapore.

Swain S, Behera N, Swain AK, Jayasingh SK, Patra KJ, Pattan ayak BK, Mohanty MN, Naik KD, Rout SS (2023). "Application of IoT Framework for prediction of heart disease using machine learning." *Int J Recent Innovation Trends Comput Commun* 11 (10s): 168–176.

Efficient FOG task Scheduling Using Humming Bird Based Task Assignment

Kodanda Dhar Naik[1], Sanjay Kumar Kuanar[1], Rashmi Ranjan Sahoo[2], Amar Ranjan Dash[2], Chiranjib Mohapatra[2], Allu Saiswari[2]

[1]Department of Computer Science & Engineering, Gandhi Institute of Engineering and Technology University, Gunupur, India, kodandadhar.naik@giet.edu, sanjay.kuanar@giet.edu
[b]Department of Computer Science & Engineering, Parala Maharaja Engineering College, Berhampur, India, rashmiranjan.cse@pmec.ac.in, amaranth_9876@yahoo.co.in, chiranjibmohapatra2001@gmail.com, allusaiswari2001@gmail.com

Abstract

With gradual evolution of computing power, the cloud is capable of solving multiple complex problem parallelly. In order to solve issues like slow response time & high latency, one intermediate network is developed, which is named FOG. FOG manages all tasks with low computing resource requirement. Generally, FOG task scheduling deals with three operations task uploading, task assignment, task management. A novel task scheduling algorithm, which integrates a hummingbird-based task assigning mechanism, has been proposed. An empirical comparison is conducted between the performance of the proposed algorithm and persisting bio-inspired algorithms typically used in FOG. The results of our approach presented in this paper demonstrate improved performance in terms of make-span and memory usage.

Keywords: Fog computing, task scheduling, humming bird, optimisation, bio-inspired,

1. Introduction

Faster evolution of IoT and tele-communication leads to exponential growth in number of task. Due to overload of task at cloud level, the whole architecture divided into three layers. Task moves from device to edge, edge to fog, fog to cloud on the basis of computing resource vs latency filter. As Cloud contains the highest resource capacity out of all and Edge need least latency. Further in each layer, computing resources of distributed system divided into mul-ti-ple nodes. It allows the fog computing to perform proper resource management by scaling up or scaling down the resources based on need of task. Several goals of fog computing are performance improvement, enha-nced scalability, reliability increment, optimal resource utilisation, archive gen-erality etc.

In fog environment, task scheduling broadly handles three operations task uploading, task assignment, task management. After receiving a set of task processing request, fog will filter them based on amount of computing resource required for their processing. Then it will upload the tasks, with high computing resource requirement, to the cloud. In the second step the scheduler will decide which task should be assigned to which fog node

Chapter 47 DOI: 10.1201/9781003581215

based on requirement of different comput-ing resource (to optimise the performance and efficiency of the system). This sched-uling in fog com-puting can be performed using a varie-ty of different algorithms, depending on the specific needs and con-straints of the system. Some common algori-thms include round-robin schedul-ing, which assigns tasks to devices in a pre-determined order, and dynamic scheduling, which adjusts the task assignments based on real-time infor-mation about the devices and tasks. In the third and final step sched-uler will perform the task management and resource allocation within each fog node during processing. This paper pro-posed a bio-inspired metaheuris-tic algorithm for fog task assignment.

2. Literature Review

Kopras *et al.* (2022) have proposed EEFFRA and LC-EEFFRA algo-rithm. Each algorithm utilises dyna-mic voltage and scaling of frequenc-ies, which helps in minimizing energy consumptions and reduces delay. Saif *et al.* (2023) have pro-posed an algori-thm named MGWO which deals with energy consumption and delays in cloud-fog computing by implemen-ta-tion of queue theory. This doesn't focus upon reducing make-span and resource utilisations. Bitam *et al.* (2018) have pro-posed a job schedul-ing algorithm based on the bee swamp algorithm to solve job scheduling problem, but the cost function is not calculated. Kui-kui *et al.* (2018) pro-posed an Improved Genetic Algorith-m (IGA) that incorporates wellness assess-ment into the primary mutation process to tackle the lack of vision of simple genetic algorithms (SGA). Ningning *et al.* (2016) proposed a load balance-ing approach for fog computing that utilises task alloca-tion via graph partitioning. The proposed approach has been proven to be effec-tive in reducing task execution time. However, because graph repartition-ing is occasion-ally required to adjust for fog variants, the technique may not be as effective for dynamic fog load balancing. Batista *et*

al. (2022) proposed a novel load bal-ancing solu-tion for Fog of Things (FoT) platfor-ms, leveraging Software-Defined Ne-tworks (SDN) to address challenges in IoT applications in remote locations. A comprehensive bio-inspired Meta-heuristic algorithm for engineering applications has been suggested by Zhao *et al.* (2022) It draws inspiration from the various forag-ing behaviours, flight skills, and memory capacities of a group of hummingbirds.

3. Proposed Methodology

The Artificial Hummingbird Algori-thm (AHA) is dependent mainly on 3 compo-nents, viz., food sources, hum-ingbirds, and visit table. A food sou-rce is a solu-tion vector, and its nectar-refilling rate is represented by the ob-jective function's fit-ness value. In our problem definition food source and humming bird (search agent) are defined as set of tuples (tasks and the fog nodes they are allocated to). Visit table is a 2d array that shows for how many iteration one food source is not visited by each humming bird.

3.1 Problem Formulation

If fog layer receives n tasks and needs to assign them to m fog nodes. T_{tn}^{fn} indicate that the tn^{th} task is executing within fn^{th} fog node.

$$Cost\_Function(FNTasks) = Min\left[\sum_{fn=1}^{m} Cost\_Function(T_{tn}^{fn}, FN_{fn})\right] \quad (1)$$

$$Cost\_Function(T_{tn}^{fn}, N_{fn}) = w1.ExecutionTime(FN_{fn}Task) + w2.Memory(FN_{fn}Task) \quad (2)$$

$$Execution\_Time(FN_{fn}Tasks) = \max_{1<=fn<=m}(FN_{fn}.Texe) \quad (3)$$

$$Memory(FN_{fn}Tasks) = Max(T_{tn}^{fn}.AllocatedMemory) \quad (4)$$

Equation 1 & 2 represent the cost-function of overall all process and all process in each fog node. Equation 3 &

4 represent the method of calculat-ing the execution time and memory requirement with respect to each fog node. Where w1, w2 are weight factors.

3.2 Mathematical Representation of AHA

We have 'n' no. of Humming birds and 'm' Food sources, and the initiali-zation done randomly as per equation 5. Where L is lower boundary and U is upper boundary of that problem and X_z is the position of zth Food source. Initialisation of visit table is as per equation 6.

$$X_z = L + r.(U-L), \forall \ 1 \le z \le n \qquad (5)$$

$$VT_z . \begin{cases} 0 & if \ z \ne fn \\ null & if \ z = fn \end{cases}, \forall 1 \le z, fn \le n \qquad (6)$$

Where for $VT_{(z, fn)}$ = null shows that the a humming bird collects food from that food source and for $VT_{(z, fn)}$ = 0 is shows that z^{th} humming bird just visited the fn^{th} food source in that iteration. There are 3 types of Flights seen in humming bird: Axial Flight, Diagonal Flight, Omnidirectional Fli-ght and those are defined mathem-ati-cally as per equation 7, 8, 9:

$$D^{(i)} \begin{cases} 1 & if \ i = rand(i)([1,d]) \\ 0 & otherwise \end{cases}, \forall 1 \le i \le d \qquad (7)$$

$$D^{(i)} \begin{cases} if(i = randperm(k)): \\ 1 & k \in [2, (\lceil r1(d-2)\rceil+1)], \\ & \forall 1 \le i \le d \\ 0 & otherwise \end{cases} \qquad (8)$$

$$D^{(i)} = 1, \forall 1 \le i \le d \qquad (9)$$

There are 3 types of Foraging strate-gi-es seen in hummingbird viz. Guided, Territorial, Migration Foraging. and those are defined mathematically as per equa-tion 10, 12, 13. The position of the zth food source at time t is deno-ted as $xz(t)$, where-as the position of the target food source that the zth hummingbird intends to visit is denot-ed as $x_{iter}(t)$. The guided factor 'a' fol-lows a normal distribution N (0, 1).

$$v_i(t+1) = x_{iter}(t) + a D (x_i(t) - x_{iter}(t)), \forall \ a \sim N(0,1) \qquad (10)$$

$$x_i(t+1) = \begin{cases} x_i(t) & f(x_i(t)) \le f(v_i(t+1)) \\ v_i(t+1) & f(x_i(t)) > f(v_i(t+1)) \end{cases} \qquad (11)$$

$$v_i(t+1) = x_i(t) + b D x_i(t), b \sim N(0,1) \qquad (12)$$

$$X_{wor}(t+1) = L + r.(U-L) \qquad (13)$$

In this case, 'b' stands for the territo-rial factor that follows a normal distribu-ti-on. *xwor* symbolises the food source in the population that has the lowest rate of refilling nectar.

4. Proposed Algorithm

AHA()

fn: fog node index
tn: task index
cl: index for column selection
r1, r2, r3: random variable within 0 to 1
t : iteration number
Z : index
tar: target food source
X_{wor}: worst position of humming bird
wor: worst index

Step 1.	z = 1
Step 2.	Repeat step 3 to 9 till z reach n
Step 3.	X_z = L + r1 (U - L)
Step 4.	fn = 1
Step 5.	Repeat step 6 to 9 till fn reach n
Step 6.	**If** (z! = fn)
Step 7.	$VT_{(z,fn)}$ = 0
Step 8.	**Else**
Step 9.	$VT_{(z,fn)}$ = null
Step 10.	Iteration= 1
Step 11.	Repeat step 11 to 54 till iteration< MaxITER
Step 12.	z=1

Step 13.   Repeat step 14 to 47 till z reach n

Step 14.   **If** r2 < 1/3

Step 15.   Execute equation (7)

Step 16.   **Else**

Step 17.   **If** r2 > 2/3

Step 18.   Execute equation (8)

Step 19.   **Else**

Step 20.   Execute equation (9)

Step 21.   **If** $(r3<=0.5)$

Step 22.   Execute equation (10)

Step 23.   **If** f $(v_z(t+1))<f(X_z(t))$

Step 24. $X_z(t+1)=v_z(t+1)$

Step 25.   fn = 1

Step 26.   Repeat step 27 till ((fn <= n) && (fn != tar))

Step 27. $VT_{(z,fn)} = VT(z,fn) + 1$

Step 28. $VT_{(z,tar)}=0$

Step 29.   fn=1

Step 30.   Repeat step 31 till fn reach n

Step 31. $VT_{(z,fn)}=max(Vt(fn,l))+1$

Step 32.   **Else**

Step 33.   fn = 1

Step 34.   Repeat step 35 till ((fn <= n) && (fn != tar))

Step 35. $VT_{(z,fn)} = VT_{(z,fn)} + 1$

Step 36. $VT_{(z,tar)}=0$

Step 37.   **Else**

Step 38.   Execute equation (12)

Step 39.   **If** f $(v_z(t+1))<f(X_z(t))$

Step 40. $X_z(t+1)= v_z(t+1)$

Step 41.   fn = 1

Step 42.   Repeat step 43 till ((fn <= n) && (fn != z))

Step 43. $VT_{(z,fn)} = VT_{(z,fn)}+1$

Step 44.   fn = 1

Step 45.   Repeat step 49 till fn reach n

Step 46. $VT_{(z,fn)} = VT_{(z,fn)}+1$

Step 47.   **Else**

Step 48.   fn = 1

Step 49.   Repeat step 53 & 54 till ((fn <= n) && (fn!=z))

Step 50. $VT_{(z,fn)} = VT_{(z,fn)}+1$

Step 51.   **If** $(mod(t,2n) == 0)$

Step 52. $X_{wor}(t+1) = L + r4 (U - L)$

Step 53. $X_z(t+1)= v_z(t+1)$

Step 54.   fn = 1

Step 55.   Repeat step 56 till ((fn <= n) && (fn!=wor))

Step 56. $VT_{(wor,j)}=VT_{(wor,j)}+1$ // row

Step 57.   fn =1

Step 58.   cl=1

Step 59.   Repeat step 56 till ((fn <= n) && (cl <= n) && (cl!= fn))

Step 60. $VT_{(fn,wor)}= max (VT_{(fn,cl)}) + 1$ //column->max+1

5. Comparative Result Analysis

Several tests were performed in the Simulation section with varying num-bers of fog node accessible. Which has Fogid, level, upbw, downbw, mbps, RAM, cost per

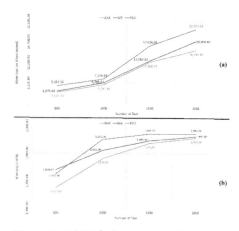

Figure 1: (a) MakeSpan comparison, (b) Memory Utilisation Comparison

second, and cost per memory as per their characterist-ics, each fog node has its own working power. Fog nodes thus exhibit a var-iety of features. Task characteristics are as follows: Task id, task's instruct-ion Length, File Size, Output Size, Memory Required. For the simulation purpose we did the evaluation by us-ing iFogSim of java v12, on windows 10 Operating System. Figures 1 and 2 re-present the makes-span and memory utilisation based comparative analy-sis of proposed algorithm with PSO and BEE algorithm.

6. Conclusion

Fog Computing task layer deals with task uploading, task assignment, and task management. This paper has implement-ed humming bird based task assignment algorithm, which results in better perfor-mance measure than traditional PSO and BEE algorithms. Furthermore, this can be extended further by implementing SLA concept in the problem domain and opti-mising the results.

References

Batista, E., Figueiredo, G., and Prazeres, C. (2022). "Load balancing between fog and cloud in fog of things-based platforms through soft-ware-defined networking." *Journal of King Saud University Computer and Information Sciences*, 34 (9): 7111-7125.

Bitam, S., Zeadally, S., and Mellouk, A. (2018). "Fog computing job scheduling optimization based on bees swarm." *Enterprise Information Systems*, 12 (4): 373-397.

Kopras, B., Bossy, B., Idzikowski, F., Kryszkiewicz, P., and Bogucka, H. (2022). "Task Allocation for energy optimization in fog computing net-works with latency constraints." *IEEE Transactions on Communicati-ons*, 70 (12): 8229-8243.

Kui-Kui, H.A.N., Zai-Peng, X.I.E. and Xin, L.V. (2018), "Fog comput-ing task scheduling strategy based on improved genetic algorithm." *Com-puter Science*, 4: 22

Ningning, S., Chao, G., Xingshuo, A., and Qiang, Z. (2016). "Fog computing dynamic load balancing mechanism based on graph repartitioning." *China Communications*, 13 (3): 156-164.

Saif, F. A., Latip, R., Hanapi, Z. M., and Shafinah, K. (2023). "Multi-objective grey wolf optimizer algo-rithm for task scheduling in cloud-fog computing." *IEEE Access*, 11: 20635-20646.

Zhao, W., Wang, L., and Mirjalili, S. (2022). "Artificial hummingbird algorithm: a new bio-inspired optimizer with its engineering applications." *Computer Methods in Applied Mechanics and Engineering*, 388: 114194.

Efficient Watermarking Framework Using Cryptography and Bit Substitution

Sanjeev Narayan Bal, Sharmila Subudhi

Department of Computer Science, Maharaja Sriram Chandra Bhanja Deo University, Baripada, India

E-mail: s_n_bal@yahoo.co.in, sharmilasubudhi1@gmail.com

Abstract

This article suggests an encryption-based digital watermarking model for securing a message inside carrier data. The watermark is ciphered initially by a symmetric key cryptography mechanism. The XOR operation modifies the encrypted watermark's pixel bits with that of the host image. The least significant bits of the original image are substituted by the encrypted watermark pixel bits to hide the watermark. The watermark is then extracted and decoded with the help of bit substitution and symmetric key decryption. Experiments are carried out to evaluate the model's efficacy concerning various performance metrics. The proposed scheme produces high-quality encrypted images and a reasonable payload factor. Further, a performance comparison with three different existing approaches establishes the supremacy of the developed system owing to the high payload.

Keywords: Watermarking, message security, pixel encryption and decryption, bit operation, substitution

1. Introduction

Nowadays, an upsurge in the usage of digital information has severely violated the copyrights of the original content creators. Thus, watermarking has become increasingly vital in preventing the exploitation of digital content. Watermarking provides authenticity and ownership to artworks, documents, and currency (Sharma et al., 2023). A digital watermark is a unique identifier inserted in a digital file, helpful in tracing the source and determining the rightful owner (Sharma et al., 2023). A survey by Kaspersky in 2020 (Hunter, 2022) mentioned that countless security attacks are designed on industrial enterprises worldwide. Malicious codes are hidden inside Excel email attachments to gain access to the systems. Despite innovating mechanisms to preserve the watermarks, the attackers are evading detection and performing novel attacks by putting malicious data in the original image (Li et al., 2023). Hence, stern action is required to handle such undesirable events.

This work presents a watermarking methodology that encrypts a watermark before embedding it in a carrier image. The highlights of the current work are mentioned as follows. (1) Symmetric key encryption enciphers the watermark image before embedding it for security. (2) Bit substitution is used for inserting the encrypted watermark in the carrier. (3) Different signal attacks are used to check the robustness of the approach.

This article is organised as follows: Section 2 highlights a few watermarking papers. Section 3 presents the proposed system extensively. Section 4 discusses the model's result. Section 5 states the synopsis of the work.

2. Related Work

This section presents a review of a few published watermarking schemes. Table 1 presents a brief description of the above-discussed works. One of the main challenges faced during the study is that the models are computationally expensive and mathematically complex. Therefore, there is a need to develop a simple yet effective watermarking model to protect the image's rightful content and ownership.

3. Proposed Watermarking Model

This section explores a watermarking mechanism for building a robust watermark. The model secures the watermark using the Symmetric Key Cryptography (SKC) and inserts it in the host using the LSB substitution. The proposed schema is categorised into three processes depicting

the encryption, embedding, and retrieval of the watermark in the host image.

3.1. Watermark Securing Process

Initially, the watermark is added with security through the SKC technique. This process uses a private key to cipher and decipher the watermark (Baldimsti et al., 2021). The encrypted watermark bits can be embedded using the proposed scheme in a much-secured manner and can also be retrieved later, thus providing dual security.

3.2. Watermark Encoding Process

After securing the watermark, the encoding process begins. A host image is used for inserting the encrypted watermark. The proposed watermarking encoding and embedding process is depicted below, whereas Figure 1 illustrates the encoder's workflow.

1. Initially, the pixels of the watermark image N undergo the SKC encryption to produce a modified watermark.

Table 1: Comparative Analysis of Some Recent Works

Work	Techniques Used	Advantage	Limitation	Image Type	Application
Deeba et al. (2020)	ANN and LSB substitution	Detect the presence of sensitive information	Not considered for high-definition information	Gray-scale	Digital Image
Liu et al. (2021)	Deep Neural Network-based watermarking	Have not used ownership indicators for verification purposes	Mathematically complex computation	Color	Content authentication
Faheem et al. (2022)	LSB and image gradient	Handles image gradient and chaotic map	Low hiding capacity	Gray-scale	Copyright Protection
Li et al. (2023)	Neural network	IP protection of NLG APIs	Computationally expensive	Color/ Gray-Scale	IP protection

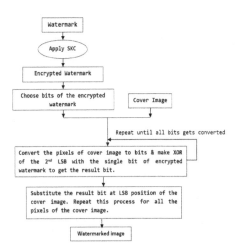

Figure 1: Watermark Encoding Process

2.　Each pixel of　and host image　are expressed in 8-bit format.
3.　An XOR operation is performed between 's encrypted bit and 's second LSB to obtain a result bit.
4.　Now, the result bit is replaced at the LSB position of .
5.　The steps are repeated until all bits of are ingrained inside the host　to produce the final watermarked image .

3.3. Watermark Decoding Process

In this section, the hidden encrypted watermark is extracted from the watermarked image and deciphered to get the original watermark. The decoding mechanism of SKC is performed on the encrypted pixel of　to get the original pixel. The proposed watermark

decoding steps are presented below, while Figure 2 depicts the decoder's workflow.

1.　Select a pixel randomly from the watermarked image　and present it in an 8-bit format.
2.　Now, take the LSB of　and do the XOR operation with its second LSB to get the encrypted watermarked bit.
3.　This process is repeated until the concealed bits are recovered from the watermarked image.
4.　Finally, the encrypted watermark can be extracted by placing these recovered bits correctly.

4. Outcomes and Discussion

This section tests the efficiency of the developed system concerning imperceptibility and robustness. Experiments are done in MATLAB 2016b on a Windows 7 Intel i5 system. A gray-scale watermark image of size 32 × 32 is used for evaluating the system's ability. Figure 3 presents the original and encrypted watermark gray-scale images.

Figure 4 depicts four gray-scale carrier images, each size 512 × 512. These host images hide the encrypted watermark (Figure 3(b)). Figure 5 presents the encrypted watermarked images.

It must be noted that embedding watermark in the carrier image changes the original image's quality. Hence, the following benchmark metrics, such as Mean Squared Error (MSE), Peak Signal-to-Noise Ratio (PSNR), Structural Similarity Index (SSIM), Universal Image Quality Index (UIQI),

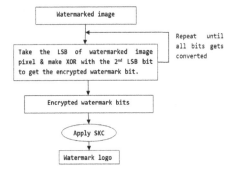

Figure 2: Watermark Decoding Process

(a) Original Image　　　(b) Encrypted Image

Figure 3: Watermark Image

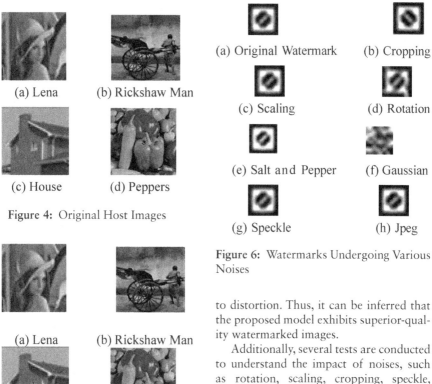

(a) Lena   (b) Rickshaw Man

(c) House   (d) Peppers

Figure 4: Original Host Images

(a) Original Watermark   (b) Cropping

(c) Scaling   (d) Rotation

(e) Salt and Pepper   (f) Gaussian

(g) Speckle   (h) Jpeg

Figure 6: Watermarks Undergoing Various Noises

(a) Lena   (b) Rickshaw Man

(c) House   (d) Peppers

Figure 5: Watermarked images

Mean of SSIM (MSSIM), are used to verify the quality (Wang and Bovik, 2002).

The higher the PSNR value, the more similar the two images are. The lesser the MSE, the better the quality. A UIQI number ≈ 1 signifies a close bond between the original cover and watermarked images. Similarly, the SSIM falls in range, where points to high similarity, and to a less likeness. Likewise, the larger the MSSIM value, the better the image (Wang and Bovik, 2002).

Table 2 depicts the outcomes of the model's imperceptibility concerning all the performance measures. The MSE, SSIM, and MSSSIM values indicate a tight similarity between the original host and watermarked images. Similarly, the UIQI values present that all the images are impervious

to distortion. Thus, it can be inferred that the proposed model exhibits superior-quality watermarked images.

Additionally, several tests are conducted to understand the impact of noises, such as rotation, scaling, cropping, speckle, salt and pepper, gaussian, and Jpeg on the watermark. This helps measure the model's robustness. Figure 6 illustrates the above attacks on the watermark image.

The following performance indicators, viz, PSNR, SSIM, Normalised Correlation (NC), and Bit Error Rate (BER), are employed to calculate the effects of attacks. The NC value ≈ 1 points to the image's stability after an attack, whereas the lesser the BER value, the fewer the error bits (Wang and Bovik, 2002). the fewer the error bits (Wang and Bovik, 2002).

Table 3 presents the model's robustness after undergoing the above attacks on the watermark. The denotes that the watermark image is highly impacted by the Gaussian noise. Besides, upon looking at all performance metrics, the watermark is shown as resistant to Rotation while plagued by the Jpeg noises.

Moreover, Figure 7 presents a performance analysis with three different contemporary works (Zhang et al. (2022), Wu et al. (2020), Duan et al. (2020)) regarding

Table 2: Imperceptibility of Proposed Model

Image	MSE	PSNR (in dB)	UIQI	SSIM	MSSIM
Lena	0.0031	51.9923	0.9932	0.9922	0.9988
Rickshaw Man	0.0032	52.2531	0.9868	0.9868	0.9998
House	0.0030	52.7215	0.9981	0.9981	0.9978
Peppers	0.0033	52.6251	0.9999	0.9999	0.9995

Table 3: Robustness of Proposed Model

Noises	PSNR (in dB)	NC	SSIM	BER
Rotation	52.324	0.9989	1	0
Scaling	50.879	0.7869	0.992	0
Cropping	48.919	1	1	0
Gaussian	45.130	0.4831	0.858	0.06
Salt and Pepper	51.355	0.7754	1	0
Speckle	50.418	0.8978	0.933	0
Jpeg	47.003	0.7923	0.888	0.02

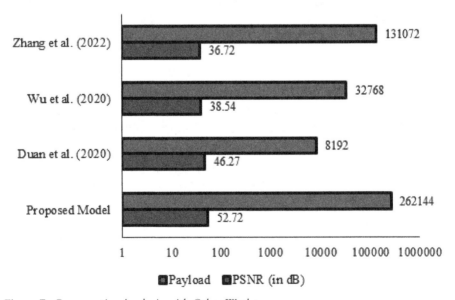

Figure 7: Comparative Analysis with Other Works

PSNR and Payload factors. The four gray-scale host images shown in Figure 4 are used for the comparison. The highest PSNR and payload values indicate the superiority of the proposed model, along with a better embedding capacity.

5. Conclusion

This article presents a cryptography-based bit substitution watermarking scheme. The model initially employs symmetric key cryptography to design a secure watermark. Tests were conducted on four gray-scale host images to evaluate the model's efficiency. Moreover, the system's robustness is also observed by analyzing the impact of seven different attacks on the watermark. The investigation also established the supremacy of the proposed mechanism with good hiding capability and imperceptibility. Additionally, the performance comparison with five other approaches illustrates the dominance of the proposed scheme.

References

Baldimtsi, F., Kiayias, A., and Samari, K. (2021). "Watermarking public-key cryptographic functionalities and implementations: the case of encryption and signatures." *IET Information Security*, 15 (3): 205-222.

Deeba, F., Kun, S., Dharejo, F. A., Memon, H. (2020) "Digital image watermarking based on ann and least significant bit." *Information Security Journal: A Global Perspective*, 29 (1): 30–39.

Duan, S., Wang, H., Liu, Y., Huang, L. and Zhou, X. (2020). "A novel comprehensive watermarking scheme for color images." *Security and Communication Networks*, 2020, pp. 1-12.

Faheem, Z. B., Ali, M., Raza, M. A., Arslan, F., Ali, J., Masud, M., Shorfuzzaman, M. (2022). "Image watermarking scheme using lsb and image gradient." *Applied Sciences*, 12 (9): 4202.

Hunter, T. (2022) "Steganography: the undetectable cybersecurity threat - builtin.com." https://builtin.com/cybersecurity/steganography, [Accessed 09-01-2024].

Liu, H., Weng, Z., Zhu, Y. (2021). "Watermarking deep neural networks with greedy residuals." In: *ICML*. pp. 6978–6988.

Li, M., Wu, H., and Zhang, X. (2023). "A novel watermarking framework for intellectual property protection of NLG APIs." *Neurocomputing*, 558: 126700.

Sharma, S., Zou, J.J., Fang, G., Shukla, P., Cai, W. (2023) "A review of image watermarking for identity protection and verification." *Multimedia Tools and Applications*, pp. 1–6.

Wang, Z., and Bovik, A. C. (2002). "A universal image quality index." *IEEE Signal Processing Letters*, 9 (3): 81-84.

Wu, H., Liu, G., Yao, Y. and Zhang, X. (2020). "Watermarking neural networks with watermarked images." *IEEE Transactions on Circuits and Systems for Video Technology*, 31 (7): 2591-2601.

Zhang, B., Wu, Y. and Chen, B. (2022). "Embedding guided end-to-end framework for robust image watermarking." *Security and Communication Networks*, 2022.

Analyzing the Effectiveness of Machine Learning Algorithms in detecting Fake News

Jaishree Jain[1], Santosh Kumar Upadhyay[1], Sanjib Kumar Nayak[2]

[1]Department of CSE, Ajay Kumar Garg Engineering College, Ghaziabad, India
[2]School of Computer Science, VSSUT, Burla, India
E-mail:jaishree3112@gmail.com,upadhyaysantosh@akgec.ac.in,sknayak_ca@vssut.ac.in

Abstract

Fake news is widely available on social media and in other media channels, which poses a serious danger to society and has the potential to cause great social and national damage. The primary objective is to create a supervised machine learning system that can tell the difference between authentic and misleading news reports. The suggested method applies natural language processing (NLP) methods for textual analysis, such as feature extraction and vectorization, and does so by using Python's scikit-learn module. The work recommends using scikit-learn's Count Vectoriser and Tfidf Vectoriser for preparing text data. The method uses a Random Forest classification algorithm to determine if a post is authentic or not, and suggestions for improvement are provided. The study piece emphasises how urgent it is to solve the problem of false news and shows how machine learning might help with this effort. From the experimental analysis Naïve Bayes and Random Forest Algorithm perform better than other algorithms.

Keywords: Machine learning, fake news classification, detection, fact-checking, semantic analysis

1. Introduction

In recent years, fake news has been widely spread online for a range of political and commercial purposes, mostly as a result of the rapid expansion of online social networks. These online bogus news reports use deceptive language, making it easy for users of online social networks to become infected. An increasing number of people are concerned about accurate information on the Internet, especially on social media, but web-scale data, the ability to identify, evaluate, and update this kind of content, which is commonly referred to as "fake news," that shows up on these platforms (Majed Alrubaian et al., 2016). Machine Learning is proved to be a very powerful tool for various real word classification problem (Upadhyay et al.,2021,2022, Kumar et al. 2022, Rukhsar et al.,2022). The integration of convolution neural networks (CNNs) with threshold segmentation in the brain tumour detection model signifies a notable development within the realm of medical imaging (Jain et al. 2023). These days, deliberate government propaganda, satire pieces, and fake news in some media are all caused by false information (Manish Gupta et al., 2012). The aim of this research is to develop a model that can precisely forecast if a certain article is fake news or not (Krzysztof Lorek et al., 2015).

Chapter 49 DOI: 10.1201/9781003581215

2. Literature Survey

Many terms with similar principles and definitions, including credibility, believability, accuracy, fairness, impartiality, and others, were used to define the creditability of information (Ballouli et al., 2017). Numerous studies compute the message's creditability using a machine learning approach (N. J. Conroy et al., 2015). Content that falsely claims to be true and occasionally contains sensitive messages is referred to as fake news. It is a fact that will barely be classified to contextualise our findings, detection systems with empirical human baseline accuracy were used (Granik, M. et al., 2017). The primary objective is to empower computers to learn on their own, devoid of human assistance or intervention, and modify their operations accordingly (Anjali Jain et al., 2018). Conroy, Rubin, and Chen offer several approaches that seem promising for precisely classifying false content (Monther Aldwairi et al., 2016). Combining n-gram approaches with Probabilistic Context Free Grammars (PCFG) for deep syntax analysis has been shown to yield very positive results (Vanya Tiwari et al., 2019).Meanwhile, search and social media firms are beginning to put defences in place, and social, cognitive, and communicative, and computer scientists are trying to provide solutions by comprehending the complex dynamics underlying the viral spread of digital misinformation (Kushal Agarwalla et al., 2016). This research suggests that restricting the usage of social bots may be a helpful strategy for lessening the online spread of misleading information.

3. Approach and Model Details

This research advances the use of Natural Language Processing (NLP) techniques to detect false news items that come from untrustworthy sources, also known as "fake news." This research used a count vectoriser- based model; alternatively, a tfidf matrix, in which a word count is determined by the frequency with which a word occurs in other dataset articles, may be utilised. Suggested framework is depicted in Figure 1.

3.1. Term frequency or TF

The number of times a term appears in a document is called TF. The more times a term appears in the search terms, the more value shows that it appears more often than other terms, suggesting that the document and the phrase are a good fit.

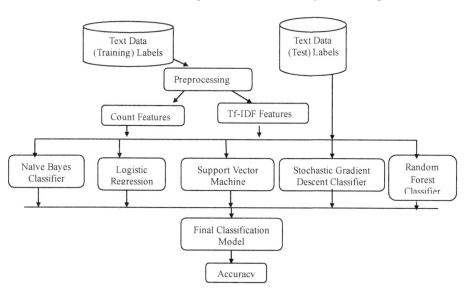

Figure 1: Suggested Framework

3.2. IDF (Inverse Document Frequency)

The IDF of a term is a measure of its significance over the whole corpus. Extracting the best features for the count vectoriser is the next stage or vectoriser for tfidf. This is accomplished by using a list of the most commonly used terms and/or phrases, regardless of whether they are lowercased.

The following approaches have been used to evaluate the news' accuracy.

1. The pre-processed Liar dataset (12.8K).
2. Following their varied configurations, the texts were manually tagged taken from www.politifact.com. The SAFAR v2 library and the NLP NLTK libraries will be used in the following step to eliminate the noise.
3. To extract features, choose lexical features such word count, average word length, article length, number count, and number of speech segments.
4. Assign 80% of the dataset to training and 20% for the Python Sklearn test.
5. After using all of the algorithms, create an ipynb file for the classification model.
6. Create a confusion matrix and evaluate the model's precision on the test subset of the dataset.

4. Empirical Findings

This section covers the algorithms that were used to clean and extract the data from the chosen dataset. Every record was provided a class label (0, 1, 2, 3) in order to train the model with it. To classify the news as Fake or real, Naïve Bayes, Logistic Regression, SVM, Stochastic Gradient Descent Classifier and Random Forest ML models are used. In result evaluation it is observed that Naïve Bayes and Random Forest works well with an accuracy of 98% and 99% respectively. Classification result of Naïve Bayes is shown in Figure 2 in form of Confusion matrix. Similarly, classification result of Random Forest is shown in Figure 3 in form of Confusion matrix.

Comparison of accuracies of all five implemented models is shown in Table 1. Bar graph of accuracy comparison is depicted in Figure 4.

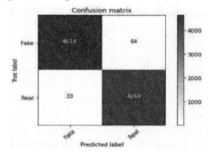

Figure 2: Classification Result of Naïve Bayes algorithm

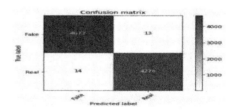

Figure 3: Classification Result of Random Forest algorithm

Table 1: Comparison of Accuracy Levels of Different Models

Different Models	Accuracy Levels
Naïve Bayes	98.91%
Logistic Regression	93%
SVM	88%
Stochastic Gradient Descent Classifier	95%
Proposed Methodology (Random Forest)	99.69%

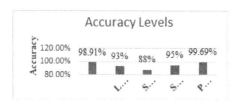

Figure 4: Comparison of Accuracy Levels of Different nod

5. Conclusion

We tackled the problem of automatically identifying bogus news in this study. We presented two new datasets of fake news: one sourced from crowd sourcing and the other from online celebrity news sources. From the experimental analysis Naïve Bayes and Random Forest Algorithm perform better than other algorithms. Instead of using Random Forest and naïve bayes to categorise the data, we can even switch to a better classifier.

References

Jain, A., Shakya, A., Khatter, H., Kumar, A. (2019). "A smart system for fake news detection using machine learning." *2019 International Conference on Issues and Challenges in Intelligent Computing Techniques (ICICT)*.

Rim, E. B., El-Hajj, W., Ghandour A., Elbassuoni S., M. Hajj H. and Shaban K. (2017). "CAT: credibility analysis of arabic content on Twitter." *Proceedings of the Third Arabic Natural Language Processing Workshop*, pp. 62–71.

Granik, M., and Mesyura, V. (2017). "Fake news detection using naive Bayes classifier." *2017 IEEE First Ukraine Conference on Electrical and Computer Engineering (UKRCON)*, pp. 900-903.

Jain J.,Sahu S. and Sahu N. (2024). "Sentiment Analysis based on NLP using Learning Techniques." *2nd Int. Con. on Adv. data driven comp.*, CRC Press Taylor & Francis.

Jain J., Sahu S., and Dixit A. (2023). "Brain Tumour Detection model based on CNN and Threshold Segmentation." *Int. J. of Exp. R. & R.*, 32: 358-364.

Kumar, A. and Upadhyay, S.K. (2022). "A novel approach for rice plant diseases classification with deep convolutional neural network." *Int. J. Inf. Tecnol.*, 14: 185–199.

Agarwalla, K., Nandan, S., Anil Nair, V., Hema, D. (2019). "Fake News Detection using Machine Learning and Natural Language Processing." *Int. J. RTE*, 6 (6).

LaPorta, R., Gupta, M., Zhao, P., Han, J. (2012). "Evaluating event credibility on Twitter." *Proceedings of the 2012 SIAM International Conference on Data Mining*, (pp. 153-164).

Conroy, N. J., Rubin, V.L., and Chen, Y. (2015). "Automatic deception detection: Methods for finding fake news." *Proceedings of the Asso. for Info. Sci. and Technology*, 52 (1): 1–4.

Rukhsar, Upadhyay, S. K. (2022). "Rice leaves disease detection and classification using transfer learning technique, ICACITE." Greater Noida, India, (pp. 2151-2156).

Rukhsar, Upadhyay S. K. (2022). Deep transfer learning-based rice leaves disease diagnosis and classification model using InceptionV3." *2022 International Conference on Computational Intelligence and Sustainable Engineering Solutions (CISES)*, Greater Noida, India, (pp. 493-499).

Upadhyay, S.K., Kumar, A. (2021). "Early-stage brown spot disease recognition in paddy using image processing and deep learning techniques. *TS*, 38 (6): 1755-1766.

Upadhyay, S. K., Kumar, A. (2022). "An Accurate and Automated plant disease detection system using transfer learning-based Inception V3Model." *ICACITE*. India. (pp. 1144-1151).

Analyzing the Machine Learning Approaches in Predicting the Crop Yield

A Decade Literature Review

Santosh Kumar Upadhyay[1], Sanjib Kumar Nayak[2]

[1]CSE Department, AKGEC, Ghaziabad, India
[2]School of Computer Science, VSSUT, Burla, India
E-mail: upadhyaysantosh@akgec.ac.in, sknayak_ca@vssut.ac.in

Abstract

Farming is a key sector that generates revenue and provides a means of livelihood in the world. The agricultural yields are influenced by a variety of financial, seasonal, and biological factors; however, unanticipated changes in these variables lead to a significant loss of crops. These risks may be evaluated when suitable statistical or mathematical methods are used to analyze data on past yield, climatic factors, and soil quality. With the development of machine learning, it will be possible to predict agricultural yields by obtaining useful data from crop fields. This would enable farmers to concentrate on the crops they wish to grow in the upcoming season in order to reap substantial advantages. This paper presents a review on the research done in field of crop yield prediction.

Keywords: Machine learning, crop yield, classification, prediction, agriculture

1. Introduction

The core of this survey is an evaluation of previous methods used in crop yield prediction systems. We categorize the methods into regression technique, Decision tree, RF, SVM, Bayesian network, ANN and deep learning, providing insights into their strengths and limitations. ML and advanced ML techniques (Upadhyay et al.,2021,2022,2023 Kumar et al. 2022, Rukhsar et al.,2022) helps a lot in precision agriculture.

2. Literature Review

The present study includes various yield prediction techniques that utilize machine learning approaches.

Ramesh et al. (2013) applied the Regression(MLR) strategy to estimate rice production in the future year based on region-specific average rainfall. The accuracy of the model ranged from 90% to 95%. Bolton et al. (2013) implemented a ridge-regression-based statistical model for the prediction of soybean and corn yields. In another experiment, Everingham et al. (2015) performed regression analysis to find the significant environmental parameters that influence the yields of sugarcane. Gandhi et al. (2016) proposed a rice crop yield prototype decision support system. Majumdar et al. (2017) uses MLR to predict yields.

Veenadhari et al. (2011) presented a model for Soybean Productivity using Decision Tree Algorithms. One of the limitations is that the model only predicts the low or high yield, but cannot predict the quantity of yield output. Upadhyay et al. (2024) introduced decision tree and random forest based yield prediction system.

A random forest model was introduced by Everingham et al.,2016 to precisely forecast sugarcane yield. Karim et al. (2023) used RF to estimate crop yields with accuracy of 97.69%. Deka et al. also used Random-forest algorithm to predict tea yields.

Su et.al. (2017) and Tripathy *et. al.* (2016) have applied SVM to predict the crop yield. Tripathy et al. achieved the precision of 83.97%, sensitivity of 68.17%, and accuracy of 78.76%.

Villordon *et. al.* (2010) developed a Bayesian belief network (BBN) model to depict the correlation between market yield and agro-climatic factors responsible to affect critical root storage initiation phases in sweet potato. Cornet et.al. (2016) have used Additive modeling technique to get an appropriate Bayesian network model for yield prediction. Huang *et. al.* (2017) and Armstrong *et. al.* (2016) also used Bayesian network to predict crop yields.

Panda et al. (2010) had developed a remote sensing technique to predict yields of corn. Fortin et al. (2011) built a neural network-based predictive model for early season region-specific potato yield prediction. Guo *et. al.* (2014) performed a comparison of spatial and temporal neural network models for forecasting of crop production. Pandey et al. (2017) and Adisa et al. (2019) applied ANN in the estimation of Potato and maize yields respectively.

Kuwata *et. al.* (2015) prepared a prediction system based on deep learning by using CNN for Fast Feature Embedding (Caffe) as a framework. You *et. al.* (2017) applied remote sensing with deep learning to forecast the crop yields. The addition of the Gaussian Process element in proposed designs attained even better efficiency with a 30 percent decrease in RMSE from the best competing techniques. (Deep Neural Network (Kuwata et al., 2015), ridge

regression (Bolton et,2014)). Khaki et al. (2019) used deep learning models to make yield predictions based on genotype and environmental data (including yield, check yield, and yield difference).

3. Results of Review

3.1. Review Statistics

Literature review statistics are shown in

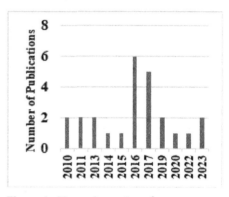

Figure 1: Year wise reviewed papers

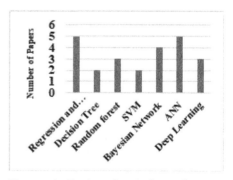

Figure 2: Reviewed Machine learning Techniques

Figures 1 and 2.

3.2. Research Gap

i.　Currently, the Maximum yield prediction techniques give individual approaches, as well as solutions and, are not properly associated with the decision making.

ii. Effective integration of temporal, spatial, and spectral domains and integration of expert knowledge into machine learning approaches for modeling and decision-making in various aspects of precision agriculture is lacking in the literature review.

iii. iii.Although research on deep learning for crop yield prediction is done in some works of literature. Furthermore, this work can be extended to create a more accurate model with few parameters and less computation time.

iv. iv.Fertilisation, quality of soil, the efficiency of the irrigation system, disease occurrence, quality of crop seeds, topographic attributes as well as changes in temperature and rainfall can be included during designing and simulation of a predictive model to get a comparatively good result.

v. The combination of image processing and various ML techniques into hybrid systems to take advantage of these techniques' strengths and balance for their limitations.

4. Conclusion

More advanced techniques and more affordable, comprehensive data sets will probably be made available in the future, allowing for more accurate assessment and decision-making on crop and environmental conditions. Numerous machine learning techniques have already been effectively implemented in a range of precision agriculture (PA) jobs. However, deep learning is currently a popular area of study for predicting crop productivity.

References

Adisa, O. M., Botai, J.O.,Adeola, A.M., Hassen, A., Botai, C.M.,Darkey, D., Tesfamariam, E. (2019). "Application of artificial neural network for predicting maize production in South Africa." *Sustainability*, 11 (4): 1145.

Armstrong, L. J., Gandhi, N., Petkar, O. (2016). "Predicting rice crop yield using Bayesian networks." In *ICACCI*, pp. 795-799. Jaipur, India.

Bolton, D. K. and Friedl, M. A. (2013). "Forecasting crop yield using remotely sensed vegetation indices and crop phenology metrics." *Agricultural and Forest Meteorology*, 173: 74–84.

Cornet, D., Sierra, J., Tournebize, R., Gabrielle, B., Lewis, F. I. (2016). "Bayesian network modeling of early growth stages explain yam interplant yield variability and allows for agronomic improvements in West Africa." *European Journal of Agronomy*, 75: 80-88.

Deka, P. C. Das, A., Saikia, U. S., and Borah, R. (2020). "Tea yield forecasting using Random Forest algorithm." *Computers and Electronics in Agriculture*, 175: 105631.

Everingham, Y. L., Sexton, J., Robson, A. (2015). "A statistical approach for identifying important climatic influences on sugarcane yields." In *Proceeding of Australian Society of Sugarcane Technologists*, pp. 8–15. Bundaberg, Australia.

Everingham, Y., Sexton, J., Skocaj, D., Inman-Bamber, G., (2016). "Accurate prediction of sugarcane yield using a random forest algorithm." *Agronomy for Sustainable Development*, 36: 27.

Fortin, J. G., Anctil, F., Parent, L. E., Bolinder, M. A. (2011). "Site-specific early season potato yield forecast by neural network in Eastern Canada." *Precision Agriculture*, 12: 905–923.

Gandhi, N., Armstrong, L.J., Petkar, O., (2016). "Proposed decision support system (DSS) for Indian rice crop yield prediction." In *TIAR*, pp. 13-18. Chennai, India.

Guo, W. W., and Xue, H. (2014). "Crop Yield Forecasting Using Artificial Neural Networks: A Comparison between Spatial and Temporal Models." *Mathematical Problems in Engineering*, 2014, Article ID 857865, 7 pages.

Huang, X., Huang, G., Yu, C., Ni, S., and Yu, L. (2017). "A multiple crop model ensemble for improving broad-scale yield prediction using Bayesian model averaging." *Field Crops Research*, 211: 114-124.

Karim, A. and Benjamin, J. (2023). "Crop prediction using random forest algorithm, NCECA,. https://doi.org/10.5281/zenodo.10147295

Khaki, S., and Wang, L. (2019). "Crop yield prediction using deep neural networks." *Frontiers in Plant Science*, 10, Article 621.

Kumar, A. and Upadhyay, S.K. (2022). "A novel approach for rice plant diseases classification with deep convolutional neural network." *Int. j. Inf. Tecnol.* 14: 185–199.

Kuwata, K. and Shibasaki, R. (2015). "Estimating crop yields with deep learning and remotely sensed data." In *IEEE IGARSS*, pp. 858–861.

Majumdar, J., Naraseeyappa, S. and Ankalaki, S. (2017). "Analysis of agriculture data using data mining techniques: application of big data." *Journal of Big Data*, 4: Article 20.

Panda, S.S., Ames, D.P., Panigrahi, S. (2010). "Application of vegetation indices for agricultural crop yield rediction using neural network techniques." *Remote Sensing*, 2: 673–696.

Pandey, A., and Mishra, A. (2017). "Application of artificial neural networks in yield prediction of potato crop." *Russian Agricultural Sciences*, 43: 266–272.

Ramesh, D., Vardhan., (2013). "Region specific crop yield Analysis: A data Mining approach." *UACEE International Journal of Advances in computer Science and its Applications*, 3: 77-80.

Rukhsar, Upadhyay, S. K. (2022). "Rice leaves disease detection and classification using transfer learning technique, ICACITE. Greater Noida, India, pp. 2151-2156

Rukhsar and Upadhyay, S. K. (2022). Deep Transfer Learning-Based Rice Leaves Disease Diagnosis and Classification model using InceptionV3", 2022 CISES, Greater Noida, India, 2022, pp. 493-499, doi: 10.1109/CISES54857.2022.9844374.

Su Y.X, Xu, H, Yan, LJ. (2017). "Support vector machine-based open crop model (SBOCM): Case of rice production in China." *Saudi Journal of Biological Sciences*, 24, 537-547.

Tripathy, A. K., Gandhi. N., Armstrong, L.J., and Petkar, O. (2016). Rice crop yield prediction in India using support vector machines. In *13th JCSSE*, pp. 1-5. KhonKaen, Thailand.

Upadhyay, S.K., Kumar, A. (2021). Early-Stage Brown Spot Disease Recognition in Paddy Using Image Processing and Deep Learning Techniques. TS.38(6), pp. 1755-1766.

Upadhyay, S. K., Kumar, A. (2022). An Accurate and Automated plant disease detection system using transfer learning-based Inception V3Model. ICACITE. India. pp. 1144-1151.

Upadhyay, S. K., Kumar, A. (2023). Automatic Recognition and Classification of Tomato Leaf Diseases Using Transfer Learning Model, Future Farming: Advancing Agriculture with Artificial Intelligence (2023) 1: 23. https://doi.org/10.2174/9789815124729123010005

Upadhyay, S.K., Kushwaha, P. and Garg,P. (2024).A Machine Learning Approach for Predicting Crop Yield in Precision Agriculture AUTOCOM, Dehradun, India, 2024, pp. 289-294.

Veenadhari, S., Mishra, D.B., Singh, C.D. (2011). Soybean Productivity Modeling using Decision Tree Algorithms. *IJCA*,27,11-15.

Villordon, A., Solis J., Labonte D., Clark C. (2010). Development of a prototype Bayesian network model representing the relationship between fresh market yield and some agroclimatic variables known to influence storage root initiation in sweet potato. *HortScience*, 45,1167- 1177.

You J., Li X., Low M., Lobell D., Ermon S. (2017). Deep gaussian process for crop yield prediction based on remote sensing data. In Proceedings of the Thirty-First Association for the Advancement of Artificial Intelligence conference, pp.4559-4565. San Francisco, California, USA.

BCViT

A Vision Transformer Enabled Deep Learning Model for Brest Cancer Identification

Debasish Swapnesh Kumar Nayak[1,2], Tejaswini Das[3], Kamalakanta Rout[4],
Sonali Priyadarshini Mohapatra[4], Suprava Ranjan Laha[1], Satyaprakash Swain[1,4],
Tripti Swarnkar[5]

[1]Department of Computer Science and Engineering, FET-ITER, Siksha 'O' Anusandhan
(Deemed to be University), Bhubaneswar, India, swapnesh.nayak@gmail.com
[2]Department of Computer Science and Engineering, SOET, Centurion University of
Technology and Management, Bhubaneswar, India
[3]Department of Statistics, Utkal University, Bhubaneswar, India
[4]Department of Computer Science and Engineering, Institute of Management and
Information Technology, Cuttack, India
[5]Department of Computer Application, FET-ITER, Siksha 'O' Anusandhan (Deemed to be
University), Bhubaneswar, India

Abstract

The complex task of BC diagnosis has sparked a revolution in the field of artificial intelligence. Clinicians can save time and effort with the use of artificial
intelligence (AI) when it comes to cancer detection. With this ground-breaking approach, healthcare systems may work more efficiently, leading to greater
results for patients. This research takes a technical approach to BC detection
by combining the architectures of Vision Transformer (ViT) and Convolutional
Neural Network (CNN) with breast-ultrasound images. To thoroughly examine
BC datasets, we have explored the possibilities of ViT and CNN architectures, as
opposed to conventional machine learning methods. The findings demonstrate
the effectiveness of our customized methodology, with the Vision Transformer
achieving an impressive 95% classification accuracy and the CNN achieving
92%. This study is a significant step forward in the field of BC detection methods. We demonstrate the efficacy of customized deep learning architectures by
avoiding conventional machine learning models, which should lead to improved
AI-driven BC detection tools in the future.

Keywords: AI, breast cancer, deep learning, CNN, vision transformers

1. Introduction

The Centres for Disease Control and
Prevention (CDC) reports that breast cancer (BC) is the most common cancer in
women. The cancer's survival rate varies
depending on the type of cancer and its
stage upon diagnosis. It originates from
breast cells and usually develops in lobules
or ducts, but it can also occur in fibrous
or adipose tissues. The goal of early detection is to stop the proliferation of cells and
avoid negative outcomes. Differentiating
between benign and malignant tumours
upon diagnosis is crucial because of how
they should be treated. Malignant cells are

Chapter 51 DOI: 10.1201/9781003581215

cancerous and have the ability to spread, whereas benign cells are not and do not spread.

The lack of accurate diagnostic methods for early identification of BC poses a challenge to prompt intervention. Since early-stage disorders can often be cured with little intervention, timely detection is essential. On the other hand, if BC is detected early on and stops from spreading to other parts of the body, it is treatable. The development of sophisticated diagnostic instruments is desperately needed to stop the growth of malignant cells or undesired cells (Bhise *et al.*, 2021).

Figure 1 provides a concise worldwide representation of the incidence of BC by showing the number of afflicted women per 100,000 people. The graph gives a thorough picture of the global prevalence of BC and sheds light on geographical and demographic differences.

Novel strategies for BC detection have been made possible by the combination of large datasets and novel clinical technologies. Classifiers based on DL have become useful tools for handling this problem. In recent years, DL techniques have been crucial in developing models for successful autonomous guidance in the healthcare business, leading to a noteworthy advancement in medical innovation (Nayak, Mahapatra et al. 2024). In our work, we demonstrate the power of DL classifiers for effectively identifying the BC. This work showcases the potential of ViT for identifying BC with time effectiveness and without any intervention of trained manpower.

2. Related Work

The lack of a comprehensive comparison between Convolutional Neural Networks (CNN) and conventional machine learning techniques for tumour classification in histopathology images represents a research gap. While the paragraph discusses varied datasets and methods, it lacks a clear evaluation of CNN against classic techniques (SVM, Random Forest, KNN, Logistic Regression, Naive Bayes) across multiple datasets. Furthermore, the emphasis on CNN's superiority calls for a more thorough investigation to determine its efficacy in various medical imaging contexts. The prevailing emphasis on datasets related to BC raises the possibility of exploring the applicability of these methods to other forms of cancer or medical imaging datasets. Closing this gap would improve our knowledge of medical image analysis using machine learning. Finally, investigating deep learning architectures that are more recent than CNN offers a potential direction for the advancement of tumor categorization studies (Panigrahi *et al.*, 2023).

3. Propose Model

Our suggested approach expedites the tumour classification process, quickly differentiating between benign and malignant tumours. Figure 2 describes the blueprint of the proposed model.

3.1. Dataset Descriptions

As part of the baseline data collection, breast ultrasound images were collected of

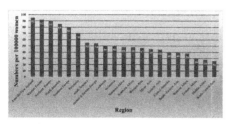

Figure 1: Number of women per 1000000 affected by breast cancer around the globe

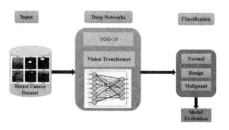

Figure 2: Blueprint of our proposed DL model

Table 1: Description of BC dataset utilized in our study

Samples	Number of images
Normal	266
Benign	421
Malignant	891
Total	1578

women aged 25 to 75. The total number of female patients is 600. The average pixel size of the 780 pictures in the sample is 500 by 500. Each image is labelled with one of three categories: normal, benign, or malignant. The dataset is sourced from Kaggle.

3.2. Data Preprocessing

To prepare raw data for analysis or DL model training, it must be cleaned, transformed, and arranged. Data preprocessing aims to improve the quality of the data, making it easier to work with and increasing machine learning algorithms' performance (Gheflati and Rivaz, 2022, Swain *et al.*, 2023).

4. Material and Methods

Our study was based on a dataset that included ultrasound images of BC. Using specialized models like VGG-19 and ViT (Garia and Hariharan 2023, Sood 2023), the major goal was to apply deep learning techniques for BC classification.

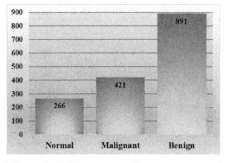

Figure 3: Different samples are presented in our studied dataset

4.1. Vision Transformer (ViT)

ViT, or Vision Transformer, is a neural network architecture intended for computer vision tasks, particularly picture classification (Li *et al.*, 2023). ViT uses a transformer architecture instead, which was first designed for natural language processing. ViT breaks away from pixel-centric processing by segmenting input images into non-overlapping patches. Each patch is then linearly embedded to create flat vectors, which are then used as the transformer's input. To fill in the gaps in spatial information, positional embeddings are included. To anticipate class labels, a standard classification head is attached. ViT demonstrates remarkable adaptability and scalability across a range of image sizes without the need for architectural modifications. (Shamshad *et al.*, 2023).

4.2. VGG-19

VGG-19 is a convolutional neural network (CNN) (Cong and Zhou 2023) with 19 layers. There are five blocks including sixteen convolutional layers, with max pooling coming after each block. A stride of 1, 3x3 filters with ReLU activation are used in the convolutional layers. After every block, stride 2 and max pooling with a 2x2 window are applied. With 4,096 neurons in the first two layers and 1,000 neurons in the third layer for ImageNet classes, three fully connected layers follow the convolutional blocks. To prevent overfitting, dropout is utilized in the first two completely linked layers. Class probabilities in the last layer are handled via SoftMax activation.

5. Result

To train and test the model, the Python 3.7 platform is employed. The performance indicators of the models under study are displayed in Table 2, which also includes important evaluation criteria for their efficacy. The models that are being examined include ViT (Vision Transformer) and VGG-19. Performance metrics include F1 score

Table 2: Performance metrics of the studied models

Models	Ac%	Pr%	Se%	Sp%	F1%
VGG-19	92	94	94	85	94
ViT	96	96	98	91	97

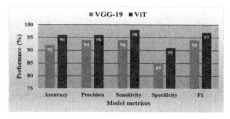

Figure 4: The performance metrics of ViT and VGG-19 on the studied BC dataset

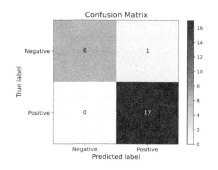

Figure 5: Confusion matrix from ViT BC classification model

(F1), sensitivity (Sen), specificity (Spe), accuracy (Acc), and precision (Pre). The VGG-19 model exhibits 92% accuracy, 94% precision and sensitivity, 85% specificity, and a 94% F1 score. ViT, on the other hand, performs better, with 96% accuracy, 96% precision, 98% sensitivity, 91% specificity and a 97% F1 score among its metrics.

In Figure 4 performance matrix, ViT performed better than VGG-19, demonstrating greater accuracy, precision, sensitivity and F1 Score. ViT's ability to manage intricate features and a global environment is emphasized by the contrast.

The confusion matrix produced by the ViT BC classification model is shown in Figure 5. The matrix indicates that there are 6 true negatives, 17 true positives, 1 false negative, and 0 false positives in this particular case.

6. Conclusion and Future Scope

The comparative study of two BC detection models, ViT and VGG-19, concludes that ViT performs significantly better, with a 96% accuracy rate. The findings highlight the potential role that transformer-based architectures, like ViT, can play in improving the accuracy and dependability of diagnostic systems, and also support its use in the diagnosis of BC. Furthermore, examining how interpretable ViT's forecasts are could

help us comprehend how it makes decisions. The robustness and clinical usability of the model may be improved by integration with cutting-edge technologies like explainable AI and continuous learning techniques.

References

Bhise, S., Gadekar, S., Gaur, A. S., Bepari, S., and Deepmala Kale, D. (2021). "Breast cancer detection using machine learning techniques." *International Journal of Engineering Research & Technology*, 10 (7): 2278-0181.

Cong, S. and Y. Zhou (2023). "A review of convolutional neural network architectures and their optimizations." Artificial Intelligence Review 56(3): 1905-1969.

Emmanuel, E. C. (2023). "Early breast cancer detection using convolutional neural networks."

Garia, L. S. and Hariharan, M. (2023). "Vision transformers for breast cancer classification from thermal images." *Robotics, Control and Computer Vision: Select Proceedings of ICRCCV 2022*, pp. 177-185. Springer.

Gheflati, B. and Rivaz, H. (2022). "Vision transformers for classification of breast ultrasound images." *2022 44th Annual International Conference of the IEEE Engineering in Medicine & Biology Society (EMBC)*. IEEE.

Li, Z., Y. Li, Q. Li, P. Wang, D. Guo, L. Lu, D. Jin, Y. Zhang and Q. Hong (2023). "Lvit: language meets vision transformer in medical image segmentation." *IEEE transactions on medical imaging* 43(1): 96-107.

Nayak, D. S. K., S. Mahapatra, S. P. Routray, S. Sahoo, S. K. Sahoo, M. M. Fouda, N. Singh, E. R. Isenovic, L. Saba and J. S. Suri (2024). "aiGeneR 1.0: An Artificial Intelligence Technique for the Revelation of Informative and Antibiotic Resistant Genes in Escherichia coli." Frontiers in Bioscience-Landmark 29(2): 82.

Panigrahi, A., A. Pati, B. Sahu, M. N. Das, D. S. K. Nayak, G. Sahoo and S. Kant (2023). "En-MinWhale: An ensemble approach based on MRMR and Whale optimization for Cancer diagnosis." *IEEE Access* 11: 113526 - 113542.

Shamshad, F., Khan, S., Zamir, S. W., Khan, M. H., Hayat, M., Khan, F. S., and Fu, H. (2023). "Transformers in medical imaging: a survey." *Medical Image Analysis*: 102802.

Sood, A. (2023). "Breast cancer detection using neural networks." NEU Journal for Artificial *Intelligence and Internet of Things* 1(1): 12-18.

Swain, R. R., Nayak, D. S. K., and Swarnkar, T. (2023). "A comparative analysis of machine learning models for colon cancer classification." *International Conference in Advances in Power, Signal, and Information Technology (APSIT)*. IEEE.

Enhancing Watermark Detection in Digital Media Ownership Protection through Hamming Coding

Priyanka Priyadarshini, Kshiramani Naik

Department of IT, VSSUT, Burla, Sambalpur, India, ppriyadarshini398@gmail.com, kshiramani@gmail.com

Abstract

Currently Digital watermarking plays a crucial role in ensuring the integrity of digital multimedia content. This paper represents a novel method for watermark embedding and extraction utilising Integer Wavelet Transform (IWT) and Hamming code encoding. The proposed scheme partitions the cover media into sub-bands using IWT, further enhancing its robustness by embedding encoded text watermarks into higher coefficients. During embedding, Hamming code is employed to encode the text, allowing for error correction and efficient extraction. The process involves converting the text into binary form, applying Hamming code. The method's effectiveness is evaluated through Bit Error Rate (BER) and Peak Signal-to-Noise Ratio (PSNR). Experimental results demonstrate the proposed technique's ability to accurately embed and extract watermarks while maintaining high image quality. Figures illustrating the application and extraction process provide a comprehensive overview of the proposed approach's efficacy and potential applications in digital content protection.

Keywords: Error correction, hamming coding, IWT, watermark detection, reversibility.

1. Introduction

Digital image watermarking is widely used for authentication, copyright protection, and ownership confirmation. Irreversible watermarking can harm the original image data but is necessary for military communication and healthcare. Reversible watermarking preserves the original data, important for online publishing and law enforcement. Researchers use techniques like SVD, DCT, DWT, and a combination of spatial and transform domain methods for watermarking. Various researchers have contributed to digital image watermarking advancements. Cox et al. emphasized fidelity and robustness, categorizing techniques into spatial and transform domains. Thapa et al. proposed an efficient SVD-based method, while Kaushik et al. suggested a DFT-based approach for enhanced security. Totla et al. compared DWT and DCT techniques, and Mohanty et al. introduced DCT-based watermarking. Singh et al. demonstrated high-resolution watermarking with DWT. Colin et al. highlighted watermarking's significance in medical imaging, and Mostafa et al. presented a DWPT-based technique for medical image security. These studies collectively enhance watermarking's security and authenticity, with this study using Hamming code for image tampering detection.

Chaper 52 DOI: 10.1201/9781003581215

The subsequent sections of this paper are structured as follows: section 2 presents the Hamming code technology, proposed watermarking scheme encompassing embedding and detection procedures are given in the Section 3, in Section 4 some experimental results illustrating the scheme's effectiveness in detecting and locating tampered regions, and finally, Section 5 summarises the conclusions

2. Hamming Codes

Hamming Codes, created by Richard Hamming, are linear block error correction codes. They use Hamming distance to measure discrepancies between check and information bits. This work focuses on binary Hamming codes, where three parity-check equations apply to any 7-bit message Craver et al.:

$$c1 \oplus c3 \oplus c5 \oplus c7 = 0 \quad (1)$$
$$c2 \oplus c3 \oplus c6 \oplus c7 = 0 \quad (2)$$
$$c4 \oplus c5 \oplus c6 \oplus c7 = 0 \quad (3)$$

where c_k denotes the 7-bit sequence's kth bit. The parity bits for the input 4-bit sequence, c1, c2, and c4, are generated using the aforementioned equations. Here is an illustration of how to expand a 4-bit sequence into a 7-bit sequence. If the Input Sequence is given as [0. 0. 1. 1.] then the Output Sequence will be [1. 0. 0. 0. 0. 1. 1.].

After reception, the Hamming code can correct up to one error in a 7-bit sequence. This is accomplished by calculating a vector known as the syndrome. The parity-check equations below are used to compute s0, s1, and s2 in the syndrome s = [s0, s1, s2]:

$$s0 = r1 \oplus r3 \oplus r5 \oplus r7 \quad (4)$$
$$s1 = r2 \oplus r3 \oplus r6 \oplus r7 \quad (5)$$
$$s2 = r4 \oplus r5 \oplus r6 \oplus r7 \quad (6)$$

where r_k denotes the kth bit in the received sequence, which is seven bits. The bits s0, s1, and s2 indicate the position of the error bit in binary form, with s0 representing the least significant bit and s2 representing the most significant bit. The error bit can then be located and edited using this information by applying the syndrome. An illustration of a recovered sequence with the wrong fourth bit. Sequence recovered: [1. 0. 0. 1. 0. 1. 1.] Error at the fourth bit: s0 = 0, s1 = 0, s2 = 1. Sequence corrected: [1. 0. 0. 0. 0. 1. 1.] [0. 0. 1. 1.] is the decoded sequence.

3. Proposed Method

This method uses IWT to partition the host image into sub-bands (LL, LH, HL, and HH), further grouped based on detail level. Sarmah et al. experimented that the text watermark is encoded with a Hamming code and inserted into second-level coefficients, enhancing robustness. Petitcolas et al. said the Hamming code, with 4 data bits and 3 parity bits, corrects a single bit error. Text is converted to binary and then encoded. PSNR assesses watermarked image quality, while BER measures watermark extraction strength. Figures 1 and 2 depict the technique's application and extraction process.

3.1. Embedding Algorithm

Step 1: Read the original image with dimensions n*n.

Step 2: Utilise Integer Wavelet Transform (IWT) to decompose original media up to the 2nd level.

Step 3: Retrieve the text intended for embedding and determine its length.

Step 4: Transform the text into its binary form.

Step 5: Utilise Hamming code for encoding the text watermark.

Step 6: Insert the encoded text into the cover image using this formula: for each coefficient (m,n) in IWT coefficients of cover image:

$$new_{coefficient} = coefficient + alpha *$$
$$coefficient * watermark[k]$$

Step 7: Execute the Inverse Integer Wavelet Transform (IIWT) to generate the watermarked image.

3.2. Extraction Algorithm

Step 1: Load the watermarked image.

Step 2: Utilise Inverse Integer Wavelet Transform (IIWT).

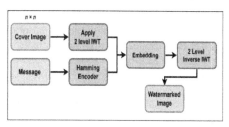

Figure 1: Embedding process of proposed method

Step 3: Extract text watermark from the watermarked media using the following equation:

for each coefficient (m,n) in IIWT coefficients of

watermarked image:

$$extracted_{watermark[k]} = \frac{(IIWT_{co} - original\_co)}{(alpha * original\_co)}$$

Step 4: Take the extracted watermark as binary values:
W[k] = sign(extracted_W[k])

Step 5: Decode the extracted text watermark.

Step 6: Return the extracted text.

4. Results and Discussions

Table 1 presents a comparison of watermark performance with and without Hamming code. The use of Hamming code results in a

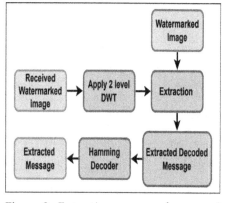

Figure 2: Extraction process of proposed method

Table 1: Robustness of watermark for different gain value

K value	Result using Hamming	
K	PSNR	BER
0.01	68.65	44.71
0.1	58.65	20.85
1	50.65	0
1.5	49.13	0
3	49.63	0
5	35.61	0

BER of 0, indicating improved text recovery quality and a reduced likelihood of errors in watermark recovery. However, the PSNR is slightly lower when using Hamming code compared to without it. This can be attributed to the increased calculations and changes to the cover image required for Hamming code application, which affects overall quality. Figure 3 shows different cover image with text messages which are experimented and corresponding watermarked images. Table 2 provides a comparison between the proposed scheme and previous work in terms of PSNR.

The following images display various cover images corresponding input text and watermarked images.

5. Conclusion

This method introduces a new color image watermarking scheme using Hamming code to enhance visual quality and security.

Table 2: Comparison of PSNR value of proposed scheme with existing work

Images	Cao et al.	Chang et al.	Proposed method
House	51.14	50.84	52.15
Couple	51.15	50.12	54.62
Baboon	51.15	50.95	53.55
Lake	50.12	50.94	52.28
Female	51.15	50.77	56.23

| Cover Image | Message | Watermarked Image |

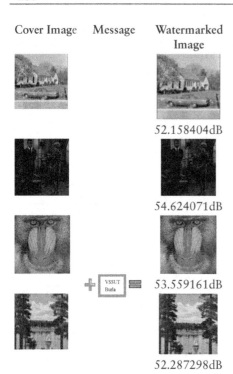

52.158404dB

54.624071dB

53.559161dB

52.287298dB

Figure 3: Experimental result of different cover image with text messages

By integrating a secret key in embedding and extraction, it improves security and tamper detection. The approach is useful for authentication, ownership identification, and copyright protection, especially in educational institutions. It uses IWT for embedding, combined with error correcting codes for better error correction. The method suggests extending the error correction code length to improve error correction capabilities, aiming to enhance both image quality and watermark robustness.

References

Cao, Z., Yin, Z., Hu, H., Gao, X., and Wang, L. (2016). "High capacity data hiding scheme based on (7, 4) Hamming code." *SpringerPlus*, 5 (1): 1-13.

Chang, C.-C., Kieu, T. D., and Chou, Y.-C. (2008). "A high payload steganographic scheme based on (7, 4) hamming code for digital images." *International Symposium on Electronic Commerce and Security*, pp. 16-21. IEEE.

Colin, R. R. and Uribe, C. F., Villanueva, J.-A. M. (2008). "Robust watermarking scheme applied to radiological medical images." *IEICE Transactions on Information and Systems*, 91 (3): 862-864.

Cox, I. J., Miller, M. L., Linnartz, J.-P. M. G., and Kalker, T. (2018). "A review of watermarking principles and practices 1." *Digital Signal Processing for Multimedia Systems*: II(VIII), 461-485.

Craver, S., Memon, N., Yeo, B.-L., and Yeung, M. M. (1997). "On the invertibility of invisible watermarking techniques." *Proceedings of International Conference on Image Processing*, vol. 1, pp. 540-543. IEEE.

Kaushik, A. K. (2012). "A novel approach for digital watermarking of an image using DFT." *International Journal of Electronic and Computer Science Engineering*, 1 (1): 35-41.

Mohanty, S. P. (1999). "Digital watermarking: a tutorial review." http://www.csee.usf.edu/~smohanty/research/Reports/WMSurvey1999Mohanty.pdf.

Petitcolas, F. A. P., Anderson, R. J., and Kuhn, M. G. (1999). "Information hiding-a survey." *Proceedings of the IEEE*, 87 (7): 1062-1078.

Sarmah, D. K. and Bajpai, N. (2010). "Proposed System for data hiding using Cryptography and Steganography Proposed System for data hiding using Cryptography and Steganography." *arXiv preprint arXiv:1009.2826*.

Singh, A. K., Dave, M., and Mohan, A. (2012). "A novel technique for digital image watermarking in frequency domain." *2nd IEEE International Conference on Parallel, Distributed and Grid Computing*, pp. 424-429. IEEE.

Thapa, M., Sood, S. K., and Sharma, A. P. M. (2011). "Digital image watermarking technique based on different attacks." *International Journal of Advanced Computer Science and Applications*, 2 (4), pp. 1-13.

Totla, R. V. and Bapat, K. S. (2013). "Comparative analysis of watermarking in digital images using DCT & DWT." *International Journal of Scientific and Research Publications*, 3 (2): 1-4.

Structural Integrity Preservation through Digital Watermarking using Dual Decomposition Techniques

Alina Dashi1, Kshiramani Naik2

1Department of CSE, VSSUT, Burla, Sambalpur, India, alinadash_cse@vssut.ac.in
2Department of IT, VSSUT, Burla, Sambalpur, India, kshiramaninaik_it@vssut.ac.in

Abstract

In In today's digital realm, where image sharing is ubiquitous, safeguarding the privacy and integrity of shared images is critical. This paper proposes advanced digital image watermarking techniques, including DCT, QSVD, and Schur Decomposition, to protect against unauthorized use and tampering. By combining these methods, the aim are to enhance image privacy and revolutionize digital watermarking, addressing challenges in the interconnected environment. Key objectives include improving security, imperceptibility, and computational efficiency, offering a robust solution for secure image sharing in the evolving digital era.

Keywords: Digital image watermarking, image privacy, Discrete Cosine Transform (DCT), Quaternion Singular Value Decomposition (QSVD), schur decomposition, imperceptibility

1. Introduction

In today's digital landscape, image sharing is integral to communication, underscoring the importance of preserving image privacy and integrity. Robust digital image watermarking techniques are essential to address this need. This work focuses on employing advanced techniques like DCT, QSVD, and Schur Decomposition to secure image privacy. DCT excels in representing images in the frequency domain, while QSVD offers a sophisticated approach using quaternion matrices, accommodating larger watermarks. Schur Decomposition enhances security and imperceptibility by capturing both magnitude and phase information. By combining these techniques, we aim to revolutionize digital image watermarking, contributing

to enhanced image privacy and protection in the evolving digital era. Zhang *et al.* (2023) introduced an innovative approach based on structure-preserving technique for color image watermarking based on a quaternion singular value decomposition. Zhang *et al.* (2023) proposed a Dual-Color QSVD based Watermarking for imperceptible embedding, robust to common attack along with enhanced computational efficiency. Tang *et al.* (2021) used hashing and QSVD based watermarking technique which extracts stable and discriminative image features from CIE Lab color space. It also utilizes Euclidean distance for compression, resulting in compact hashes. Chen *et al.* (2020) introduced blind color image watermarking using quaternion QR decomposition which utilizes strong correlations in quaternion matrix for high capacity watermarking. Computational load minimized with structure preserving method. Sun *et al.* (2022) proposed dual color image watermarking which offers strong robustness against common and geometric attacks enhancing invisibility compared to other methods. Chen *et al.* (XXXX) introduces novel real structure preserving QSVD algorithm which Incorporates coefficient pair selection and adaptive embedding.

2. Background Study

2.1. Discrete Cosine Transfer

A commonly employed technique for incorporating image watermarking within the frequency domain involves utilising the Discrete Cosine Transform. The procedure involves partitioning the original image into smaller, non-overlapping blocks. Discrete Cosine Transform is mathematically defined as shown below in Equation 1.

$$D = \int_{a=0}^{N-1} x_a \cos\left[\frac{\pi}{N}\left(a+\frac{1}{2}\right)k\right] (1)$$

Where, k = 0,1,2....N-1

2.2. Schur Decomposition

Schur Decomposition is a matrix factorisation technique that, similar to Eigenvalue Decomposition (EVD), breaks down a square matrix into simpler components. Specifically, for a square matrix A,
The Schur Decomposition expresses it as the product of mentioned in equation 2.

$$A = QTQ^* \qquad (2)$$

Where Q is unitary matrix, T is an upper triangular matrix. The conjugate transpose of Q is noted as Q^*.

2.3. Quaternion Singular Value Decomposition

In QSVD, matrix A, which is quaternion in nature, is decomposed into three components: R, T, and V. The decomposition equation can be expressed as in equation 3

$$A = RTV^\circ \qquad (3)$$

Where R is an orthogonal quaternion matrix, T is a diagonal quaternion matrix containing the singular values, and V^* denotes the conjugate transpose of the orthogonal quaternion matrix V.

3. Proposed Methodology

The proposed watermark technique is based on DCT, Schur Decomposition and Quaternion Singular Value Decomposition for achieving a robust and imperceptible watermarked image preserving the structural integrity and authenticity. Overall method is divided into two phase of Embedding and Extraction.

3.1. Steps of Embedding Phase

Step-1: DCT is applied to the cover image, producing the DC Coefficient Matrix.
Step-2: Proceed with Schur Decomposition on the result of the DCT of the cover image, yielding a matrix that captures both magnitude and phase information.
Step-3: Pass the watermark image through QSVD, obtaining a quaternion matrix representation.
Step-4: Perform Schur Decomposition on the QSVD watermark matrix, resulting in a quaternion diagonal matrix.
Step - 5: Utilise the Xor embedding technique to combine the matrices derived from the original image and the watermark image.
Step - 6: Produce the watermarked image by implementing the ISD on the embedded image.
Step - 7: Conclude the embedding process by employing inverse DCT on the outcome of the inverse Schur Decomposition, leading to the generation of the watermarked image.
The total process of embedding is illustrated in Figure 1 as below.

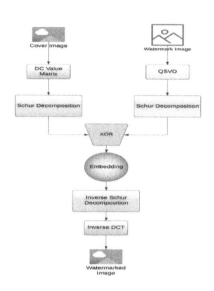

Figure 1: Embedding procedure for proposed model

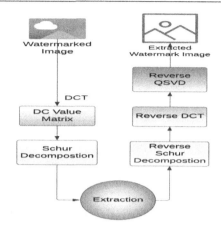

Figure 2: Extraction procedure for proposed model

3.2. Steps for Extraction Phase

Step 1: Applying Discrete Cosine Transform (DCT) on watermarked image to extract DC value matrix.

Step 2: Inverse Schur Decomposition: Apply the inverse of the Schur Decomposition performed during embedding to retrieve the combined matrix containing the watermark.

Step 3: Xor Extraction: Reverse the XOR operation used for embedding by applying the same operation to the matrices derived from both images.

Step 4: Inverse Schur Decomposition on QSVD Watermark Matrix:

Step 5: Inverse QSVD: Reverse the QSVD transformation performed on the watermark image to obtain the original watermark.

Step 6: Inverse DCT: Finally, apply the inverse DCT to the outcome of the inverse Schur Decomposition to recover the original cover image.

4. Experimental Analysis

To evaluate the efficacy of the proposed watermarking technique in the context of secure image sharing systems, a comprehensive experimental analysis was conducted. The dataset includes cover images such as mandril, ship Lena, cameraman, and tulip. These images are commonly employed in image processing and computer vision research due to their diverse visual characteristics and complexity. 5 cover images of size 256 x 256 and an Elephant image as watermark image, displayed in Figure 3

Figure 3: Cover images and watermark

Table 1: Proposed method parameters

Image	PSNR	MSE	NCC	SSIM
Mandril	49.7163	0.0521	0.9971	0.9989
Ship	48.1531	0.0252	0.9992	0.9999
Lena	49.3814	0.8741	0.9999	0.9997
Cameraman	49.6123	0.0492	0.9996	0.9697
Tulip	48.8238	0.0321	0.9999	0.9988

Table 2: Comparision of proposed method with other state of arts

Methods	Av.PSNR	Av.NCC	Av.SSIM
Chen *et al.* (XXXX)	42.1634	0.9218	0.9431
Tang *et al.* (2021)	41.8312	0.9611	0.9261
Yong *et al.* (XXXX)	45.3184	0.9883	0.9514
Sun *et al.* (2022)	46.4714	0.9971	0.9902
Proposed	**49.1373**	**0.9991**	**0.9934**

4.2. Performance Parameters

For evaluating the performance of proposed work a set of key evaluation metrics is employed. These metrics include Mean Squared Error (MSE), Peak Signal-to-Noise Ratio (PSNR), Structural Similarity Index (SSIM), and Normalised Cross Correlation (NCC) as shown in Table 1.

4.3. Result Analysis and Discussion

In performance result we have taken different performance evaluation matrices like MSE, PSNR, NCC, SSIM and results and also Histogram of Embedded Watermarked image is also considered in Table 1. We have got average PSNR of 49.1373 db
which is a quite good value for imperceptibility. Again NCC and SSIM are 0.9991 and 9934 respectively, which are nearly equal to 1 indicating preservation of structural integrity is very high in our proposed work. We have compared our proposed methodology with other existing work in Table 2 and it has been noted that it outperforms others in Average PSNR, NCC, and SSIM.

5. Conclusion

The proposed methodology has delved into the domain of securing image privacy through advanced digital image watermarking techniques, specifically employing the Discrete Cosine Transform (DCT), Quaternion Singular Value Decomposition (QSVD), and Schur Decomposition. The combination of these techniques represents a novel approach in addressing the escalating challenges associated with the widespread sharing of digital images in today's interconnected world.

References

Chen, Y., Jia, Z., Peng, Y., and Peng, Y. (2020). "Robust dual-colorwatermarking based on quaternionsingular value decomposition." *IEEE Access*, 8: 30628-30642.

Chen, Y., Jia, Z.-G., Peng, Y., Peng, Y.-X., and Zhang, D. (2021). "A new structure-preserving quaternion QR decomposition method for color image blind watermarking." *Signal Processing*, 185: 108088.

Chen, Y., Jia, Z.-G., Peng, Y.-X., and Peng, Y. (2023). "Efficient robust watermarking based on structure-preserving quaternion singular value eecomposition." *IEEE Transactions on Image Processing.*

Sun, Y., Su, Q., Wang, H., and Wang, G. (2022). "A blind dual color images watermarking based on quaternion singular value decomposition." *Multimedia Tools and Applications*: 1-23.

Tang, Z., Yu, M., Yao, H., Zhang, H., Yu, C., and Zhang, X. (2021). "Robust image hashing with singular values of quaternion SVD." *The Computer Journal*, 64 (11): 1656-1671.

Zhang, M., Ding, W., Li, Y., Sun, J., and Liu, Z. (2023). "Color image watermarking based on a fast structure-preserving algorithm of quaternion singular value decomposition." *Signal Processing*, 208: 108971.

Use of Machine Learning Models and Ayurveda in Early Malignancy Prediction

Ritadrik Chowdhury, Suparna Das Gupta, Prolay Ghosh, Kallol Acharjee, Anunay Ghosh

Department of IT, JIS College of Engineering, West Bengal, India

Abstract

Cancer is a leading cause of death in India, necessitating early diagnosis and prediction. Medical teams use deep learning and machine learning models to simulate disease emergence and management. Various methods, including Support Vector Machines, Bayesian Networks, Decision Trees, and Artificial Neural Networks, are used for feature identification from large datasets.

Keywords: Dhatu, granthi, cancer, DT, SVM, AMA

1. Introduction

The detection rate of cancer prediction models is around 88.89%, indicating their effectiveness and precision. Deep learning and machine learning approaches can accelerate cancer prediction, but they must be accurate and consider medical procedures. The study categorises methods, datasets, and situations, evaluating performance indicators. Ayurveda, the oldest plant-based therapy in India, has been used to prevent and reduce various cancers. Researchers are now studying complementary and alternative therapies for cancer treatment. Early detection and treatment are expensive, and late diagnosis can lead to poor outcomes. Early detection and treatment are crucial for preventing malignancy and improving overall health outcomes.

Here, Table 1 shows the modern and ayurvedic views on cancer differ, with recent molecular data supporting ayurvedic theories. Ayurveda posits that nutrition and environmental factors influence Agni and immunity, potentially increasing cancer risk. The "shared pathology" between cancer and metabolic syndrome resembles Ayurvedic theory, involving systemic malfunctions in weak tissues. This article provides an overview of cancer detection using machine learning and traditional Ayurvedic methods, highlighting the potential of combining the Humanoid Brain with ancient wisdom for cost-effective early detection.

2. Literature Review

Studies on malignancy prediction, prevention, and curative approaches have been conducted in various fields. In India, 14,61,427 new cases of cancer are expected in 2022, with fatigue, nausea, vomiting, decreased appetite, altered tastes, hair loss, dry mouth, and constipation being the most common side effects. Deep Neural Networks (CNNs) have been used to identify or label medical images, with a multi-scalar two-layer CNN confirming lung cancer in 2015 with 86.84% accuracy (Sathishkumar et al., 2022).

SVM using Artificial Neural Networks and Decision-making Trees (SVM) has been used to predict prostate cancer survival

Table 1: Difference between modern and ayurvedic view of cancer

Earlier Medical concepts of Cancer	Ayurvedic concepts of Arbuda	Supporting Concept of Cancer in Ayurveda
Cancer is the outcome of a series of genetic events that cause unchecked cell proliferation and resistance.	Arbuda develops when the Doshas get vitiated and immunity is compromised as a result of aberrant interactions between Prakriti (genotype) and environmental variables.	Cancer is caused by abnormalities other than abnormal cell proliferation and cell death. Phenotypes are influenced by epigenetic regulation, food, environmental variables, and immunity.
The majorities of malignancies start as spontaneous mutations in particular tissues and progress to other organs.	Organ arbuda result from the interaction of vitiated doshas and weak tissues (dhatus).	Metabolic syndrome and cancer share a common molecular pathogenesis.
Diets high in fibre are linked to a lower risk of cancer and heart disease. Cancer and the inflammatory process were not connected.	Links between diseases and poor diet, digestion, metabolism, and inflammation. The biomarker "Ama" may be a new one for early inflammation.	All phases of carcinogenesis are actively promoted by chronic inflammation.
Radiotherapy and chemotherapy do not target specific cancer tissue. Additionally, these treatments damage healthy tissue and have negative side effects.	Through the eradication of vitiated Doshas, the rejuvenation of Dhatus, and the restoration of immunity, therapies indirectly target granthi.	Drugs that treat diabetes and inflammation inadvertently kill cancerous tissue. Immunotherapy. Vaccinations for cancer.

with 92.85% accuracy. However, discrete time models may not be suitable for situations where imaging observations are made at strange frequencies. A continuous time model could better handle this problem. A more advanced method could involve learning the state space from the data, which is planned to be studied in the future.

Figure 1 describes the process how cancer cells are developed as per Ayurvedic scriptures.

CAM (Cellular Mechanics) is establishing itself as a powerful modality to bridge the gap in cancer treatment. Swarna (gold) and Bhasma (ash) are used as the main form of treatment, with gold nano-particles showing pharmacological effects. Nanoshell-Assisted Photothermal Therapy (NAPT) uses gold-silica nanoshells to kill tumor cells using near-infrared radiation and has shown

analgesic qualities. Gold-EGFR conjugates have been applied to the detection of cancer cells by using their ability to scatter light (Altun and Sonakya, 2018).

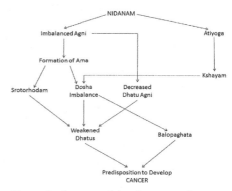

Figure 1: Process of development of cancer cells

However, the study's sample size is insufficient to produce statistically meaningful results, requiring more balanced, blinded research with a larger sample size.

The study investigates the harmonisation of Ayurvedic and biological principles in oncology, focusing on the pathophysiology of cancer. This is the first study to use rigorous research approaches to look for connections between principles from biological oncology and Ayurvedic medicine. This study's samprapti of cancer, which combines historical and contemporary biomedical frameworks, is the best Ayurvedic explanation of cancer. [Dhruva A et.al] Nuclear receptors that sense cholesterol, called LXRs, control lipid transport and metabolism, reducing inflammatory signaling in macrophages. Adiponectin synthesis is stimulated by PPARγ ligands' anti-inflammatory activities, counteracting the effects of pro-inflammatory cytokines. PPARγ nuclear receptors are considered a crucial part of the molecular pathways linking metabolic syndrome and cancer formation (Wochna Loerzel , 2015).

FXR, another nuclear receptor, can also have anti-tumor effects. Mice lacking the FXR gene are more prone to developing intestinal tumors and inflammation caused by endotoxin lipopolysaccharide (LPS). FXR serves as a major therapeutic target for colorectal cancer prevention due to increased excretion of secondary bile acids. Guggulsterone may be a beneficial treatment for colon cancer due to its effectiveness against hyperlipidemia and its capacity to bind FXR. These findings collectively suggest that FXR expression and activity at normal levels have significant anti-inflammatory and anti-tumor effects (F.J. Shaikh, D.S. Rao (2022)).

The term "Ama" in Ayurveda refers to a waste product of digestion and metabolism that is poisonous, heavy, unctuous, and sticky. It accumulates in individuals with underdeveloped or overburdened digestion with incorrect nutrients. Ayurvedic medical professionals identify "signs and symptoms" of "Ama," such as nausea, vomiting, diarrhea, lethargy, indigestion, constipation, inflammation, sudden fatigue, heaviness, and pain.

The main objectives of Ayurvedic treatment are Agni strengthening and "Ama" digestion in its whole. Therapies like purgation, enema, or therapeutic emesis are used to fully digest "Ama" and separate the vitiated Dosha from different channels. Extra "Ama" can be found in urine and on the tongue, allowing comparison and quantification biochemically between healthy individuals and patients with various malignancies, metabolic syndrome, or both disorders.

Table 2: Review-based comparison study of accuracy in different methodologies

Publication	Method	Cancer type	No of Patients	Type of data	Accuracy	Validation method	Important features
Ayer T et al.[11]	ANN	Breast cancer	62,219	Mammographic, demographic	69.55%	10-fold cross validation	Age, mammography findings
Waddell M et al.[12]	SVM	Multiple myeloma	80	SNPs	71%	Leave-one-out cross validation	snp739514, snp521522, snp994532
Listgarten J et al.[13]	SVM	Breast cancer	174	SNPs	69%	20-fold cross validation	snpCY11B2 (+) 4536 T/C snpCYP1B1 (+) 4328 C/G
Stajadinovic et al [14]	BN	Colon carcinomatosis	53	Clinical, pathologic	71%	Cross-validation	Primary tumor histology, nodal staging, extent of peritoneal cancer
I3SET2k22	LG, BN	Multiple Myeloma	10	Clinical	78.9%	Cross Validation	Age, Habits, Pathological Findings, Genetic History

Clinical research may discover statistically significant associations between the biochemical make-up and concentrations of "Ama" and the degree of metabolic syndrome and/or malignancies. Ama should be detected in its early stages because it results from incorrect digestion and metabolism in healthy, non-obese persons.

Pre-clinical and clinical studies can establish whether "Ama" is an accurate biomarker of early inflammation. Although well-known inflammation markers like C- reactive protein, TNF-α, and IL-6 are linked to high-fat diets, these markers only show immunological dysfunction in its later phases. The ratio of omega-6 to omega-3 fatty acids is a highly accurate indicator of oxidative stress and diet-related inflammation (M. Waddell et.al (2005)).

In conclusion, "Ama" could be a reliable biomarker of early inflammation, connecting inflammation with poor food and digestion.

Table 2 shows that various machine learning techniques have varying detection accuracy for diagnosing diseases. The combination of LG and BN may offer a slight advantage, while SVM, ANN, BN, and LG are the most popular. Predictive patient criteria vary and depend on physical factors, pathological findings, past internal problems, tissue disorganisation, genetic data, and habits. The patient may have ignored the issue due to a belief it was a typical discomfort (Acharya SS (2018)).

3. Conclusion

Despite great advances in applying machine learning for cancer detection, numerous knowledge gaps remain. Some of the significant challenges are:

Limited access to high-quality data: Machine learning algorithms use vast datasets to find patterns and generate accurate predictions. However, there is often a shortage of high-quality cancer data available for analysis, particularly for rare or understudied cancer types.

Lack of standardisation in data collection and analysis: Cancer data is often collected from a variety of sources using different protocols and standards, making it difficult to compare and analyze data across studies. This lack of standardisation can limit the accuracy and reliability of machine learning models.

Difficulty in accounting for tumor heterogeneity: Tumor cells can vary significantly in their genetic makeup and behavior, even within the same cancer type. Machine learning algorithms need to account for this heterogeneity to accurately predict cancer outcomes and guide treatment decisions (Ayer. T. et. al.(2010)).

Machine learning models have limited interpretability: Many machine learning models are called "black boxes," which makes it difficult to understand how the model makes its predictions. This lack of interpretability can make it challenging for clinicians to accept and apply the models in clinical practice. Addressing these knowledge gaps will require continued collaboration between researchers, clinicians, and data scientists. Efforts to standardise data collection and analysis develop more sophisticated machine learning algo, and increase the availability of high-quality cancer data will is critical to improving the accuracy and reliability of cancer detection and treatment using ML.

References

Acharya, S. S. (2018) *Sushruta Samhita.* AVR publication.

Altun, I. and Sonkaya, A. (2018). "The most common side effects experienced by patients were receiving first cycle of chemotherapy." *Iranian Journal of Public Health*, 47 (8): 1218-1219.

Ayer, T., Chhatwal, J., Alagoz, O., Kahn, C. E., Jr, Woods, R. W., and Burnside, E. S. (2010). "Comparison of logistic regression and artificial neural network models in breast cancer risk estimation." *RadioGraphics*, 30 (1): 13-22.

Das, S., Das, M. C., and Paul, R. (2022). "Swarna Bhasma in cancer: a prospective clinical study." https://doi.org/10.4103/0974-8520.108823.

Loerzel, V. W. (2015). "Symptom experience in older adults undergoing treatment for cancer." *OncolNurs Forum*, 42 (3): E269-E278. https://doi.org/10.1188/15.ONF.E269-E278. PMID: 25901389.

Petousis, P., Winter, A., Speier, W., Aberle, D. R., Hsu, W., and Bui, A. A. T. (2019). "Using sequential decision making to improve lung cancer screening performance." *IEEE Access*, 7: 119403-119419. https://doi.org/10.1109/ACCESS.2019.2935763.

Sathishkumar, K., Chaturvedi, M., Das, P., Stephen, S., and Mathur, P. (2022). "Cancer incidence estimates for 2022 & projection for 2025: result from National Cancer Registry Programme, India." *Indian Journal of Medical Research*, 156 (4&5): 598-607. https://doi.org/10.4103/ijmr.ijmr_1821_22.

Shaikh, F. J. and Rao, D. S. (2022). "Prediction of cancer diase using machine learning approach." *Materials Today: Proceedings*, 50, (Part 1), 40-47, ISSN 2214-7853, https://doi.org/10.1016/j.matpr.2021.03.625.

Waddell, M., Page, D., and Shaughnessy, J. (2005). "Predicting cancer susceptibility from single-nucleotide polymorphism data: a case study in multiple myeloma." *Proceedings of the 5th international workshop on Bioinformatics (BIOKDD '05)*. Association for Computing Machinery, New York, NY, USA, pp. 21–28. https://doi.org/10.1145/1134030.1134035.

Predictive Modelling of Disease-Gene Association through Machine Learning Approaches

Riyam Patel1, Aishani Rachakonda2, T. S. Shiny Angel2

1Department of Computational Intelligence, College of Engineering and Technology, SRM Institute of Science & Technology, Chennai, India, rp1950@srmist.edu.in
2Department of CINTEL, COET, SRM Institute of Science & Technology, Chennai, India, sr2665@srm ist.edu.in, shinyant@srmi st.edu.in

Abstract

The study of genetic factors and disease susceptibility is crucial for biomedical research, as it helps understand molecular mechanisms and develop targeted therapeutic interventions, despite traditional methods being time-consuming. In response to this challenge, the use of machine learning (ML) with genomics research has arisen as an innovative method, offering a rapid and efficient means to predict disease-gene-associations. Here, in this paper, a framework has been proposed using ensemble learning to predict disease-gene association which is producing good results with high accuracy as compared to existing methods.

Keywords: Convolutional Neural Network (CNN), Disease Gene Prediction (DGP), Decision Tree (DT), Machine Learning (ML), Random Forest (RF), Disease Gene Database (DGDB), Ensemble Learning (EL)

1. Introduction

Machine learning is a powerful tool for analyzing large-scale genetic and biological datasets, enabling researchers to make more accurate predictions about genes involved in specific disorders. This is particularly useful in disease-gene association prediction due to its efficiency in handling the growing volume of genomic information. Algorithms like supervised learning and network-based methods provide a flexible tool for integrating data sources and revealing hidden connections between genes and illnesses. This study explores and proposes a framework for predicting disease-gene associations, highlighting its potential for precision medicine and tailored treatment interventions. It also explores challenges, approaches, and outcomes, highlighting the potential for significant healthcare advancements.

1.1. Contribution

It has been developed a comprehensive ensemble framework that integrates diverse machine learning algorithms such as Convolutional Neural Network, Stacked Regression, Decision Tree, Random Forest, and Ensemble Learning to obtain the high accuracy mode.

2. Literature Review

Many literatures are available in the biomedical related to predicting the disease-gene association. A few of them have been described in this section.

Chapter 55 DOI: 10.1201/9781003581215

This study (Asif's *et al.*, 2018) proposed a supervised algorithm to forecast Genes linked to ASD (Autism Spectrum Disorder). The classifiers, trained on ASD and non-ASD genes, performed better than previous ASD classifiers.

The authors (Luo *et al.*, 2019) introduce a multimodal DBN (dgMDL) technique which uses two Deep Belief Networks to acquire Encoded representations of protein-protein interaction-networks and based on gene-ontology, then combines them to create cross-modality representations that outperform other algorithms.

The study by (Wu *et al.*, 2019) uses RENET to identify gene-disease connections, improving accuracy and recall rates to 85.2% and 81.8%, respectively and (Han *et al.*, 2019) introduces GCN-MF, a new structure for establishing connections between diseases and genes is proposed by merging matrix factorisation with Graph Convolutional Network.

The author (Le, 2020) has presented a strategy for disease gene prediction using machine learning methodologies. Advanced techniques like deep learning, ensemble learning, and matrix factorisation are used to identify non-disease-causing genes. The authors (Madeddu *et al.*, 2020) uses a technique use machine learning algorithms to analyze the functional-and-connectivity patterns of surrounding proteins. Its purpose is to predict the links between diseases and genes inside the human interaction network.

The use Knowledge-Graph Embedding tech. Vilela *et al.*, (2023) on a biological Knowledge Graph to enable logical deduction based on publically available databases. It has been introduced a machine-learning framework based on network (Yang *et al.*, 2021) for identifying functional modules and illness candidate genes. MapGene, a disease gene-prioritising technique, outperforms current algorithms and generates highly dependable rankings of candidate genes.

GediNET is a machine learning model that uses the G-S-M method to detect meaningful connections among diseases

stated by (Qumsiyeh *et al.*, 2022). The authors (Li *et al.*, 2023) developed a Prediction method with Parallel Graph Transformer Network (DGP-PGTN), that combines information from illnesses, genes, ontologies, and phenotypes. The study uses a gene-disease network, gene ontology, and protein interaction networks, and hypothesises the correlation between genes and illness classes.

2.1. Proposed Model

The proposed model shows in Figure 1 uses a disease gene dataset and pre-processing techniques like quality control, data cleaning, and filtering. It uses 60% data for training and 40% for testing, with five algorithms applied in two phases: MLP, CNN, Decision Tree, and Random Forest. The best algorithm is selected for the next phase, and the results are stored separately. In the second phase, random forest and gradient-boosting are trained parallelly during ensemble learning, and logistic regression is used as a meta-classifier in the second layer of stacking regression.

2.2. Performance Measures

The confusion matrix used in disease-gene association prediction are as follows. It helps identify classes mistaken by model as other classes. Here, P=True-Positive(TP), Q=False-Positive (FP), R=False-Negative(FN), and S=True-Negative(TN).

The formula used here are;

$$Precision_{disease} = \frac{P}{(P+Q+R)} \qquad (1)$$

$$Recall_{disease} = \frac{P}{(P+X+A)} \qquad (2)$$

$$Accuracy =$$
$$\left(\frac{P+Y+C}{P+Q+R+X+Y+Z+A+B+C} \right) \times 100 \qquad (3)$$

$$F_1\,Score_{disease} =$$
$$\left(\frac{2 \times Precision \times Recall}{Precision + Recall} \right) \times 100 \qquad (4)$$

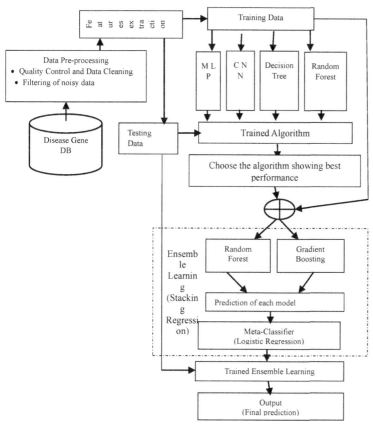

Figure 1: Disease Gene Association Prediction Model (Proposed)

Figure 1: Disease Gene Association Ensemble Prediction Model

2.3. Experimental Analysis and Results

The experiment uses data from the DisGeNET dataset, a discovery platform with a large collection of genes and variants related to human diseases. Ensemble methods, including sequential and parallel techniques, are used. Parallel ensemble techniques generate base learners in parallel formats, like random forest, promoting independence between learners. This reduces error due to averages. Stacking, a parallel ensemble technique, uses predictions for multiple nodes, such as Random Forest and Gradient Boost in layer-1, and Logistic Regression in the output layer as a meta classifier. AUC, AUPR, precision, recall, and F1-Score are

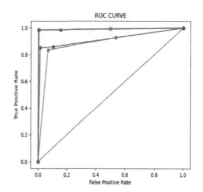

Figure 2: ROC Curve of all ML algorithms

evaluation measures. Figure 2 shows the Receiver-Operating-Characteristics-curve (ROC) with the horizontal & vertical axes

Table 1: Performance of classifiers

Algorithm	Accuracy	Precision	Recall	F1 - Score
RF Classifier	0.988	0.988	0.988	0.988
AdaBoost	0.742	0.744	0.742	0.743
Decision Tree	0.984	0.984	0.984	0.984
CNN	0.689	0.699	0.698	0.569
MLP Classifier	0.887	0.877	0.886	0.876
Stacking Regresso	0.998	0.99	0.992	0.991

set to the false-positive rate (FPR) and true-positive rate (TPR). Set the precision ratio as the horizontal-axis and the recall ratio as the vertical-axis to construct the precision-recall-curve.

Table 1 compares classifier performance and it reveals that the suggested technique has high accuracy, F1 score, and recall.

3. Conclusion

Successful prediction of disease-gene associations holds significant implications for personalised medicine, drug discovery, and therapeutic target identification. The application of machine learning in predicting disease-gene associations represents a promising avenue for accelerating biomedical-research & improving complicated disease understanding. The integration of computational approaches with experimental validation will pave the way for transformative breakthroughs in personalised medicine and drug development.

References

Asif, M., Martiniano, H. F., Vicente, A. M., & Couto, F. M. (2018). Identifying disease genes using machine learning and gene functional similarities, assessed through Gene Ontology. PloS one, 13(12), e0208626.

Han, P. *et al.* (2019). "GCN-MF: disease-gene association identification by graph convolutional networks and matrix factorization." *Proceedings of the 25th ACM SIGKDD International Conference on Knowledge Discovery & Data Mining*, pp. 705-713, ACM Digital Library

Le, D. H. (2020). "Machine learning-based approaches for disease gene prediction." *Briefings in Functional Genomics*, 19 (5-6): 350-363.

Li, Y., Guo, Z., Wang, K., Gao, X., and Wang, G. (2023). "End-to-end interpretable disease–gene association prediction." *Briefings in Bioinformatics*, 24 (3): bbad118.

Luo, P. *et al.* (2019). "Enhancing the prediction of disease–gene associations with multimodal deep learning." *Bioinformatics*, 35 (19): 3735-3742.

Madeddu, L., Stilo, G., and Velardi, P. (2020). "A feature-learning-based method for the disease-gene prediction problem." International *Journal of Data Mining and Bioinformatics*, 24 (1): 16-37.

Qumsiyeh, E., Showe, L., and Yousef, M. (2022). "GediNET for discovering gene associations across diseases using knowledge based machine learning approach." *Scientific Reports*, 12 (1): 19955.

Vilela, J., Asif, M., Marques, A. R., Santos, J. X., Rasga, C., Vicente, A., and Martiniano, H. (2023). "Biomedical knowledge graph embeddings for personalized medicine: predicting disease-gene associations." *Expert Systems*, 40 (5): e13181.

Wu, Y., Luo, *et al.* (2019). "Renet: a deep learning approach for extracting gene-disease associations from literature." *RECOMB, USA, May 5-8, 2019, Proceedings*, pp. 272-284. Springer International Publihser.

Yang, K., Lu, K., Wu, Y., Yu, J., Liu, B., Zhao, Y., and Zhou, X. (2021). A network-based machine-learning framework to identify both functional modules and disease genes." *Human Genetics*, 140, 897-913.

Design of Wireless Sensor Network-based IoT platform for Precision Agricultural Applications

Sanjib Kumar Nayak[1], Sohan Kumar Pande[2], Suresh Kumar Srichandan[1]

[1]Department of CA, Veer Surendra Sai University of Technology, Odisha, India, sknayak_ca@vssut.ac.in, suresh.vssut15@gmail.com
[2]Department of CSE, Silicon Institute of Technology, Odisha, India, [b]ersohanpande@gmail.com

Abstract

This research work provides a brief description of wireless sensor network based IoT application for precision agriculture application. The WSN comprises of different sensors which measures ambient temperature, humidity and soil moisture level and sends the measured value via internet to an android mobile device where the user has the option to do some corrective actions. The complete system overview along with the related instrumentation have been discussed in the paper. Experimental analysis has been provided for the application.

Keywords: Precision agriculture, IoT, wireless sensor network, environmental impact, artificial intelligence, machine learning

1. Introduction

According to United Nations 2030 agenda for sustainable development, global food security is second most important objective which is being tackled by world leaders and well-known scientists. Tackling and reducing the negative environmental impacts due to food growth such as greenhouse gas and tackling the climate change due to global warming needs to be tackled before the objective can be achieved. To rapidly increase the food production, scientists have included different technical devices in agriculture. Modernising the agriculture and making it more productive is one of the basic steps on achieving the long term goals Bodirsky *et al.* (2015). Due to global warming, the productivity of the crop is affected and there is a shortage of water due to the climate change. In different climatic condition and in rural areas, production of food is one of the major challenges. Precision agriculture aims to optimise the outcome of the agricultural system by introducing technology for monitoring and controlling different stages of crop growth. The basic technology used in precision agriculture is wireless sensor network. The wireless sensor network (WSN) comprises of N number of wire- less sensor nodes and a connecting network Chong and Kumar (2003). Wireless sensor node comprises of four different units such as sensing unit, processing unit, communication unit and power management unit as shown in Figure 1.

IoT is a multidisciplinary concept that encompasses wide range of technologies, devices, application domains. It is widely acknowledged that the WSN is the basic building block of IoT system Chi, Yan, Zhang,

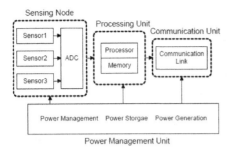

Figure 1: Schematic Diagram of WSN nodes

Pang, and Da Xu (2014). A lot of work has been carried out on how to implement IoT in agricultural field where apart from classical sensors and micro- controller node, researchers have adapted low altitude remote sensing, geographical mapping, machine vision, artificial intelligence and machine learning approaches to address certain aspects of the problem Shadrin *et al.* (2019.

A review of WSN deployment in precision agriculture has been discussed in Abbasi *et al.* (2014); Garcia-Sanchez *et al.* (2011). There are different commercial IoT solutions for agriculture such as onFarm, PhyTech, Semios, EZfarm, CropX, farmlogs, KAA and Mbeguchoice which provide precision IoT solution for agricul- tural requirements.

This paper provides an experimental prototype of WSN node comprising of different sensors and wi-fi connectivity which is then connected to the WSN hub with the help of wireless communication and the measured value is provided in the android application of the mobile.

2. Literature Review

In this tech-driven world, the thriving demand for food is the primary concern due to the exponential growth of the global population. It is expected that by the year 2050, the global population will be 9.6 billion, which will present a scenario where humans will struggle to feed themselves Maroli *et al.* (2021). The traditional agricultural system cannot feed such a large population, which enforces the integration of IoT, big data, edge computing, and wireless sensor networks in the agricultural sectors to increase productivity, minimise the usage of pesticides, and lower water wastage. Monitoring the environment and soil moisture, analysing factors affecting plant growth, plant management, product traceability, food safety applications, better food supply management, reducing food wastage are the objectives of IoT in the field of agriculture. In the agricultural domain, IoT is primary used to forecast various environmental parameters like humidity, temperature, moisture of soil. In different researches authors have proposed innovative IoT systems which uses machine learning to predict temperature and weather of upcoming days. Various WSN and IoT systems were pro- posed by authors and researchers to estimate various parameters of soil to improve productivity. These proposed methods can be adopted for a large variety of crops and can be implemented in diverse geographical locations. In Kiani and Seyyedabbasi (2018), authors have designed a smart system which predict the necessary requirements of a irrigation land by collecting data of soil moisture and temperature, UV, humidity, precipitation and weather data from the web. The proposed system also uses smart methods to avoid water wastage.

Energy constraint is another major challenge in IoT system. In order to reduce the power consumption and for optimal use of energy, various authors have proposed innovative and smart methods and algorithms. In other works authors have proposed an agricultural method to reduce the energy consumption by various IoT sensors and nodes. In some works authors have developed a network model in which the sensors and nodes uses minimum energy and the implementation cost is also low. Proposed model enhances the agricultural productivity and optimises the energy consumption. Zhang *et al.* (2015) have designed a smart irrigation system to inform and guide the farmers to properly utilise the resources such that wastage of water would be minimised. The proposed system uses data collected from various sensors of the farm and are processed using the Microsoft Azure to predict the correct time to start the irrigation process.

3. Problem Formulation

Figure 2 provides the problem statement of the system where a wireless sensor node is placed in an independent farm area and the sensor node comprises of different sensors and a transceiver. The transceiver is used for wireless communication purposes. The sensor nodes are connected to the hub using either wi-fi or IEEE 802.15.4 protocol. The hub is connected to other devices via internet.

Figure 3 provides the different layers of wireless communication of IoT. The different layers are as such

- Physical layer
- Data link layer
- Network layer
- Transportation layer
- Application layer

Figure 4 presents the architectural layout of physical layer of IEEE 802.15.4 protocol which comprises of different headers such as synchronisation header, physical layer header and physical payload.

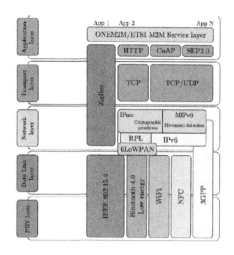

Figure 3: Different layers of IoT

Preamble 32 bits	SFD 8 bits	Frame length 7 bits	Reserved bit 1 bit	PSDU (0-127) bytes
Synchronization Header		Physical layer Header		PHY Payload

Figure 4: Physical layer of IEEE 802.15.4

4. Results

For experimental prototype development, a Node MCU is used as microcontroller. For measuring temperature and moisture, DHT 11 is used. For soil moisture measurement, a probe with LM393 operational amplifier has been used. Details of data acquisition and data logging can be found. Node MCU is chosen for this application because its a low cost and open source IoT platform and it runs on ESP8266 Wi-Fi SoC.

Figure 5(a) illustrates the circuit diagram of the hardware setup, Figure 5(b) presents

the photograph of the developed hardware setup (WSN node) and Figure 5(c) presents the snapshot of android application.

Figure 6 presents the bit-error-rate of IEEE 802.15.4 protocol in different modulation scheme. Different physical layer supports different modulation scheme and it supports different data rates. Different variations of IEEE 802.15.4 system use different modulation schemes and different data rates. A comparative analysis has been provided in Figure 6. The bit-error-rate analysis of the developed WSN system is useful to analyse the system performance.

5. Conclusion

This paper provides a brief description of the use of WSN and IoT in precision agricultural system. The overall system design schematic and different wireless communication system has been analysed for the system. An experimental setup has been developed which acts as a WSN node and wi-fi communication between WSN node and WSN

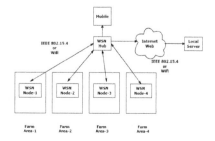

Figure 2: Problem Formulation

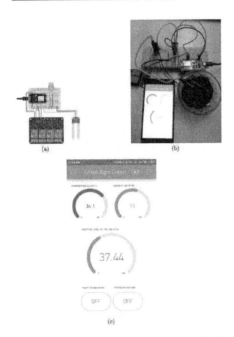

Figure 5: a) Circuit diagram of the hardware setup, (b) photograph of hardware setup, (c) snapshot of android application

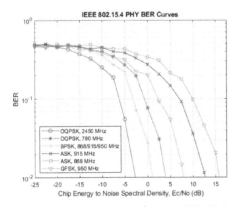

Figure 6: Bit-error-rate of IEEE 802.15.4 physical layer in different modulation scheme

hub has been established along with the development of android application.

References

Abbasi, A. Z., Islam, N., Shaikh, Z. A., *et al.* (2014). "A review of wireless sensors and networks' applications in agriculture." *Computer Standards & Interfaces*, 36 (2): 263-270.

Bodirsky, B. L., Rolinski, S., Biewald, A., Weindl, I., Popp, A., & Lotze-Campen, H. Global food demand scenarios for the 21 st century. PloS one, 10(11):e0139201, 2015. https://journals.plos.org/plosone/article?id=10.1371/journal.pone.0139201

Chi, Q., Yan, H., Zhang, C., Pang, Z., and Da Xu, L. (2014). "A reconfigurable smart sensor inter- face for industrial wsn in iot environment." *IEEE Transactions on Industrial Informatics*, 10 (2): 1417-1425.

Chong, C.-Y. and Kumar, S. P. (2003). "Sensor networks: evolution, opportunities, and challenges." *Proceedings of the IEEE*, 91 (8): 1247-1256.

Garcia-Sanchez, A.-J., Garcia-Sanchez, F., and Garcia-Haro, J. (2011). "Wireless sensor network deployment for integrating video-surveillance and data-monitoring in precision agriculture over distributed crops." *Computers and Electronics in Agriculture*, 75 (2): 288-303.

Kiani, F., & Seyyedabbasi, A. (2018). Wireless sensor network and internet of things in precision agriculture. International Journal of Advanced Computer Science and Applications, 6(9). https://openaccess.izu.edu.tr/xmlui/handle/20.500.12436/1841

Maroli, A., Narwane, V. S., and Gardas, B. B. (2021). "Applications of iot for achieving sus- tainability in agricultural sector: a comprehensive review." *Journal of Environmental Management*, 298: 113488.

Shadrin, D., Menshchikov, A., Somov, A., Bornemann, G., Hauslage, J., and Fedorov, M. (2019). "Enabling precision agriculture through embedded sensing with artificial intelligence." *IEEE Transactions on Instrumentation and Measurement*, 69 (7): 4103-4113.

Zhang, Z., Yu, X., Wu, P., and Han, W. (2015). "Survey on water-saving agricultural inter- net of things based on wireless sensor network." *International Journal of Control and Automation*, 8 (4): 229-240.

Optimization of Optical Communication Using Optical Fiber at Tri-Optical Windows

Sangram Keshari Nayak[1], Pratap Kumar Panigrahi[1], Gopinath Palai[2], Rabinarayan Satpathy[2]

[1]Department of Electrical Engineering, GIET University, Gunupur, India
[2]Faculty of Engineering and Technology Sri Sri University, Cuttack, India,
E-mail: gpalai28@gmail.com

Abstract

The current study focuses on optimising polymer-based optical fibers to maximise efficiency across the three optical communication windows by varying the core region diameter and V-number. Through simulations, the research aims to enhance the electric field through the fiber for signals at 850 nm, 1310 nm, and 1550 nm. Results show that increasing the fiber diameter improves efficiency. For instance, diameters of 300 nm, 460 nm, and 580 nm yield efficiencies of 0.9976, 0.9991, and 0.9993, respectively, across the communication windows, with a fiber length of 1000 meters. Loss, measured in decibels, is minimal at 0.02 dB, 0.01 dB, and 0.01 dB for 850 nm, 1310 nm, and 1550 nm signals, respectively. The proposed fiber demonstrates potential suitability for local area network applications.

Keywords: Optical fiber, photonics, polymer

1. Introduction

Optical communication revolutionises the way we transmit information, leveraging light signals for high-speed, long-distance data transfer. At its core, optical communication harnesses the principles of optics, employing lasers or LEDs as light sources, modulation techniques to encode data, optical fibers as transmission mediums, and receivers to convert light signals back into electrical ones. Optical amplifiers ensure signal strength over extended distances, contributing to reliable communication.

Central to this technology are the optical communication windows, critical wavelength ranges optimised for minimal signal loss in optical fibers. The first window at 850 nm, historically significant, offers low attenuation suitable for long-distance communication. The second window at 1310 nm boasts even lower attenuation, enhancing signal transmission, particularly in metropolitan networks. The third window, at 1550 nm, stands out for its minimal attenuation, ideal for long-distance communication with minimal loss.

Efficient waveguide designs prioritise these optical communication windows, optimising signal transmission. This optimization journey involves enhancing component designs, minimising signal loss, increasing data rates, and overall system reliability.

In academic discourse, a structured paper delves into optical fiber's significance and operational intricacies. It explores proposed structures, operational mechanisms, mathematical models, and discusses

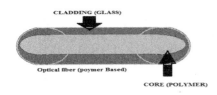

Figure 1: Schematic of optical fiber

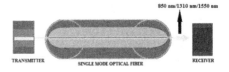

Figure 2: Oprational Mechanism of polymer based optical fiber

findings. Through conclusive remarks, it encapsulates essential insights, contributing to the ongoing evolution of optical communication systems.

In essence, the optimization and understanding of optical communication are paramount, ensuring efficient, reliable, and high-performance data transmission in our interconnected world.

1.1. Proposed Structure

The Figure 1 represents as a conventional optical fiber where polymer materials is at at core position, which is called as polymer optical fiber. In essence, Polymer Optical Fiber (POF) is a type of optical fiber made from polymer materials, as opposed to traditional optical fibers that are typically made from glass. Polymer Optical Fibers (POFs) are utilised for optical communication and data transmission, offering unique characteristics tailored for specific applications. Key parameters like material, core diameter, numerical aperture, and attenuation determine fiber efficiency. In this study, polymer material is the core, with diameters ranging from 300 nm to 600 nm across a proposed 1000-meter length, ensuring V-parameter stays below its cut-off of 2.45.

1.2. Operational Mechanism

After discussing the proposed structure of the fiber, Figure 2 focuses on the opeartional mechnism to eealise an its efficeincy pertaining to the signal of 850 nm, 1310 nm and 1550 nm, which is considered as 1st, 2nd and 3rd optical communication widows.

The present mechanism consists of transmitter which is laser source and photodioden which converts optical to electrical conter part. However the present research concentrates on the wavegude section only. Here the waveguide is considered as optical fiber which is discussed in the previos section. As far as mechnism is concerned; we want to chose a proper refractive indices of materails, diamter of core region such that high efficeincy of signal (850 nm ,1310 nm and 1550 nm) can propgate it . Alternately, we can say that the loss in the fiber would be feeable.

2. Mathematical Treatment

In the context of optical fibers, which are commonly used for transmitting optical signals in telecommunications, the electric field plays a crucial role in guiding and transmitting light signals through the fiber. To realise the same, the mathematical treatment involves solving the relevant partial differential equations that govern the behavior of the electric field in the optical fiber. These equations are obtained from Maxwells electromagnetic equations.

The solution of electric filed equation can be written as

$$E\left(x,y,z,t\right)=E_0\left(x,y\right)\cos\left(\beta z-\omega t\right) \text{ 1(a)}$$

Where E0(x,y) is the field distribution in the transverse plane , ω is the optical angular frequency and β is called as propagation constant which is defined as the ration of $\frac{2\pi n_{eff}}{\lambda}$

Further equation 1(a) can written in terms of propagation mode as

$$E_{lm}\left(x,y,z,t\right)=e^{i\left(\beta_{lm}-\omega t\right)}E_{0lm}\left(x,y\right) \text{ 1(b)}$$

If we consider E_{in} be the input incident with respect to the transverse field distribution, it can be expressed as of filed inside the optical fiber. It can be written as

$$E_{in}(x,y,0) = \sum_{l,m} c_{lm} E_{0lm}(x,y)$$
$$+ \int\int_0^\infty C_{\mu\vartheta} E_{0\mu\vartheta}(x,y) d\mu \, d\vartheta \qquad 1(c)$$

Where C_{lm} are complex weighting coefficient and the indices l, m represents the guided modes.

Finally, the coefficient can be expressed as [9]

$$|C_{l,m}|^2 = \frac{\left|\int\int |E_{in}(x,y,0) E_{0,l,m} dxdy\right|^2}{\int\int |E_{in}(x,y,0)|^2 dxdy \int\int |E_{0,l,m}|^2 dxdy} \qquad 1(d)$$

The equation 1(d) represents the coefficient to find out the field distribution

3. Result and Interpretation

Basically, Optical communication is a crucial technology with several important aspects and applications that contribute to various fields. It provides lies in its ability to provide high-speed, high-bandwidth, secure, and reliable transmission of data over long distances. Further, in the context of communication, the term "optical communication windows" typically refers to specific wavelength ranges in the electromagnetic spectrum where optical fibers exhibit low attenuation, allowing for efficient transmission of signals [10, 11]. The three main optical communication windows are commonly referred to as the first (850 nm), second (1310 nm), and third (1550 nm) windows. These windows are essential for the development and implementation of fiber optic communication systems (Manoharan et al., 2023). Considering the importance of optical fiber at the signal of 850 nm, 1310 nm and 1550 nm, the present paper design optical fibers where one can find high efficiency. To do so, simulation is made with the help of equation (4 to 6). The outcomes of the same is indicated in the Figures 3–5 for allowing the signal of 850 nm, 1310 nm and 1550 nm respectively.

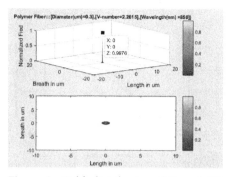

Figure 3: Field distribution (3D and 2D variation) through optical fiber at the signal of 850 nm

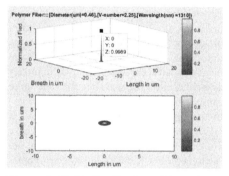

Figure 4: Field distribution (3D and 2D variation) through optical fiber at the signal of 1310 nm

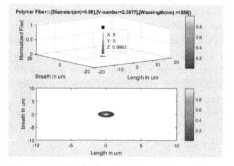

Figure 5: Field distribution (3D and 2D variation) through optical fiber at the signal of 1550 nm

The Figures 3–5 represents the variation of electric filed with respect to diameter , wavelength and V- numbers. Here two figures are inset in each diagram; for

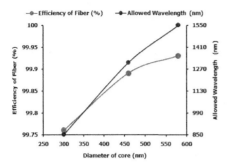

Figure 6: Variation of efficiency of the fiber with respect to the diameter and allowed wavelength

example; the upper part is represented as 3 dimensional representations where lower is indicated as 2 dimensional representation of electric field distribution. Form these figures; it is observed that the efficiency increases with the increasing of diameter as well as allowing wavelengths. The same can be understood from the Figure 6.

In the Figure 6; efficiency of fiber (%) is taken along primary vertical axis, where allowed wavelength (nm) is taken along secondary vertical axis, where diameter in nm is taken along horizontal axis. After going though the above said graph it is found that the efficiency of fiber is a function of wavelength and diameter ; For example; the frequency of 99.76 %, 99.89 % and 99.93 % of the fiber for the combination of (diameter, allowed wavelengths) of (300 nm , 850 nm), (460 nm, 1310 nm) and ((580 nm, 1550 nm) respectively. However, the V-number varies nonlinearly as it depends on not only numerical parameter but also materials properties (refractive indices of material) because refractive indices of the core and cladding are differed from different wavelengths. To sum up, the present research deals with three optical fiber having different core diameter which allows first , second and third optical communication windows with high efficiency

4. Conclusion

This study optimises polymer-based optical fibers across three optical communication windows by varying core diameter and V-number. Simulations aim to increase electric field efficiency for 850 nm, 1310 nm, and 1550 nm signals. Results show efficiency improves with fiber diameter increase, demonstrating suitability for local area network applications with minimal signal loss.

References

Brockman, P., French, D., and Tamm, C. (2014). "REIT organizational structure, institutional ownership, and stock performance." Journal of Real Estate Portfolio Management, 20 (1): 21-36.

Hecht, J. (2004). *City of Light: The Story of Fiber Optics (revised ed.). Oxford University, pp. 55-70.*

Catania, B., Michetti, L., Tosco, F., Occhini, E., and Silvestri, L. (1976). "First Italian experiment with a buried optical cable." *Proceedings of 2nd European Conference on Optical Communication (II ECOC), pp. 315–322.* Accessed August 18, 2022.

Pearsall, T. (2010). *Photonics Essentials,* 2nd ed. *McGraw-Hill.* ISBN 978-0-07-162935-5. *Archived from the original on August 17, 2021.* Retrieved February 24, 2021.

Gloge, D. (1971). "Weakly Guiding Fibers." *Applied Optics, 10 (10): 2252-2258.* Bibcode:1971ApOpt.10.2252G. *https://doi.org/10.1364/AO.10.002252.* PMID 20111311. Retrieved December 21, 2023.

Single-mode fibre. Retrieved November 26, 2021. *"Single-mode fibre (also referred to as fundamental or mono-mode fibre) Maxwell's Equations." Engineering and Technology History Wiki. 29 October 2019.* Retrieved December 24, 2021.

Hasanin, T., Manoharan, H., Alterazi, H. A., Srivastava, G., Selvarajan, S., and Lin, J. C.-W. (2023). "Mathematical approach of fiber optics for renewable energy sources using general adversarial networks, Front. Ecol. Evol." Frontiers in Ecology and Evolution, 11, pp 1-12. https://doi.org/10.3389/fevo.2023.1132678.

A Swarm Optimised Deep Learning Model for Financial Time Series Forecasting

Sudersan Behera[1], Sarat Chandra Nayak[2], Sanjib Kumar Nayak[3], Sung-Bae Cho[4]

[1]Department of AIML, Sphoorthy Engineering College, Hyderabad, India, sbehera.04@gmail.com
[2]Department of CSE, School of Technology, GITAM University, Hyderabad, India, saratnayak234@gmail.com
[3]Department of CA, Veer Surendra Sai University of Technology, Burla, India, sknayak_ca@vssut.ac.in
[4]Department of Computer Science, Yonsei University, Seoul, South Korea, sbcho@yonsei.ac.kr

Abstract

Prediction of data on financial time series is extremely difficult due to the inherent complexity and constantly changing dynamics of financial market. Deep learning (DL) methods are found efficient in capturing such underlying dynamism. This article uses a Particle Swarm Optimisation (PSO) to fine-tune the biases and weights of a popular DL method, i.e., Long Short-Term Memory (LSTM) network which resulted in the creation of the ground-breaking hybrid model LSTM+PSO. To evaluate LSTM+PSO, it is used to forecast the values of two widely tracked currency exchange rates. Our concurrent use of the conventional LSTM for the same prediction task serves as a benchmark for comparison. The MSE statistic is used for performance evaluation. According to the results, LSTM+PSO significantly outperform basic LSTM in terms of making correct predictions. Exhaustive experimental outcomes show that the combined forecast is good at dealing with the complex problems of currency exchange rate time series forecasting.

Keywords: LSTM, PSO, exchange rate prediction, financial time series prediction

1. Introduction

Time series forecasting, which encompasses challenges in stock price prediction, and weather forecasting, plays a crucial role in various aspects of human life (Kargar, 2021). The abundance of time series data can be transformed into valuable information for forecasting purposes Mishra *et al.* (2017). This predictive approach, aiming to anticipate future occurrences Wulandari *et al.* (2021), has witnessed accelerated advancements through cutting-edge computer technology, particularly in the realm of artificial intelligence. ANNs, characterised as

connectionist agent networks, emulate the information transmission between artificial neurons, drawing inspiration from biological neurons. According to Abiodun *et al.* (2018), ANNs serve as effective models for information management, mirroring the function of the human brain's biological nervous system in problem-solving. Deep learning (DL), or deep representation learning, involves studying hierarchies of representations as inputs traverse neurons, leading to enhanced performance accuracy Varghese and Kandasamy (2021). Given the dynamic and multi-dimensional nature of time series data, nonlinear systems are well-suited for

Chapter 58 DOI: 10.1201/9781003581215

their analysis. Recurrent Neural Networks (RNN), a DL technique tailored for sequential data, are particularly effective for nonlinear predictions Gyamerah and Korda (2021). However, RNN faces challenges in handling long sequences, leading to gradient loss and long-term memory failure (Wu *et al.*, 2021). In contrast, LSTM addresses these issues by utilising memory cells and gate units Madireddy (2018). Additionally, He *et al.* (2023) have observed the integration of LSTM in evolutionary algorithms, highlighting the effectiveness of such approaches in training machine learning models as compared to conventional methods. This reflects the evolving landscape of methodologies for addressing the challenges of financial time series prediction. Considering evolutionary infusion into DL techniques like LSTM, various hybrid models such as FA-PSO-LSTM (Chen and Long, 2023), and CNN-LSTM-GA (Baek, 2023) have been proposed for financial time series predictions and achieved success. We started an integration project by combining LSTM and PSO, which were both inspired by the findings from the aforementioned research. In this innovative fusion, PSO takes charge of fine-tuning the weights and biases of LSTM, culminating in the creation of the hybrid model LSTM+PSO. This novel amalgamation aims to leverage the strengths of both techniques, enhancing the overall predictive capabilities of the model. Our application of the LSTM+PSO hybrid model is directed towards forecasting the closing prices of two widely monitored currency exchange rates, such as EUR/USD and INR/USD. To provide a benchmark for evaluation, we concurrently employed the conventional LSTM model. The comparison of these models hinges on the Mean Squared Error (MSE) metric, allowing for a quantitative assessment of their performance. We want to find out how well the LSTM+PSO hybrid model can predict closing prices compared to the traditional LSTM method by applying a comparison test. The MSE error metrics serve as a valuable tool for objectively measuring the accuracy and efficacy of both models in capturing the dynamics of currency exchange rates.

2. Materials and Methods

2.1. Long Short-Term Memory

Long Short-Term Memory (LSTM) is a type of recurrent neural network designed to address the challenges of capturing long-term dependencies in sequential data. The key components of an LSTM cell include the cell state; forget gate, and output gate. Perspective readers can refer (He *et al.*, 2022) to know about the working principles of LSTM and how LSTM contributes to financial time series forecasting. The architecture of a LSTM model is given in Figure 1.

2.2. Particle Swarm Optimisation

The inception of the Particle Swarm Optimisation (PSO) algorithm can be attributed to Kennedy and Eberhart (1995). In this study, we leverage PSO as a well-established optimisation tool known for its versatility and effectiveness. Perspective readers can refer (Chen and Long, 2023) to know more about the working principles of PSO and how PSO integrates with LSTM for financial time series prediction purpose.

3. PSO –based Training

Parameters, especially weights (W) and biases (b), play a crucial role in determining the performance of a LSTM model, which in turn affects the accuracy of classifications. The learning step of a Long Short-Term Memory (LSTM) model trained using Particle Swarm Optimisation (PSO) aims to decrease training error through the use of Structural Risk Minimisation. The values of 'W' and 'b' are

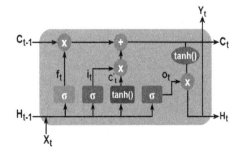

Figure 1: LSTM architecture

adjusted dynamically using PSO as the training error improves, with the aim of identifying the parameters that result in the lowest error. The best LSTM settings will then be determined through an optimisation procedure. After the optimum parameters have been determined, they are used to retrain the LSTM model so that it can identify fresh samples in the testing phase. Feature selection from the initial dataset is used to pick a testing set during the testing phase. We then feed the learned LSTM model testing patterns. Here are the main steps involved in optimising the parameters of the LSTM model:

1. First set the parameters for 'W' and 'b' to their initial values.
2. Structural Risk Minimisation should be used to train the PSO-based LSTM model in order to decrease training error.
3. Adjust the values of 'W' and 'b' using PSO during the training stage for optimal parameter adjustment.
4. Finding the optimal parameters that minimise training error is the goal of optimisation.
5. It is necessary to retrain the LSTM model using the optimal parameters.
6. Feature selection from the source dataset is used to select a testing set.
7. "Imputation" means to insert test patterns into the LSTM that has been trained.
8. Use test data and conduct testing.

4. About the Data

From January 24, 2023, through January 24, 2024, we collected daily currency data on the conversion rate of Indian Rupees to US Dollars and Euros to US Dollars from Yahoo Finance. Our method involved using a 3-window sliding window (Behera *et al.*, 2023) with a 50/50 split between the training and testing datasets. We were able to separate the attributes that were chosen for training and testing purposes in this way. Using this methodology simplified both the training and evaluation processes of our prediction model. The proposed LSTM metaheuristic relies on a novel learning mechanism known as LSTM. A

basic model with one kernel layer is utilised. Using PSO to find the weights associated with hidden-input and hidden-output neurons and the bias vectors between them led to the development of PSO.

5. Simulations and Result Analysis

A Windows 10 environment was used to conduct the experiments for both models on Google Colab. After conducting lengthy iterative experiments, the maximum number of iterations was determined to be 100 for the PSO model, which utilised an iterative technique. The population size was initially set at 50. Following the guidelines provided by Kennedy and Eberhart (1995), all other algorithmic parameters for PSO were chosen. High, Low, Open, and Adj Close data from all datasets were the input features for all models. After that, exchange rate closing prices for both 1-day and 10-day ahead intervals were predicted using these models. The MSE metric, as described by (Nayak *et al.*, 2022), was used to evaluate the accuracy of each model's forecasts. Table 2 displays the testing errors, which were recorded with great care and are shown as mean squared errors (MSE) over 10 runs for both datasets and time frames.

6. Conclusion

In this study, PSO served as a metaheuristic for LSTM training, optimising weights and biases to create the LSTM+PSO hybrid model. This innovative model effectively predicted closing prices for two widely observed currency exchange rates, outperforming the traditional LSTM in forecasting accuracy, as evidenced by the MSE metric. Future efforts will focus on enhancing network topology and implementing more efficient hybrid training techniques to further refine prediction accuracy. While LSTM+PSO excel in currency exchange rate forecasting, acknowledging limitations is crucial. The model hasn't been tested on other financial time series, lacks hyperparameter tuning, and does not incorporate technical indicators. Addressing these limitations will be a

Table 1: MSE score comparison

Model	Period	INR-USD	EUR-USD
LSTM+PSO	1-day	2.00246	2.29881
	10-day	3.18214	3.99849
LSTM	1-day	8.08157	9.37792
	10-day	9.06924	11.36559

Source: Author's compilation.

key aspect of our future research to ensure a more comprehensive and robust approach to financial time series forecasting.

References

Abiodun, O. I., Jantan, A., Omolara, A. E., Dada, K. V., Mohamed, N. A., and Arshad, H. (2018). "State-of-the-art in artificial neural network applications: a survey." *Heliyon*, 4 (11): e00938.

Baek, H. (2024). A CNN-LSTM Stock Prediction Model Based on Genetic Algorithm Optimization. *Asia-Pacific Financial Markets*, 31, 205-220.

Behera, S., Kumar, A. V. S., & Nayak, S. C. (2024). Improved Set Algebra-Based Heuristic Technique for Training Multiplicative Functional Link Artificial Neural Networks for Financial Time Series Forecasting. SN *Computer Science*, 5(5), 1-34.

Chen, X. and Long, Z. (2023). "E-commerce enterprises financial risk prediction based on FA-PSO-LSTM neural network deep learning model." *Sustainability*, 15 (7): 5882.

Gyamerah, S. and Korda, D. R. (2021). "Prediction of stock market returns using LSTM model and traditional statistical model." *International Journal of Computer Applications*, 183 (37): 57-61.

He, Y., Zeng, X., Li, H., and Wei, W. (2022). "Application of LSTM model optimized by individual-ordering-based adaptive genetic algorithm in stock forecasting." *International Journal of Intelligent Computing and Cybernetics*, 16 (2): 277-294.

Kargar, N. (2021). "Generalized autoregressive conditional heteroscedasticity (GARCH) for predicting volatility in Stock Market." *International Journal of Multidisciplinary Research and Growth Evaluation*, 2 (3): 73-75.

Kennedy, J., & Eberhart, R. (1995, November). Particle swarm optimization. In Proceedings of ICNN'95-international conference on neural networks 4, pp. 1942-1948). IEEE.

Madireddy, V. R. (2018). "Stock market prediction in BSE using Long short-term memory (LSTM) algorithm." *SSRN Electronic Journal*.

Mishra, N., Soni, H. K., Sharma, S., and Upadhyay, A. (2017). "A comprehensive survey of data mining techniques on time series data for rainfall prediction." *Journal of ICT Research and Applications*, 11 (2): 168.

Nayak, S. C., Dehuri, S., and Cho, S. B. (2022). "Intelligent financial forecasting with an improved chemical reaction optimization algorithm based dendritic neuron model." *IEEE Access*, 10, 130921-130943.

Varghese, L. R. and Kandasamy, V. (2021). "Convolution and recurrent hybrid neural network for hevea yield prediction." *Journal of ICT Research and Applications*, 15 (2): 188-203.

Wu, J. M. T., Sun, L., Srivastava, G., and Lin, J. C. W. (2021). "A novel synergetic LSTM-GA stock trading suggestion system in internet of things." *Mobile Information Systems*, 2021: 1-15.

Wulandari, S., Sufri, S., and Yurinanda, S. (2021). "Implementation of ARIMA method in predicting stock price fluctuations of PT Bank Central Asia Tbk." *Buana Matematika : Jurnal Ilmiah Matematika Dan Pendidikan Matematika*, 11 (1): 53-68.

IoT Enabled Green House Environment Automation

Subham Barik[1], Sandeep Sahani[1], Satyabrat Sahoo[1], Manas Ranjan Kabat[2]

[1]Department of Computer Science & Engineering, Silicon Institute of Technology, Sambalpur, India, bariksubham159@gmail.com, sandeepsahani76j@gmail.com, satyabrat.sahoo@silicon.ac.in

[2]Department of Computer Science & Engineering, Veer Surendra Sai University of Technology, Burla, Sambalpur, India, manas_kabat@yahoo.com

Abstract

Agriculture is crucial for national development, but global warming and climate change challenge crop yields. Greenhouses offer a solution, providing an optimal growing environment. However, declining crop production persists due to poor monitoring and human error in greenhouse management. To address these issues, we propose an automated greenhouse monitoring system using IoT technology. It collects data on temperature, soil moisture, humidity, and light intensity via sensors, uploading it to the Arduino Cloud. Data is compared against thresholds, triggering actuators to maintain ideal conditions. The Arduino Cloud enables real-time monitoring through the Arduino Remote IoT app or website. Our system includes an attack detection mechanism to ensure secure data transfer. By integrating technology and automation, we aim to enhance crop production and safeguard data transmission security amid climate change and human error.

Keywords: Arduino Cloud, IoT, agriculture, ESP32 wroom, sustainability, DoS

1. Introduction

A greenhouse optimizes environmental conditions for plant growth. IoT technology in agriculture boosts productivity and resource management. Our system uses sensors for essential greenhouse variables (temperature, humidity, soil moisture, light). The ESP32 Wroom 32U transmits data to the Arduino Cloud platform for real-time visualisation. Actuators respond to data, comparing with optimal values. Historical data is downloadable from the Arduino Cloud in CSV format. To secure data transmission, an attack detection layer is incorporated. The ESP32 Wroom collects sensor data; Arduino Cloud stores it. The DHT11 sensor measures humidity and temperature; a soil moisture sensor assesses water content. An LDR module gauges light intensity, and an LCD display monitors environmental parameters onsite.

2. Literature Survey

The studies referenced introduce various technologies for greenhouse monitoring and control. In Akkaş, and Sokullu (2017), a Wireless Sensor Network (WSN) using MicaZ nodes tracks temperature, light, pressure, and humidity, allowing farmers to manage conditions remotely via IoT-enabled devices. Study Dan et al. (2015) utilizes CC2530 and ZigBee for greenhouse monitoring, with nodes managing

Chaper 59 DOI: 10.1201/9781003581215

environmental data and transmitting readings in real-time. In Modani *et al.* (2017), the S3C2440 microprocessor manages data processing and control via ZigBee, directing greenhouse equipment based on sensor inputs. System Danita *et al.* (2018) employs Raspberry Pi 3 with sensors to autonomously regulate greenhouse conditions, storing and presenting data via ThingSpeak. Lastly, Soheli *et al.* (2022) utilizes Arduino Uno with sensors and ANFIS for data processing and control, transmitting data wirelessly using MQTT and connecting to mobile devices for remote monitoring and control. These systems showcase diverse approaches to smart greenhouse management, leveraging IoT and wireless technologies for efficient agriculture.

3. Problem Identification

Contemporary greenhouse monitoring systems offer remote monitoring and control of temperature, humidity, and light. However, they often lack robust security, user-friendly interfaces, and uninterrupted power supply, posing challenges for data privacy, cost, and reliability. The initial and ongoing costs can be prohibitive for smaller farms. Inconsistent power supply can disrupt system operations. This project aims to address these limitations by developing a secure, user-friendly, cost-effective, and reliable greenhouse monitoring system.

4. Methodology

4.1. Basic Work Flow

4.1.1. Sensor Data Acquisition

The system employs strategically positioned sensors in the greenhouse to monitor environmental conditions such as temperature, humidity, soil moisture, and light intensity.

4.1.2. Data Transmission

Sensor data is transmitted to the Arduino Cloud via the ESP32 Wroom connected to a network.

4.1.3. Arduino Cloud

Data is stored and processed on the Arduino Cloud, a platform designed for IoT project development, implementation, and management.

4.1.4. Decision-Making

Analytics algorithms interpret sensor data to optimize growing conditions.

4.1.5. Automated Control (Actuators)

Actuators adjust temperature, activate irrigation systems, and control shading based on data analysis.

4.1.6. User Interface

The system offers a user-friendly interface accessible through a web portal or mobile app, featuring intuitive monitoring dashboards for web and mobile (iOS/Android) platforms.

4.2. Hardware Description

Figure 1 depicts the components of a smart agriculture system, centered around the ESP32 WROOM microcontroller.

4.3. Software Description

4.3.1. Arduino Cloud

The Arduino Cloud is an Arduino service enabling users to connect their devices and third-party Wi-Fi boards to the internet for remote management. It integrates IoT capabilities into projects, enabling monitoring and control from anywhere with internet access.

Figure 1: All hardware devices used in the project

Figure 2 illustrates the interface of Arduino Cloud, serving as a centralized platform for configuring and coding various IoT devices.

Figure 2: Home page of Arduino Cloud

4.4. Flow Chat of the System

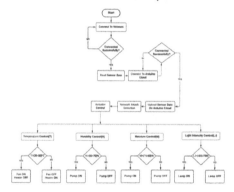

Figure 3: Flow chart of the IoT-driven automation system

4.5. Block Diagram of the System

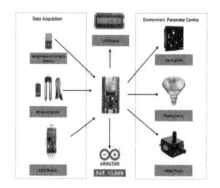

Figure 4: Block Diagram of the IoT driven automation system

Figure 5: Network attack detection algorithm

4.6. Network Attack Detection Algorithm

Figure 5 illustrates the network attack detection algorithm, which continuously monitors network activity by sniffing the connected network. Anomalies detected, such as de-authentication and disassociation attacks, are reflected in the Arduino Cloud Dashboard.

4.7 Threshold Values for Healthy Plant Growth

Table 1: Temperature controlling values

Temperature (°C)	Fan	Pump	Threshold
20-30°C	OFF	OFF	20-30°C
<20°C	OFF	ON	
>30°C	ON	OFF	

Table 2: Humidity controlling values

Humidity (%RH)	Fan	Pump	Threshold
65-70	OFF	OFF	65-70
<65	OFF	ON	
>70	ON	OFF	

Table 3: Soil moisture controlling values

Soil Moisture	Pump	Threshold
70-80	OFF	70-80
<70	ON	
>80	OFF	

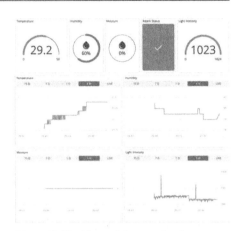

Table 4: Light intensity controlling value

LDR	Lamp	Threshold
400-700	OFF	400-700
<400	ON	
>700	OFF	

Figure 7: Dashboard for monitoring sensor data

5. Result and Discussion

Our Figure 6 model includes an LCD for real-time display of environmental parameters: Temperature (T), Moisture (M), Soil Humidity (H), and Light Intensity (LI), alongside the internet connection status. Sensors connect via screw terminals for the LDR module, moisture sensor, and DHT11 module. Each sensor has three terminals: VCC ("+"), GND ("-"), and an analog output ("S"). The system uses three relay channels for the cooling fan, heating lamp, and water pump. Operation requires a 12V DC power supply, with a 12V to 5V converter for powering the microcontroller, modules, and sensors at 5V DC.

6. Conclusion and Future Prospects

In our agricultural IoT study, we used the ESP32 Wroom 32U to store environmental data in the cloud. The DHT11 sensor measured humidity and temperature, a soil moisture sensor assessed soil water content, and an LDR module measured light the existing monitoring system: data security and privacy concerns, project costs, complex user interface, and power supply fluctuations. Our focus is on improving data security, reducing costs, simplifying the user interface, and ensuring reliable power supply. These enhancements will address current drawbacks and promote wider adoption of IoT in agriculture, supporting sustainable and efficient greenhouse management.

References

Akkaş, M. A. and Sokullu, R. (2017). An "IoT-based greenhouse monitoring system with Micaz motes." *Procedia Computer Science, 113*: 603-608.

Dan, L. I. U., Xin, C., Chongwei, H., and Liangliang, J. I. (2015). "Intelligent agriculture greenhouse environment monitoring system based on IOT technology." *2015 International Conference on Intelligent Transportation,*

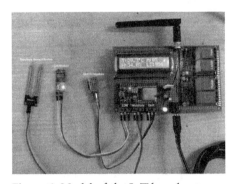

Figure 6: Model of the IoT-based automation system

Big Data and Smart City, pp. 487-490. IEEE.

Modani, V., Patil, R., Puri, P., and Kapse, N. (2017). "IoT based greenhouse monitoring system: technical review." *International Research Journal of Engineering and Technology*, 10: 2395-2456.

Danita, M., Mathew, B., Shereen, N., Sharon, N., and Paul, J. J. (2018). "IoT based automated greenhouse monitoring system." *2018 Second International Conference on Intelligent Computing and Control Systems (ICICCS)*, 1933-1937. IEEE.

Soheli, S. J., Jahan, N., Hossain, M. B., Adhikary, A., Khan, A. R., and Wahiduzzaman, M. (2022). "Smart greenhouse monitoring system using internet of things and artificial intelligence." *Wireless Personal Communications*, 124 (4): 3603-3634.

Restaurant Recommendation System Utilising User Preferences with Content-Based Filtering

Arnold Anthony[1], Kshitij Satpathy[1], Satyabrat Sahoo[1], Manas Ranjan Kabat[2]

[1]Department of Computer Science & Engineering, Silicon Institute of Technology, Sambalpur, India, st.arnold2003@gmail.com, kshitijsatpathy@gmail.com, satyabrat.sahoo@silicon.ac.in

[2]Department of Computer Science & Engineering, Veer Surendra Sai University of Technology, Burla, Sambalpur, India, manas_kabat@yahoo.com

Abstract

In the rapidly evolving landscape of the restaurant industry, we introduce a innovative restaurant recommendation system based on user preferences, which highlights the fast changing restaurant business. The system uses advanced content filtering algorithms, semantic analysis, and state-of-the-art vectorisation techniques such as TF-IDF Vectorizer in an effort to revolutionize the restaurant finding experience. The primary objective is to enhance the precision and personalisation of recommendations through the seamless integration of online evaluations and local preferences. In order to provide precise, contextually relevant restaurant suggestions, semantic analysis is utilized to identify establishments with comparable features and ratings. Improved customer satisfaction and user experience in restaurant recommendation systems are promised by its seamless integration of regional preferences with semantic analysis.

Keywords: Machine learning algorithm, recommendation system, NLP, content-based filtering, cosine similarity

1. Introduction

In living everyday life, people have to make decisions about various situations in their lives. Our decision-making process is frequently complicated by the abundance of options. We frequently seek advice or opinions from others in order to get over this. These kinds of issues have been solved in the current day by putting in place a recommendation system (Munaji et al., 2021). Recommendation systems, which are essential in a variety of industries including social networking, e-commerce, and streaming, evaluate user data to offer tailored content recommendations (Nandan et al., 2023). Public opinion has a big impact on restaurant selections; unfavorable reviews can turn away customers and emphasize how difficult it can be to locate good options in a sea of possibilities (Mahajan et al., 2020).

Our restaurant recommendation system transforms culinary discovery with context-aware recommendations by leveraging Tf-Idf Vectorizer, cosine similarity metrics, content-based filtering, and user locations to provide tailored choices based on review text analysis.

Chapter 60 DOI: 10.1201/9781003581215

2. Literature Review

2.1. Collaborative Filtering Approaches

In collaborative filtering, Munaji et al. (2021) employ Pearson correlation for user-based suggestions, whereas Nandan *et al.* (2023) recommend hybrid filtering with sentiment analysis. Mahajan et al. (2020) provide a machine learning-based method for cold start circumstances. Koetphrom *et al.* (2018) compare filtering algorithms, emphasising hybrid benefits. Choenyi *et al.* (2021) confirm the hybrid's effectiveness.

2.2. Sentiment Analysis and personalized Suggestions

Koetphrom *et al.* (2018) emphasize the benefits of hybrid filtering in sentiment analysis and tailored recommendations, whereas Choenyi *et al.* (2021) demonstrate its usefulness with various filtering modalities. Afoudi *et al.* (2019) investigate the impact of feature selection on content-based systems in order to improve performance.

2.3. NLP for Preference Prioritisation

Gomathi *et al.* (2020) use NLP to extract customer preferences from textual data, whereas Zhang *et al.* (2018) use sentiment analysis to propose ethnic restaurants based on user sentiment.

2.4. Opinion Mining and Decision Enhancement

Jayasri *et al.* (2021) refine suggestions using sentiment analysis and opinion mining, whereas Eidul *et al.* (2022) use machine learning-based review systems to predict consumer happiness.

3. Methodology and Model Specifications

The proposed restaurant recommendation system employs content-based filtering algorithms, focusing on enhancing user experience through personalised suggestions. Users input their location, and the system recommends restaurants based on that location, ensuring local relevance. Leveraging cosine similarity and correlation features, the system analyses user reviews to identify similarities between different restaurants. Cosine similarity measures the cosine of the angle between two vectors, determining their similarity, while correlation features provide insights into the correlation between various restaurant attributes. By combining methods, the model improves user experience by suggesting eateries based on correlation ratings or location.

We will outline the specifics of the three modules in this section. We will give a quick explanation of the TF-IDF and the two feature selection models in the subsection.

3.1. Feature Selection Module

In machine learning, feature selection automatically selects pertinent features to minimise dimensionality, improve accuracy, and maximise computing efficiency for the purpose of generalising the model.

3.1.1. Correlation

A statistical metric called correlation denotes the closeness and linear link between two variables. In order to improve model efficiency and interpretation, characteristics that are highly correlated should be eliminated since they have a strong relationship with the dependent variable.

3.1.2. RFE

Recursive Feature Elimination (RFE) improves the efficiency and interpretability of models by removing features repeatedly. For refining, we used RFE in conjunction with Logistic Regression.

3.2. Vector Space Model

Collaborative filtering is impacted by data sparsity. To tackle this, we use in the similarity estimate process restaurant attributes like pricing, cuisine type, and geography. TF-IDF weighting, which consists of TF (term frequency) and IDF (inverse document frequency), is frequently employed in the vector space model.

3.2.1. Term Frequency

The frequency of term Ti in document D_j is represented by TF(T_{ij}), which increases term importance by removing stop words.

This has the following definition

$$TF\left(T_{ij}\right) = log\left(T_i, D_j\right) / N\left(D_j\right) \quad (1)$$

3.2.2. Inverse Document Frequency

Ti's recurrence in documents U, N(U,T_i), which is decreased by IDF, has an inverse effect on its uniqueness, increasing relevance by reducing common terms. This has the following definition

$$IDF\left(T_i\right) = log\left(N\left(U\right) / N\left(U, T_j\right)\right) \quad (2)$$

3.3. Similarity Measure Model

3.3.1. Cosine Similarity

Cosine similarity compares texts according to their TF-IDF word count and computes the angle between vectors to show the angle of their relationship.

3.3.2. Euclidean Similarity

The Euclidean distance serves as the foundation for several similarity metrics. This defines the separation between two restaurants.

4. Results and Discussions

4.1. The Dataset

A dataset comprising more than 13,000 eateries until 2022 was chosen from Kaggle.com. It was produced by Chirag Samal for document term matrices and analysis.

4.2. Implementation

4.2.1. Experimental Steps

In order to demonstrate how features selection affects our outcome, we use cosine similarity to propose all restaurant attributes. Then, using RFE and correlation, we choose the first seven characteristics once more and evaluate the results.

4.2.2. Experiment Results

After Building our model which is shown in Figure 1, we have tested all features and the RFE and correlation, and get similar restaurants to a restaurant using cosine similarity, the results of the experiment is shown in Table 1. We get better overall result by using the Cosine Similarity over Euclidean Similarity.

4.2.3. Performance Evaluation

In the Recommendation System, the top-N restaurant recommendations are prioritised by user choice. RMSE analysis is performed using a variety of models, including cosine similarity, Euclidean similarity, correlation features, and RFE.

Cosine Similarity: 0.19578900244
Euclidean Similarity: 0.479583152

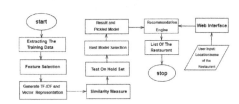

Figure 1: Working model of content-based filtering approach

Table 1: Similar restaurant to restaurant "San Churro Cafe"

All features	RFE Features	Correlation Features
Community	Ciclo Cafe	Ciclo Cafe
Dyu Cafe	Hoot	Crawl Street

(Source: Author's compilation)

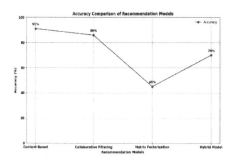

Figure 2: Accuracy comparison of various recommendation models

Correlation Features: 0.1870828693
RFE Features: 0.740913207435

When we computed the accuracy as a percentage (out of 100), the model with cosine similarity and correlation combined with RFE characteristics obtained 91% accuracy, whereas the model with Euclidean similarity and RFE characteristics scored 70.89%. Our recommendation method is 91% accurate, outperforming Collaborative Filtering (86%), Hybrid Models (70%), and Matrix Factorisation (45%) as shown in Figure 2.

5. Conclusion and Future

our restaurant recommendation system has demonstrated promising results. To improve precision and dependability, cosine similarity and correlation measures are used. Flask's integration has made it easier to install on a web interface and given consumers quick access to customised restaurant recommendations. Our objective is to investigate additional improvements, such adding sophist-icated machine learning algorithms and user feedback systems, in order to keep enhancing the system's performance. Future updates and development of the web interface platform may be easily accomplished due to its versatile nature.

References

Munaji, A. A. and Emanuel, A. W. R. (2021). "Restaurant recommendation system based on user ratings with collaborative filtering." *IOP Conference Series: Materials Science and Engineering*, vol. 1077, p. 012026. IOP Publishing.

Nandan, M. and Gupta, P. (2023). "Designing an efficient restaurant recommendation system based on customer review comments by augmenting hybrid filtering techniques." *Universal Journal of Operations and Management*, 59–79.

Mahajan, K., Joshi, V., Khedkar, M., Galani, J., and Kulkarni, M. (2020). "Restaurant recommendation system using machine learning." *International Educational Applied Research Journal*, 4 (3): 1-4.

Koetphrom, N., Charusangvittaya, P., and Sutivong, D. (2018). "Comparing filtering techniques in restaurant recommendation system." *2nd International Conference on Engineering Innovation (ICEI)*, pp. 46-51. IEEE.

Choenyi, T., Tseyang, T., Choikyong, S., Tsering, P., and Gurme, T. (2021). "A review on filtering techniques used in restaurant recommendation system." *International Journal of Computer Science and Mobile Computing*, 10 (4): 113-117.

Afoudi, Y., Lazaar, M., and Al Achhab, M. (2019). "Impact of feature selection on content-based recommendation system." *International Conference on Wireless Technologies, Embedded and Intelligent Systems (WITS)*, pp. 1-6. IEEE.

Gomathi, R. M., Ajitha, P., Krishna, G. H. S., and Pranay, I. H. (2019). "Restaurant recommendation system for user preference and services based on rating and amenities." *International Conference on Computational Intelligence in Data Science (ICCIDS)*, pp. 1-6. IEEE.

Zhang, S, Salehan, M., Leung, A., Cabral, I., and Aghakhani, N. (2018). "A recommender system for cultural restaurants based on review factors and review sentiment."

Jayasri, T., Vaneesha, V., Srinivas, U. M., Venkataramana, K., and Rala, S. C. (2021). "Best restaurant review and opinion mining rating." *Mathematical Statistician and Engineering Applications*, 70 (2):470-478.

Eidul, T. S., Imran, M. A., and Das, A. K. (2022). "Restaurant review prediction using machine learning and neural network."

A Logistic Regression Approach for Cancer Prediction

Smriti Shikha Behera[1], Rojalin Biswal[1], Neha Lakra[1], Satyabrat Sahoo[1], Manas Ranjan Kabat[2]

[1]Department of Computer Science and Engineering, Silicon Institute of Technology, Sambalpur, India, smritishikha19@gmail.com, rojalinbiswal507@gmail.com, nehalakra48@gmail.com, satyabrat.sahoo@silicon.ac.in
[2]Department of Computer Science and Engineering, Veer Surendra Sai University of Technology, Burla, Sambalpur, India, manas_kabat@yahoo.com

Abstract

The goal is to apply logistic regression to mature a cancer prediction system. The use of data mining methods for a specially created dataset is the primary goal. Improving the accuracy of the dataset's percentage utilised to forecast breast, lung, and cervical cancer is the main goal of this study. This model's workflow incorporates data mining methodologies including decision trees and logistic regression with a proprietary dataset that has been rigorously chosen to contain a wide range of symptoms and behaviours. The training procedure uses neural network architecture to improve performance and achieve high face detection accuracy. An application has been developed to demonstrate the system's functionality and usefulness.

Keywords: Machine learning algorithm, recommendation system, cosine similarity, content-based filtering, NLP

1. Introduction

The development of abnormal cells that can invade and harm healthy body components is what characterises cancer. These cells multiply sporadically as well. This article examines the features and effects of cancer, with a special emphasis on lung, breast, and cervical cancer. It underlines the need of early identification and treatment, particularly considering the increased incidence of breast cancer (Chaurasia *et al.*, 2018). The research intends to create a predictive model for breast cancer revealing utilizing data mining approaches, namely support vector machine (SVM) classification (Ramachandran *et al.*, 2018). Globally, lung-cancer is acknowledged as one of the primary reasons for death, with difficulty in early identification due to metastases. Cervical cancer is also discussed, including its link to HPV infection and other risk factors. The paper emphasises the significance of early diagnosis for appropriate medical management. However, it recognises limitations such as inadequate infrastructure and public awareness, particularly in underdeveloped nations (Arutchelvan and Periyasamy, 2015).

2. Literature Review

2.1. Breast Cancer

Breast-cancer diagnosis employs a range of data mining techniques, including ANOVA

Chapter 61 DOI: 10.1201/9781003581215

for verification and regression methods for discovery (Sarvestani *et al.*, 2010). Neural networks, fuzzy systems, and SVM classifiers are prominent in predictive modelling, as evidenced by various studies (Gupta *et al.*, 2020, Ojha and Goel, 2017).

2.2. Lungs Cancer

The study delves into AI algorithms for identifying risk groups in radiotherapy-induced radiation pneumonitis (RP) in individuals with lung cancer. Various techniques like hybridised K-means, HAC, and EACCD clustering are explored, aiming to propose targeted treatments for those at high RP risk (Alam *et al.*, 2019, Chauhan and Jaiswal, 2016, Chandra *et al.*, 2019, Dass *et al.*, 2014).

2.3. Cervical Cancer

Previous research using the same dataset evaluated classification techniques such as Decision Tree, Naive Bayes, Random Forest and Sequential Minimal Optimisation in WEKA. SVM combined with RFE and PCA successfully decreased cervical cancer risk factors (Choudhury *et al.*, 2016, Delen, 2009). Other studies found that SVM outperformed neural-networks and decision-trees in the prediction of heart disease.

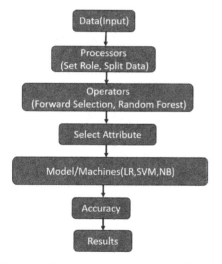

Figure 1: Comparative analysis of predictions for cervical-cancer

3. Methodology and Model Specifications

4. Results and Discussions

WEKA was the software program used in the majority of the research papers that we read and examined, while we employed RapidMiner as our software tool for the forecasting of various types of cancer viz., breast, lung, and cervical cancer; with their previous percentage of accuracy and our rating higher comparing all of them with their previous processes and operations.

4.1. Breast Cancer

After Building our model which is shown in Figure 1, we have tested results of every possible course of action in the case of breast cancer. A supervised learning method used in regression modelling and classification is the decision tree. Regression is a predictive modelling technique; therefore, these trees are used to classify data or anticipate what will happen next. A decision tree diagram is a form of flowchart that simplifies decision making by categorising the various pathways of action accessible. To find the correct and proper accuracy of the predictions of benign and malignant tumours with all the symptoms occurring in Breast Cancer, the algorithm of Logistic Regression with Neural Net and Performance is used to calculate the accuracy of the particular dataset, which differs from the result of the decision-tree model. It counts the occurence of positive and negative observations that were correctly categorised.

Accuracy level = (TruePositive + TrueNegetive) / (TruePositive + TrueNegetive + FalsePositive + FalseNegetive),   (1)

4.2. Lungs Cancer

K-NN and ANN, together with Naive Bayes, are commonly employed in lung cancer prediction to forecast symptoms and damage to the patient's lungs. In this project, we used Forward Selection, Cross Validation, Neural Network Architecture, and Model and Performance to predict cancer cell damage and stage, allowing for appropriate treatment. As a result, we employed a dataset with different outcomes than the prior techniques, and we obtained the following accuracy percentage.

4.3. Cervical Cancer

Women are disproportionately affected by cervical cancer, so in order to predict the cancer cells and the specific stage of the patient undergoing treatment, a variety of methods were employed, such as Naive Bayes and ANN, to determine the accurate percentage of cancer damage that has

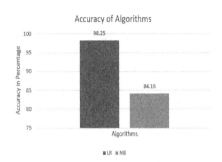

Figure 3: Comparative analysis of predictions for lungs-cancer

the results of the affected area. In contrast, we used different methods to determine the accuracy, such as apply model and logistic regression, in order to obtain

5. Conclusion and Future Scope

In order to address cancer prediction and the detection of the presence of symptoms and tumour, we have implemented a comprehensive and extensive Global Cancer Prediction System of Breast, Lungs, and Cervix. This system uses data mining techniques, such as the Logistic and Linear Regression, Artificial Neural Network (ANN), and Naive Bayes algorithms. First, the ability of Logistic Regression to yield an accurate percentage of cancer-affected areas is influenced and dominated by our Cancer Prediction System. By customising and monitoring the patients' symptoms and behaviours, such as yellowing of the skin and coughing continually with spits of blood, the system may forecast if the patient has cancer, what type of cancer, and what stage they are in. In such cases, people who have been diagnosed with cancer are made aware of the malignancy and begin treatment quickly, despite the risks of reaching the end stage of the cancer and losing their lives.

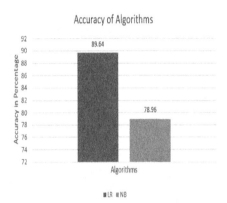

Figure 2: Comparative analysis of predictions for breast-cancer

affected each individual in order to obtain

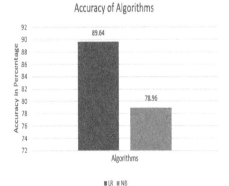

Figure 2: Process flow diagram of cancer prediction using data mining

References

Alam, T. M., Khan, M. M. A., Iqbal, M. A., Abdul, W., and Mushtaq, M. (2019). "Cervical cancer prediction through different screening methods using data mining." *International Journal of Advanced Computer Science and Applications (IJACSA)*, 10: 2.

Arutchelvan, K. and Periyasamy, R. (2015). "Cancer prediction system using datamining techniques." *International Research Journal of Engineering and Technology (IRJET)*, 2: 1179-1183.

Chandra, T. K. R. P. M., Venkata, E Y., and Ismail, B. M. (2019). "Lung cancer prediction using data mining techniques." *International Journal of Recent Technology and Engineering (IJRTE)*, 8: 4.

Chauhan, D. and Jaiswal, V. (2016). "An efficient data mining classification approach for detecting lung cancer disease." *International Conference on Communication and Electronics Systems (ICCES)*, pp. 1-8. IEEE.

Chaurasia, V., Pal, S., and Tiwari, B. (2018). "Prediction of benign and malignant breast cancer using data mining techniques." *Journal of Algorithms & Computational Technology*, 12: 119-126.

Choudhury, T., Kumar, V., Nigam, D., and Mandal, B. (2016). "Intelligent classification of lung & oral cancer through diverse data mining algorithms," *International Conference on Micro-Electronics and Telecommunication Engineering (ICMETE)*, pp. 133-138. IEEE.

Dass, M. V., Rasheed, M. A., and Ali, M. M. (2014). "Classification of lung cancer subtypes by data mining technique." *Proceedings of the 2014 international conference on control, instrumentation, energy and communication (CIEC)*, pp. 558-562. IEEE.

Delen, D. (2009). "Analysis of cancer data: a data mining approach." *Expert Systems*, 26 (1): 100-112.

Gupta, K. D., Shelly, and Sharma, A. (2020). "Data mining R. Patra, Prediction of lung cancer using machine learning classifier." *Computing Science, Communication and Security: First International Conference, COMS2 2020, Gujarat, India, March 26–27, 2020, Revised Selected Papers 1*, pp 132-142. Springer.

Ojha, U. and Goel, S. (2017). "A study on prediction of breast cancer recurrence using data mining techniques." *7th International Conference on Cloud Computing, Data Science and Engineering Convergence*, pp. 527-530. IEEE.

Ramachandran, P., Girija, N., and Bhuvaneswari, T. (2014). "Early detection and prevention of cancer using data mining techniques." *International Journal of Computer Applications*, 97: 13.

Sarvestani, S. A. P. N., Soltani, A., and Salehi, M. (2010). "Predicting breast cancer survivability using data mining techniques." *2nd International Conference on Software Technology and Engineering*, 2, pp. V2-227. IEEE.

Exploring the Underlying Patterns and Relationships Between Temperature and Heavy Rainfall Events over Western Japan

Sridhara Nayak[1], Joško Trošelj[2], Kanhu Charan Pattnayak[3], Khagendra P. Bharambe[4]

[1]Research and Development Center, Japan Meteorological Corporation, Japan, nayak.sridhara@n-kishou.co.jp
[2]Ruđer Bošković Institute, Zagreb, Croatia
[3]School of Earth and Environment, University of Leeds, United Kingdom
[4]Water Resources Research Center, DPRI, Kyoto University, Japan

Abstract

In this study, we explore the underlying patterns and relationships between temperature and heavy rainfall events using exploratory data analysis techniques to anticipate the rate provided by the Clausius-Clapeyron (CC) relationship, which is defined as approximately a 7% increase in the water vapour per degree rise in temperature. Despite the absence of a universal trend of increasing heavy rainfall events globally, we focus on specific events in western Japan. Our analysis involves collecting data on rainfall and temperatures at each impacted observation station locations during two flood events. Our results confirm a CC-like relationship pattern between temperature and precipitation during the flood events, notably up to a temperature threshold of 22°C. Interestingly, beyond this threshold, precipitation intensity starts to diminish. We hypothesise that this downtrend could be attributed to insufficient moisture availability at higher temperatures. Overall, our analysis offers valuable insights into the intricate dynamics of heavy rainfall events within the framework of climate change relationships, underscoring the nuanced influence of temperature on rainfall trends.

Keywords: Clausius-Clapeyron relationship, heavy rainfall event, temperature-precipitation relationship

1. Introduction

Recent studies have revealed that heavy rainfall events, such as floods, are becoming more intense due to global warming (O'Gorman, 2015). This is because the atmosphere can hold more water vapor in a warmer environment, as predicted by the Clausius-Clapeyron (CC) equation (Trenberth et al., 2003). Numerous studies from around the world confirm the increase in heavy rainfall events attributed to global warming showing a pattern consistent with the CC relationship (Linderink and Van Meijgaard, 2008, Nayak and Dairaku 2016, Nayak and Takemi 2019, Prein et al., 2017, Westra et al., 2014). Utsumi et al. (2011) and Nayak et al. (2018) documented the relationship between heavy rainfall events and temperature over Japan; Hardwick et al. (2010) investigated this relationship over Austraila, Similar studies

Chapter 62 DOI: 10.1201/9781003581215

are also conducted over Europe, India, and the United States (Berg *et al.*, 2009, Mishra *et al.*, 2012, Nayak, 2018).

Most of these studies have examined the underlying patterns and relationship between heavy rainfall events and temperature using daily datasets. However, this relationship is not well documented specifically for specific heavy rainfall events. For example, Nayak *et al.* (2020) reported that the relationship between these two quantities during two heavy rainfall events follow the CC relationship up to a certain temperature (22°C) and stressed the requirement for additional analysis with a more number of such events to conclude the robustness of their finding.

In this study, we examined the relationship between heavy rainfall events and temperature during the occurrence of two another heavy rainfall periods in Western Japan. We aim to assess the robustness of previously documented results of holding CC relationship and to discuss the major contributing factors prompting the presence or absence of the CC relationship during these events.

2. Overview of the Heavy Rainfall Events and the Study Area

Western Japan, particularly Kyushu Island, experiences major heavy rainfall events, typically in July, throughout the year. These events bring excessive rainfall to the region, resulting in significant property damage and loss of lives. In 2022, a heavy rainfall event occurred around July 14th in the Kyushu Island (Figure 1) and carried unusual heavy rainfall to the region. Similarly, in 2023, a heavy rainfall event occurred in around July 10th that flooded 6 rivers in this region. (Figure 1) leading to widespread soci0-economic damages.

3. Data and Variables

We used the weather datasets including rainfall, temperature, dew point temperature

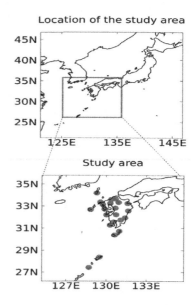

Figure 1: Location map of Kyushu Island of Western Japan (upper); AMeDAS stations (lower). AMeDAS refers to automated meteorological data acquisition system

and relative humidity as obtained at each hour from the Automated Meteorological Data Acquisition System (AMeDAS, https://www.data.jma.go.jp/) which is run by the Japan Meteorological Agency (JMA). We collected these datasets from 24 AMeDAS stations (Figure 1) each for the July month of the years 2022 and 2023.

4. Methodology

We first explored the characteristics of the hourly temperature and rainfall data through the normal probability distribution function (defined in Eq. 1) with various thresholds of no-rain and 1 mm rainfall, 5 mm rainfall, and 10 mm rainfall.

$$f(x) = \frac{1}{\sigma\sqrt{2\pi}} e^{-\frac{1}{2}\left(\frac{x-\mu}{\sigma}\right)^2} Eq \qquad (2)$$

Where, $f(x)$ is the normal probability function, x is the random variable with rainfall amount, μ is the mean, σ is the standard deviation.

We then paired the hourly rainfall amounts at each rainfall thresholds (i.e. no-rain to 10 mm) with their corresponding location's and time's temperature at 24 stations for 31 days in 2 years resulting a total of 35,712 pairs (31 days x 24 hours x 2 years x 24 stations). We stratified the rainfall amounts at each threshold into different temperature bins at 1-degree intervals (Lenderink and Van Meijgaard, 2008, Nayak *et al.*, 2018). Finally, for each bin, the 90th percentile was computed, which is considered the top 10% rainfall amounts to be in extreme situations. These extreme values were plotted against the temperature bins to find out the underlying relationship between heavy rainfall events and temperature and compared with the CC equation (defined in Eq. 2).

$$e_s = 6.11 \times \exp\left(\frac{17.62 \times T}{243.04 + T}\right) Eq \qquad (2)$$

Where, e_s is the saturated vapor pressure and T is temperature.

To understand the contributing factors to retain the CC relationship, we further analyzed the relative humidity and dew point temperature during these events in a similar way against temperature.

5. Results and Discussion

In this section, we first discussed the intrinsic flexibility and statistical properties of the temperature and rainfall datasets by using equation (1). Subsequently, we discuss the underlying patterns and relationships between these two variables, together with CC relationship as outlined in equation (2). Finally, we explore the factors influencing the maintenance of this relationship.

5.1. Temperature and Rainfall Distribution

Figure 2 presents the probability distribution of hourly rainfall and temperature datasets by using Eq. (1) under various

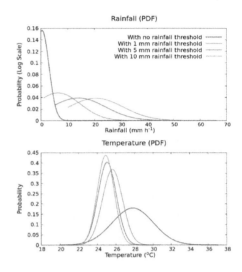

Figure 2: PDF of rainfall (upper) and temperature (lower)

rainfall thresholds. We find that the most frequent occurrences of hourly rainfall and temperature values are 1 mm and 25°C respectively when all datasets are considered (i.e., without any rainfall threshold). However, considering 1mm rainfall threshold, the most occurrences of hourly rainfall and temperature shift to higher values of 9 mm and 25.2°C respectively. This tendency continued with increasing thresholds and shows the values of 15 mm and 26°C respectively with 5 mm rainfall threshold; and 22mm and 28°C respectively with 10 mm rainfall threshold. This indicates that the weather remains relatively warmer during heavy rainfall, implying to have some relationship with temperatures. To explore this relationship, we paired the rainfall amounts with temperature and stratified at each threshold of no-rain to 10 mm rainfall, as outlined in the methodology section, and discussed their connection with the CC equation.

5.2. Underlying Relationship and Pattern

Figure 3 presents the underlying relationship between heavy rainfall events (99th percentile) and temperature with the

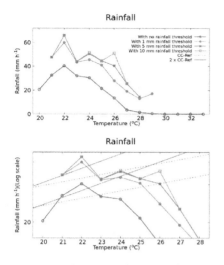

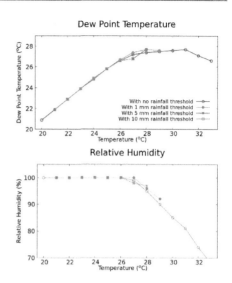

Figure 3: Relationship between the heavy rainfall events with temperature (upper); and their connection with CC equation (lower)

Figure 4: Dew point temperature (upper) and relative humidity (lower) as a function of temperature

consideration of different rainfall thresholds, ranging from no rain to 10 mm. We compared this relationship with the CC equation at log scale. We find that the relationship curve for all rainfall thresholds exhibits a peak at 22°C, with a surprising decrease in intensity for temperatures above this threshold. Furthermore, the peak intensity of rainfall events varies towards the higher side as the rainfall threshold increases. For instance, the 99th percentile of rainfall events shows peak values of 40 mm, 60 mm, 70 mm, and 70 mm at rainfall thresholds of no rain, 1 mm, 5 mm, and 10 mm, respectively. The relationship curve follows the CC reference up to the peak, indicating that the CC equation can explain heavy rainfall events concerning temperature, but only up to a certain temperature threshold. These findings are consistent with other results highlighted in Japanese regions, indicating robust features in the relationship curve. To understand why the relationship curve between rainfall and temperature diverges from the CC relationship above 22°C, we further analyzed and discussed other weather variables during heavy rainfall events.

5.3. Contributing Factors to Hold the Relationship

Figure 4 illustrates the relationships of dew point temperature and relative humidity with temperature for all rainfall thresholds. Interestingly, we find that both dew point temperature and relative humidity decrease at higher temperatures, suggesting a possible insufficient amount of water vapor available at elevated temperatures. Specifically, for heavy rainfall events, a significant amount of water vapor is necessary as temperature rises to maintain the Clausius-Clapeyron (CC) relationship. Numerous previous studies have emphasized this phenomenon and documented that water vapor becomes inadequate at higher temperatures to uphold the CC relationship.

6. Conclusion

This study explored the underlying patterns and relationships between temperature and heavy rainfall events using exploratory data analysis techniques to anticipate the rate provided by the Clausius-Clapeyron relationship. We confirm a CC-like relationship pattern between temperature and precipitation during two flood events over western Japan, notably up to a temperature threshold of 22°C. Interestingly, beyond this threshold, rainfall intensity starts to decline. We hypothesize that this downtrend could be attributed to insufficient moisture availability at higher temperatures. Overall, our analysis offers valuable insights into the intricate dynamics of heavy rainfall events within the framework of climate change. In our observational dataset, water vapor data is not available. Thus, further research with observed water vapor is necessary, which is a focus of our future studies.

References

Berg, P., Haerter, J. O., Thejll, P., Piani, C., Hagemann, S., and Christensen, J. H. (2009). "Seasonal characteristics of the relationship between daily precipitation intensity and surface temperature." *Journal of Geophysical Research: Atmospheres*, 114 (D18), D18102, 1-9.

Hardwick Jones, R., Westra, S., and Sharma, A. (2010). "Observed relationships between extreme sub-daily precipitation, surface temperature, and relative humidity." *Geophysical Research Letters*, 37 (22), L22805, 1-5.

Lenderink, G. and Van Meijgaard, E. (2008). "Increase in hourly precipitation extremes beyond expectations from temperature changes." *Nature Geoscience*, 1 (8): 511-514.

Mishra, V., Wallace, J. M., and Lettenmaier, D. P. (2012). "Relationship between hourly extreme precipitation and local air temperature in the United States." *Geophysical Research Letters*, 39 (16), L16403, 1-7.

Nayak, S. (2018). "Do extreme precipitation intensities linked to temperature over India follow the Clausius-Clapeyron relationship?" *Current Science*, 115 (3): 391-392.

Nayak, S. and Dairaku, K. (2016). "Future changes in extreme precipitation intensities associated with temperature under SRES A1B scenario." *Hydrological Research Letters*, 10 (4): 139-144.

Nayak, S. and Takemi, T. (2019). "Dependence of extreme precipitable water events on temperature." *Atmósfera*, 32 (2): 159-165.

Nayak, S. and Takemi, T. (2020). "Clausius-Clapeyron scaling of extremely heavy precipitations: Case studies of the July 2017 and July 2018 heavy rainfall events over Japan." *Journal of the Meteorological Society of Japan. Ser. II*, 98 (6): 1147-1162.

Nayak, S., Dairaku, K., Takayabu, I., Suzuki-Parker, A., and Ishizaki, N. N. (2018). "Extreme precipitation linked to temperature over Japan: current evaluation and projected changes with multi-model ensemble downscaling." *Climate Dynamics*, 51 (11): 4385-4401.

O'Gorman, P. A. (2015). "Precipitation extremes under climate change." *Current Climate Change Reports*, 1: 49-59.

Prein, A. F., Rasmussen, R. M., Ikeda, K., Liu, C., Clark, M. P., and Holland, G. J. (2017). "The future intensification of hourly precipitation extremes." *Nature Climate Change*, 7 (1): 48-52.

Trenberth, K. E., Dai, A., Rasmussen, R. M., and Parsons, D. B. (2003). "The changing character of precipitation." *Bulletin of the American Meteorological Society*, 84 (9), 1205-1218.

Utsumi, N., Seto, S., Kanae, S., Maeda, E. E., and Oki, T. (2011). "Does higher surface temperature intensify extreme precipitation?" *Geophysical Research Letters*, 38 (16), L16708, 1-5.

Westra, S., Fowler, H. J., Evans, J. P., Alexander, L. V., Berg, P., Johnson, F., ... Roberts, N. (2014). "Future changes to the intensity and frequency of short-duration extreme rainfall." *Reviews of Geophysics*, 52 (3), 522-555.

Water Quality Analysis Using Intelligent Framework

Alok Kumar Pati[1], Alok Ranjan Tripathy[1,2], Alakananda Tripathy[3]

[1]Department of Computer Science, Ravenshaw University, Cuttack, India, alokpati6065@gmail.com
[2]Department of Computer Science & Engineering, IMIT, Cuttack, BPUT, India, tripathyalok@gmail.com
[3]Centre for Artificial Intelligence and Machine Learning, Siksha 'O' Anusandhan Deemed to be University, Bhubaneswar, India, tripathy.alka@gmail.com

Abstract

Water, a fundamental component on Earth for life, faces significant challenges due to increase in population and industrialisation leading to increased pollution of water resources. This research accentuate on machine learning approaches to estimate the quality of water, with different parameters. Various supervised classification algorithms and Regression were implemented for the study. The study aims to identify the optimal algorithm for water quality prediction. Through analysis and comparison of historical data, machine learning models demonstrate the ability to predict future pollutant levels with high accuracy.

Keywords: Machine learning, water quality, water pollution, prediction, data analytics, decision making

1. Introduction

Groundwater, along with surface water, serves as a vital source of fresh water for numerous communities worldwide. For maintaining good health clean water is necessary, as it is use for a multitude of purposes ranging from drinking to agricultural and industrial activities (Ahmed et al., 2019). Consequently, ensuring the suitability and safety of groundwater before its consumption or usage is paramount. Traditional methods of water quality monitoring involve manual collection of samples from various locations within water distribution networks, followed by laboratory testing for contamination. While these methods provide valuable insights into water quality, they are often time-consuming, labour

intensive, and expensive. Moreover, the complexity of water quality parameters necessitates the development of simpler yet effective approaches for evaluation and prediction. The popular tool for assessing the water quality Water Quality Index (WQI) is use, which provides a simplified classification system based on key chemical, physical, and biological characteristics of water. By providing a single value indicative of water quality, the WQI facilitates easy interpretation and communication of complex data to the general public. Leveraging machine learning algorithms for predicting water quality based on WQI values presents a promising avenue for streamlining the assessment process (Ahmed et al., 2019).

Dataset that includes different parameters like pH, total dissolved solids (TDS), potassium (K), total hardness (T.H.), electrical conductivity (E.C.), nitrate (NO3), sulphate (SO4), magnesium (Mg), sodium (Na), calcium (Ca) among other chemical components. The paper highlights an analysis of groundwater quality.

2. Literature Review

Nair and Vijaya (2021) highlights the importance of water quality prediction and evaluation, emphasizing challenges in assessing running water and proposing solutions like real-time monitoring and hybrid modelling. Ooko et al. (2023) evaluates various ML algorithms to determine water safety based on physio-chemical properties. It explores different classification algorithms and artificial neural networks. Jalagam et al. (2023) analyse water quality parameters using different machine learning model by, focusing on total suspended solids in urban streams. Various regression models were evaluated, with random forest showing the best performance. Anita et al. (2023) evaluates machine learning algorithms to predict water quality indices, empha- sizing metrics like accuracy, recall, and precision. SVM outperforms other models, followed by XGBoost, K-NN, and Decision Tree. Brindha et al. (2023) compares machine learning classification algorithms for predicting water quality, with the Random Forest Classifier outperforming others with 91.97% accuracy.

3. Methodology and Model Specifications

The workflow begins with data collection on groundwater, followed by preprocessing and splitting for training/testing sets. Model selection involves considering different classification model like K nearest neighbour, support vector machine, logistic regression and random forest. The chosen model undergoes testing, with results documented before concluding the process, as indicated by the "End" marker in the workflow diagram.

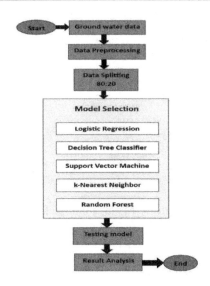

Figure 1: Workflow diagram

3.1. Data Description

The dataset comprises groundwater quality measurements from various districts in Telangana State, India, spanning three years: 2018, 2019, and 2020. Serial number, district, mandal, village, latitude, longitude, chemical compositions, total hardness, total dissolved solids, SAR, RSC, and two target variables ('Classification' along with 'Classification1') are among the 26 columns that make up each dataset.

3.2. Data Preprocessing

Data preparation is the process of organizing, modifying, and cleansing raw data in order to make it ready for analysis. This includes scaling numerical properties, removing duplicates, addressing missing values, and dividing the data into training and testing sets.

3.3. Data Cleaning

Boxplots were utilised to visualise the distribution of each numerical variable in the dataset. Outliers, which are data points lying significantly beyond the whiskers of the boxplots, were identified for each variable. Figure 2 address outlier using Interquartile Range (IQR) method.

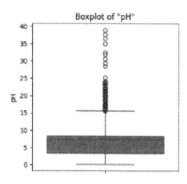

Figure 2: Address the outliers

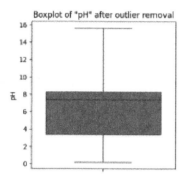

Figure 3: After removing the outliers

The outliers were removed using remove outliers' functions applied to all numerical columns in the dataset. Figure 3 shows the boxplots after removing outliers for column PH given in the dataset, this process is repeated for all columns in the dataset.

3.3.1. Correlation Analysis

In Figure 4, heat maps simplify the interpretation of large datasets, enabling analysts to discern relationships and anomalies at a glance. This graphical representation facilitates efficient data analysis and decision-making processes, particularly in field of scientific research, where understanding patterns and trends is crucial for informed decision-making (Uddin *et al.*, 2022).

3.4. Machine Learning Models

3.4.1. Logistic Regression

The model calculates the likelihood that an outcome variable will fall into a specific category by considering predictor factors. The logistic function is used to transform linear regression outputs into probabilities, enabling classification. Eq. (1) displays the logical regression equation. In this case, the bias is b0, the predicted output is y, for the single value x input the coefficient is b1.

$$y = \frac{e^{(b0+b1\times x)}}{1 + e^{(b0+b1\times x)}} \tag{1}$$

3.4.2. Decision Tree

To generate a tree-like structure and a set of if-else decision rules, it recursively divides the dataset into subsets based on features. While Eq. (3) computes the entropy using two features, Eq. (2) computes the entropy of each class.

$$H(S) = \sum_{i=1}^{c} -P_i \log_2 P_i \tag{2}$$

$$H(S,D) = \sum_{c \in D} P(C).E(C) \tag{3}$$

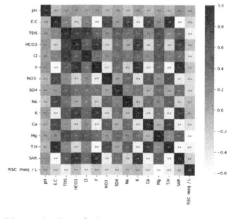

Figure 4: Correlation matrix

3.4.3. Support Vector Machine (SVM)

It determines the ideal hyperplane in the feature space that maximally divides data points belonging to various classes. The equation in eq. (4) is a p-by-q matrix X where the training data for both the classes are put together. X^i is denoted as the ith row of X. A diagonal p × p matrix Y with -l and +l determines the class of X^i either in +1 or 1 (Liu et al., 2022).

$$D = \left\{ \left(x^1, y^1 \right), \ldots, \left(x^l, y^l \right) \right\},$$
$$x \in R, y \in \left\{ -l, l \right\} \tag{4}$$

3.4.4. K – Nearest Neighbors (KNN)

This is a supervised learning non parametric technique applied to resolve classification problems. The data points are classified using the majority vote where each data point's is a k nearest neighbors in the feature space. This classifier uses a new formula to assess the distance. A widely used formula for distance is provided in Eq. (5).

$$Euclidean = \sqrt{\sum_{i=1}^{k} \left(x_i - y_i \right)^2} \tag{5}$$

3.4.5. Random Forest

It is applied to re-gression and classification problems. In order to increase accuracy and decrease overfitting, it constructs several decision trees and merges their predictions shows in Eq. (6).

$$P_{jk} = m_j T_k - m_{left(k)} T_{left(k)}$$
$$- m_{right(k)} T_{right(k)} \tag{6}$$

Where, P_{jk} is node k's significance. m_j indicating samples that reach node k. T_k indicating the impurity value of nodes k. left(k) being a child node of node k is split off from the left. right(k) is a child node from the right is separated by node k.

4. Result Analysis

4.1. Logistic Regression

The logistic regression model achieved a commendable accuracy of around 90.5%. It exhibited strong performance for certain classes such as C2S1 and C3S1, indicating its capability to accurately classify water quality into these categories.

4.2. Decision Tree Classifier

With a maximum depth of 5, the decision tree classifier attained an accuracy of approximately 95.0%. It demonstrated robust performance for classes like C2S1 and C3S1, showcasing its ability to effectively classify groundwater quality.

4.3. Vector Machine (SVM)

The SVM classifier, utilizing an RBF kernel, exhibited exceptional performance with an accuracy score of about 97.7%. It showcased strong predictive capabilities across various classes, particularly excelling in classes like C3S1 and C4S1.

4.4. K – Nearest Neighbors (KNN)

The KNN classifier, configured with 31 neighbors, achieved an accuracy score of approximately 88.3%. While it demonstrated acceptable precision and recall for certain classes such as C2S1 and C3S1, its performance may benefit from further optimisation, especially for classes with lower scores.

4.5. Random Forest Classifier

The Random Forest classifier delivered a high accuracy score of around 96.8%, indicating its effectiveness in predicting groundwater quality classes. It demonstrated robust precision and recall for several classes, suggesting its suitability for accurate classification across diverse water quality categories.

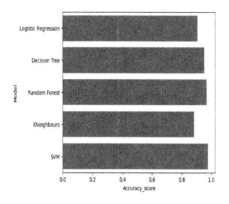

Figure 5: Result analysis

5. Conclusion

The study highlights the effectiveness of classification algorithm of machine learning to predict the groundwater quality, essential for safeguarding public health and environmental sustainability. With the use of supervised classification models and there are evaluated for their performance in classifying water quality classes based on key parameters. While each model exhibited varying degrees of accuracy and predictive capability across different classes, Support Vector Machine emerged as the top-performing model, demonstrating exceptional accuracy and robust predictive capabilities. Other models such as Logistic Regression and Random Forest also showed promising results, albeit with varying levels of accuracy for different classes. These findings offer valuable insights for stakeholders.

References

Ahmed, A. N., Othman, F. B., Afan, H. A., Ibrahim, R. K., Fai, C. M., Hossain, M. S., Ehteram, M., and Elshafie, A. (2019). "Machine learning methods for better water quality prediction." *Journal of Hydrology.*

Ahmed, U., Mumtaz, R., Anwar, H., Shah, A.A., Irfan, R., García-Nieto, J.(2019). Efficient Water Quality Prediction Using Supervised Machine Learning. Water, 11, 2210. https://doi.org/10.3390/w11112210

Nair, J. P., & Vijaya, M. S. (2021). Predictive models for river water quality using machine learning and big data techniques-a Survey. In 2021 *International conference on artificial intelligence and smart systems (ICAIS)* (pp. 1747-1753). IEEE.

Ooko, S. O., Pamela, E. K., & Kwagalakwe, G. (2023). Use of Machine Learning for Realtime Water Quality Prediction. 2023 *IEEE AFRICON*, 1-6.

Jalagam, L., Shepherd, N., Qi, J., Barclay, N., & Smith, M. (2023). Water quality predictions for urban streams using machine learning. In *SoutheastCon* 2023 (pp. 217-223). *IEEE*

Anita, M., Dinesh, M. G., Lakshmipriya, C., Sharmila, V., & Muthuram, A. (2023). Water Quality Prediction using Machine Learning: A Comparative Study. In 2023 *Second International Conference on Augmented Intelligence and Sustainable Systems (ICAISS)* (pp. 348-353). IEEE.

Brindha, D., Puli, V., NVSS, B. K. S., Mittakandala, V. S., & Nanneboina, G. D. (2023, February). Water quality analysis and prediction using machine learning. In 2023 *7th International Conference on Computing Methodologies and Communication (ICCMC)* (pp. 175-180). IEEE.

Uddin, K. M. M., Biswas, N., Rikta, S. T., & Dey, S. K. (2023). Machine learning-based diagnosis of breast cancer utilizing feature optimization technique. *Computer Methods and Programs in Biomedicine Update*, 3, 100098.

Liu, X., Lu, D., Zhang, A., Liu, Q., & Jiang, G. (2022). Data-driven machine learning in environmental pollution: gains and problems. *Environmental science & technology*, 56(4), 2124-2133.

Printed in the United States
by Baker & Taylor Publisher Services